Handbook of Experimental Pharmacology

Volume 137

Springer
Berlin
Heidelberg
New York
Barcelona
Hong Kong
London
Milan
Paris
Singapore
Tokyo

Novel Therapeutics from Modern Biotechnology

From Laboratory to Human Testing

Contributors

A. Abuchowski, R.D. Armstrong, C.F. Bennett,
J.A. Cavagnaro, D.L. Cole, S.B. Dillion, J.C. Erickson,
D.E. Everitt, S.G. Griego, G.E. Hardee, T.K. Hart,
S.P. Henry, H.E.J. Hofland, L. Huang, M.R. Koller,
H. Kotani, R.J. Kreitman, A.A. Levin, A.S. Lubiniecki,
J. Maluta, G.J. McGarrity, C.R. Middaugh, C.L. Nolan,
M.L. Nucci, W.C. Ogier, I. Pastan, R. Pearlman, T.G. Porter,
R.G. Scott, M.A. Serabian, P.J. Shadle, R.G.L. Shorr,
T.M. Smith, G.S. Srivatsa, D.W. Zabriskie

Editors

Dale L. Oxender and Leonard E. Post

Springer

Dale L. Oxender, Ph.D.
Cell Biology

Leonard E. Post, Ph.D.
Vice President, Discovery Research

Parke-Davis Pharmaceutical Research
2800 Plymouth Road
Ann Arbor, MI 48105
USA

With 34 Figures and 19 Tables

ISBN 3-540-65025-3 Springer-Verlag Berlin Heidelberg New York

Library of Congress Cataloging-in-Publication Data
Novel therapeutics from modern biotechnology: from laboratory to humantesting / contributors, A. Abuchowski . . . [et al.]; editors, Dale L. Oxender and Leonard E. Post.
p. cm. – (Handbook of experimental pharmacology; v. 137)
Includes bibliographical references and index.
ISBN 3-540-65025-3 (hardcover: alk. paper)
1. Pharmaceutical biotechnology. I. Abuchowski, A. II. Oxender, Dale L. III. Post, Leonard E., 1952– . IV. Series.
[DNLM: 1. Drug Design. 2. Drug Evaluation. 3. Biotechnology. 4. Technology, Pharmaceutical. W1HA51L v. 137 1999]
QP905.H3 vol. 137
[RS380]
615′.1s – dc21
[615′.19]
DNLM/DLC
for Library of Congress 98-41690
CIP

Cover design: *design & production* GmbH, Heidelberg

Typesetting: Best-set Typesetter Ltd., Hong Kong

Production Editor: Angélique Gcouta

SPIN: 10689474 27/3020 – 5 4 3 2 1 0 – Printed on acid-free paper

Preface

A cover story of *Business Week Magazine* in January 1984 stated "Biotech Comes of Age". In February 1986, *Venture Magazine* had a cover article entitled "The Biotech Revolution is Here". This article went on to say "New Genetic Technologies Will Transform Our Lives". These announcements were made many years after the first biotechnology companies, such as Genentech, Cetus, Amgen and Biogen, were formed to commercialize the "New Biology".

At the time of writing this book, there are over 1300 biotech companies developing new technologies or identifying potential biotech drugs. Most of these companies were started in the height of the "high-technology hype", although companies are still forming as the technology advances.

A more recent survey showed only a relatively small number of Food and Drug Administration (FDA) approvals among over several hundred biotechnology products now in clinical trial. One could ask why it has taken so long to produce biotechnology products. Part of the reason is that each new class of biotech products brings with it a set of problems that need to be solved before they enter clinical trials. These problems are often unique to biotechnology products, such as peptides, proteins, monoclonal antibodies, nucleic acids and cellular therapies.

Although the "biotech" therapeutics represent unique scientific issues, there are now numerous examples that show that the principles of drug development apply to the newer systems as well as to development of novel small synthetic compounds. For example, there needs to be a well-controlled process for making the therapeutics, appropriate analytical methodologies to characterize the product, appropriate formulation to store and deliver the therapy to the patient, and a thorough characterization of the safety issues surrounding the new entity.

Our goal in this volume is to address the particular problems associated with several classes of biotechnology products and, at the same time, demonstrate that the principles are the same as in development of small new chemical entities. We have invited several scientists at biotechnology or pharmaceutical companies to share with us some of the unique problems that they encountered in getting biotech products into clinical trials and how they solved them.

The first chapter addresses FDA regulatory expectations for biotech products. The next several chapters discuss general issues common to each class of biotech drug, such as proteins, peptides and nucleic acids. The balance of the chapters deals with specific biotech drugs that have successfully made it into clinical trials.

We would like to express our appreciation to the authors and the companies in which they work. The patience of the staff of Springer-Verlag is also appreciated. Special thanks goes to Joyce Peplinski for secretarial and organizational assistance.

D.L. Oxender
L.E. Post

List of Contributors

ABUCHOWSKI, A., NPC Inc., 62 Bunnvale Road, Califon, NJ 01830, USA

ARMSTRONG, R.D., AASTROM Biosciences, Inc., P.O. Box 376, Ann Arbor, MI 48106, USA

BENNETT, C.F., Isis Pharmaceuticals, Inc., 2292 Faraday Avenue, Carlsbad, CA 92008, USA

CAVAGNARO, J.A., Human Genome Sciences, 9410 Key West Avenue, Rockville, MD 20850, USA

COLE, D.L., Isis Pharmaceuticals, Inc., 2292 Faraday Avenue, Carlsbad, CA 92008, USA

DILLON, S.B., Department of Molecular Virology and Host Defense, SmithKline Beecham Pharmaceuticals, Philadelphia, USA

ERICKSON, J.C., Biopharmaceutical Development, SmithKline Beecham Pharmaceuticals, King of Prussia, PA 19406, USA

EVERITT, D.E., Clinical Pharmacology Unit, SmithKline Beecham Pharmaceuticals, Philadelphia, USA

GRIEGO, S.G., Department of Molecular Virology and Host Defense, SmithKline Beecham Pharmaceuticals, Philadelphia, USA

HARDEE, G.E., Isis Pharmaceuticals, Inc., 2292 Faraday Avenue, Carlsbad, CA 92008, USA

HART, T.K., Department of Toxicology, SmithKline Beecham Pharmaceuticals, Philadelphia, USA

HENRY, S.P., Isis Pharmaceuticals, Inc., 2292 Faraday Avenue, Carlsbad, CA 92008, USA

HOFLAND, H.E.J., RPR Gencell Corporation, 3825 Bay Center Place, Hayward, CA 94545, USA

Huang, L., Laboratory of Drug Targeting, Department of Pharmacology, University of Pittsburgh School of Medicine, Pittsburgh, PA 15261, USA

Koller, M.R., AASTROM Biosciences, Inc., P.O. Box 376, Ann Arbor, MI 48106, USA

Kotani, H., Genetic Therapy, Inc. A Novartis Company, 938 Clopper Road, Gaithersburg, MD 20878, USA

Kreitman, R.J., Laboratory of Molecular Biology, National Cancer Institute, National Institutes of Health, 9000 Rockville Pike, 37/4E16, Bethesda, MD 20892, USA

Levin, A.A., Isis Pharmaceuticals, Inc., 2292 Faraday Avenue, Carlsbad, CA 92008, USA

Lubiniecki, A.S., Biopharmaceutical Development, SmithKline Beecham Pharmaceuticals, King of Prussia, PA 19406, USA

Maluta, J., AASTROM Biosciences, Inc., P.O. Box 376, Ann Arbor, MI 48106, USA

McGarrity, G.J., Genetic Therapy, Inc. A Novartis Company, 938 Clopper Road, Gaithersburg, MD 20878, USA

Middaugh, C.R., Department of Pharmaceutical Chemistry University of Kansas, 2095 Constant Ave., Lawrence, KS 66047, USA

Nolan, C.L., Biopharmaceutical Development, SmithKline Beecham Pharmaceuticals, King of Prussia, PA 19406, USA

Nucci, M.L., NPC Inc., 62 Bunnvale Road, Califon, NJ 01830, USA

Ogier, W.C., AASTROM Biosciences, Inc., P.O. Box 376, Ann Arbor, MI 48106, USA

Pastan, I., Laboratory of Molecular Biology, National Cancer Institute, National Institutes of Health, 9000 Rockville Pike, 37/4E16, Bethesda, MD 20892, USA

Pearlman, R., Megabios Corp., 863A Mitten Ave., Burlingame, CA 94010, USA

Porter, T. G., Department of Protein Biochemistry, SmithKline Beecham Pharmaceuticals, Philadelphia, USA

Scott, R.G., Biopharmaceutical Development, SmithKline Beecham Pharmaceuticals, King of Prussia, PA 19406, USA

SERABIAN, M.A., Center for Biologics Evaluation and Research (CBER), FDA, Rockville, MD 20852–1448, USA

SHADLE, P.J., Biopharmaceutical Development, SmithKline Beecham Pharmaceuticals, King of Prussia, PA 19406, USA

SHORR, R.G.L., Enzon Inc. 40 Kingsbridge Road, Piscataway, NJ, USA

SMITH, T.M., Biopharmaceutical Development, SmithKline Beecham Pharmaceuticals, King of Prussia, PA 19406, USA

SRIVATSA, G.S., Isis Pharmaceuticals, Inc., 2292 Faraday Avenue, Carlsbad, CA 92008, USA

ZABRISKIE, D.W., Biopharmaceutical Development, SmithKline Beecham Pharmaceuticals, King of Prussia, PA 19406, USA
Present address: Biogen Inc., Cambridge, MA 02139, USA

Contents

CHAPTER 10

Clinical Systems for the Production of Cells and Tissues for Human Therapy
R.D. Armstrong, M.R. Koller, J. Maluta, and W.C. Ogier.
With 4 Figures 221

CHAPTER 1

Overview of Regulatory Expectations for Introducing Novel Therapies into Clinical Trials

J.A. CAVAGNARO and M.A. SERABIAN

A. Introduction

Early introduction of novel therapies into clinical trials for diagnosis, treatment or prevention of disease has relied on the accumulation of adequate scientific data. For the most part, the data to support "first in humans" studies have been derived from preclinical studies designed to ask specific questions of the new product and/or product class. This product-specific approach to (pre)clinical program planning and evaluation has been commonly referred to as the "case-by-case" approach. A combination of early dialogue, to identify the critical questions unique to an emerging technology, and a science-based rational approach, to answer these questions, has resulted not only in facilitation of clinical development, but in providing the appropriate data to support marketing applications.

Most of the novel therapies developed over the past decade have been realized through advances in biotechnology. As such, the resultant products have been referred to as biotechnology-derived pharmaceuticals, "biotech" products or biopharmaceuticals. To help identify and focus on the productspecific issues, clinical development strategies and regulatory-needs products are often grouped in one or more categories. Examples of product classifications/categories include: 1) method of production, e.g., expression system, parameters of production, purification, 2) type of product, e.g., recombinant DNA-derived proteins, such as hormones, cytokines, clotting factors or vaccines, 3) monoclonal antibodies, 4) cellular or gene therapies, 5) similarity to human molecule, e.g., nature identical, nature non-identical, 6) general clinical indication, e.g., replacement product, response regulator, and/or 7) specific clinical use, e.g., acute life-threatening, chronic.

Many of the currently approved biotechnology-derived pharmaceuticals represent significant improvements on previously available therapies. Many more novel therapies are in various phases of clinical development. The "biotech" environment has evolved rapidly in a relatively short period of time and continues to evolve. Products continue to become more complex, as do business arrangements. Clinical development is also becoming more complex; in part due to changing standards, competing products, resource constraints and consumer activism.

As experience accumulates, regulatory practices for emerging technologies have evolved with the science. Systematic review of product development successes and failures has provided the data to revise regulatory practices in order to optimize the guiding principles for regulating the products derived from the next generation of novel technologies.

B. Roles of Regulatory Scientists

The roles of Food and Drug Administration (FDA) regulatory scientists in facilitating early introduction of novel technologies into the clinic include: 1) timely review of investigational new drug applications (INDs), 2) the use of targeted research programs to identify and assess potential issues that may prevent, delay or otherwise interrupt a clinical development plan, and 3) the development of regulatory standards and policy to provide a framework to guide development plans (U.S. Government Printing Office; Food and Drug Administration 1990).

Strategic planning and project management are key factors, not only in the commercial development of a product, but also in ensuring the timely review of applications by regulatory authorities. The establishment of a managed review process in regulatory agencies with attendant timelines, negotiated deadlines and milestones has been an important factor in improving review times of investigational agents.

Coordinated research programs in academia, industry and government, which address specific issues related to novel technologies, have also been important in introducing novel therapies into the clinic, as they have provided the fundamental information used in developing the principles for regulation. FDA's responsibility for performing research relating to foods, drugs, cosmetics and devices is explicitly stated in the Food Drug and Cosmetic Act (Food and Drug Administration 1990).

The regulatory standards for each new product and/or product class are developed, as needed, based on assessment of available data derived from relevant models using appropriate methodologies and reagents. In assessing the adequacy of the data and establishing appropriate standards, the validity and feasibility of the test methods are also considered. Significant progress has been made in harmonization of product requirements to support international public health issues and complement the emergence of multinational companies and global development strategies (Office of Communication, Training and Manufacturers Assistance; International Conference on Harmonisation and related documents).

C. Product Development and Availability

In 1995, the FDA introduced reform proposals related to INDs in efforts to expedite entry of new drug and biological products into clinical testing. The intent was to reduce the amount of data needed for initial review, e.g., acceptance of summary reports of preclinical toxicology studies and abbreviation of product stability requirements, without compromising traditional safety standards. Advances in biotechnology have also provided the data to support major streamlining initiatives in the regulation and approval of biological products. These initiatives have included: 1) elimination of lot-release testing for specified biotechnology and synthetic biological products, 2) the establishment of a single application (the biologics licensing application, or BLA) to replace the establishment licensing application (ELA) and product licensing application (PLA) for specified biological products, and 3) the new definition of manufacturer, which makes it easier for companies to use outside contract manufacturers and not have to build or purchase their own manufacturing facility before they gain approval to market a product (OFFICE OF COMMUNICATION, TRAINING AND MANUFACTURERS ASSISTANCE).

Regulatory authorities are also committed to helping provide broad patient access to potentially effective treatments during clinical development. Under FDA's "expanded access" mechanisms, including the Treatment-IND protocol and the "compassionate use" single-patient protocol, patients may receive promising, but not yet fully studied or approved new therapies that are undergoing clinical testing when no other satisfactory clinical options exist.

The FDA utilizes the "accelerated approval" process to allow marketing of therapeutics for patients with serious and life-threatening diseases. Under existing regulations, a new drug or biologic agent that is intended to provide a meaningful therapeutic benefit over existing therapies may be approved on the basis of adequate and well-controlled clinical trials establishing that the drug product has an effect on a surrogate endpoint that is reasonably reflective of clinical outcome. Epidemiologic, therapeutic, pathophysiologic or other evidence is used to predict clinical benefit or on the basis of an effect on a clinical endpoint other than survival or irreversible morbidity, e.g., CD4 count on viral load effects to support effectiveness of therapies for human immunodeficiency virus (HIV) infections. Approval under this section, however, requires that the applicant study the drug further to verify and describe its clinical benefit, notably where there is uncertainty about the relation of the surrogate endpoint to clinical benefit, or of the observed clinical benefit to ultimate outcome (21 CFR 314 Subpart H and 21 CFR 601 Subpart E). Alternative endpoints have been proposed and validated for a variety of products and disease indications in efforts to make new therapies available as soon as possible. For example, in 1996, an accelerated approval program was established for cancer drugs, which allows the FDA to approve drugs based upon well-established surrogate endpoints that reasonably predict clinical benefit. Examples of this would include evidence of tumor shrinkage in solid tumor disease and meaningful remission in hematologic disease, rather than requiring demonstration of actual clinical benefit, such as improved survival or quality of life (OFFICE OF COMMUNICATION, TRAINING AND MANUFACTURERS ASSISTANCE).

D. Data Requirements

It is difficult to determine a priori the complete list of studies that will provide the essential data for all new products to be introduced and subsequently advanced in the clinical setting. This is especially true for novel therapies where there is little or no existing precedent. As with traditional therapies, there are different levels of regulatory control, based upon product class, patient population, disease state and available alternative therapies. Thus, the analysis of risk vs benefit often differs for each new product.

The critical parameters in developing and assessing novel biopharmaceutical therapies are similar to those of traditional biological products. These include demonstration of safety, purity, potency, identity and effectiveness (U.S. Government Printing Office; Food and Drug Administration 1990). Product-specific concerns and data requirements are addressed in regulations or product-specific guidelines, e.g., Points to Consider Documents (Office of Communication, Training and Manufacturers Assistance). In addition, an international body of experts representing both industry and the regulatory authorities, under the aegis of the International Conference on Harmonization of Technical Requirements for Pharmaceuticals (ICH), has developed guidelines in the areas of quality, safety and efficacy for pharmaceuticals and biotechnology-derived biopharmaceuticals that are accepted worldwide, by both industry and the regulatory authorities. These guidelines serve the purpose of supporting consistent testing and evaluation in order to streamline the approval process (International Conference on Harmonisation and related documents).

E. Manufacturing

Product development is a continuum which parallels clinical development. The primary goal of the manufacturing process is to produce the desired active product. Control of the manufacturing process is essential to ensuring a consistent final product (Siegel et al. 1995) The level of process control has been particularly important for development of traditional biological products and is also important for novel biological products.

The quality of biological products and the manufacturing process are assessed by several criteria, including purity, biological potency, stability and consistency of production. In addition to final product testing, validation of steps and equipment is needed throughout the manufacturing process (Stromberg et al. 1994).

The product should be sufficiently characterized, with well-defined specification, e.g., assessment of compatible components, adventitious agent testing, examination for extraneous materials and a knowledge of product stability, as early as possible. In order to provide an overall evaluation of the product, it is necessary to use a combination of physicochemical analyses, as no single assay can provide enough information to characterize a product. To date, biological assays, i.e., in vitro assays and/or use of one or more animal models, remain the only valid and reliable analytical tool to assess biological potency. As data accumulate, however, information requirements may change, e.g., for growth hormone and insulin, decades of research are available correlating biological activity and physicochemical analysis.

While early test material need not strictly comply to good manufacturing practice regulations (GMPs), adherence to GMP concepts insures minimum compliance for safety, purity and potency. The product intended for use in the clinical trials is expected to be in sufficient supply in order to use it in the appropriate preclinical studies. Since an understanding of exposure in relationship to dose administered is important, development of assays for measuring the product in the appropriate medium, e.g., tissue, plasma, cerebrospinal fluid, etc., should be assigned a high priority in early clinical development.

F. Preclinical Safety Testing

Timing is critically important to any clinical development plan, but especially for the economic fate of small companies involved in developing novel therapies. While there may be many reasons why development plans may be delayed, preclinical toxicology studies are generally viewed by most applicants as the "rate-limiting step" in clinical development. The challenge is to provide guidelines for testing programs designed to increase the predictive value of data for extrapolation to humans, therefore, decreasing the unnecessary use of animals and avoiding redundant testing.

The amount of preclinical safety testing necessary to support the clinical protocol for a biopharmaceutical is dependent on the stage of product development. Prior to initiation of the first human clinical study, the FDA requires demonstration through review of scientific literature, preclinical in vitro data, and/or in vivo animal data, that the product is safe at the proposed starting clinical dose and in the proposed study population. While the data needed to support initiation of phase-I trials focuses on safety, some demonstration of activity is expected, i.e., in support of a rationale, especially in those cases where the initial human studies are performed on patients.

While preclinical studies used to support safety should ideally conform to good laboratory practice (GLP) regulations (21 CFR 58), non-GLP studies may be acceptable, providing they are performed "in the spirit of GLPs" and statements are made as to specific aspects of non-compliance. Such studies are carefully scrutinized with respect to study design, including the incorporation of one or more internal controls. These studies are generally performed very early during the discovery and development phases for screening potential lead compounds or supporting initial proof-of-concept studies in humans.

G. Case-By-Case Approach

The case-by-case approach to preclinical safety assessment need not be more resource intensive than the more standardized approach used to assess conventional pharmaceuticals. However, to apply the case-by-case approach, scientists responsible for designing studies and those responsible for reviewing their adequacy must have a basic knowledge of the product in order to ask product-specific questions. In addition, they must have a common understanding of the limitations of the various models that are available to answer the questions and, more importantly, have a willingness to depart from past practices when appropriate. Although regulatory guidelines exist for sponsors to follow, if certain concerns outlined in specific guidelines are inapplicable or inappropriate for a particular product, then sponsors are encouraged to provide explanatory information and justify alternative approaches if necessary. With respect to preclinical safety assessment, the paradigm shift for evaluation of novel therapies is one of ritual to rationale. This may be a consequence of a shift in development of therapies in search of disease to deliberately developed therapies ("rational drug design").

H. Testing Goals

The goals of preclinical safety evaluation are to provide data to support initial studies in humans that will ensure a safe and uninterrupted clinical development. Early in clinical development, the data are used to support the recommendation of: 1) an initial safe starting dose and dose escalation scheme in humans, 2) the identification of potential target organs of toxicity, 3) the identification of appropriate clinical parameters to monitor in humans, and 4) the identification of individuals "at risk" for toxicity, i.e., refinement of patient inclusion/exclusion criteria (Cavagnaro 1993).

Preclinical studies should be designed to answer specific questions in order to assess the adequacy of the human study design and determine whether it is reasonably safe to proceed to the clinical trial. Questions to be considered include: 1) whether sufficient knowledge of the dose/toxicity relationship exists, 2) the dose/activity relationship, and 3) a basic understanding of the potential risks for toxicity. In general, preclinical findings which may modify, but not necessarily delay, clinical development plans include: 1) similar adverse findings in more than one species, 2) irreversible findings, 3) delayed effects, 4) evidence of toxicity not associated with pharmacological effects, 5) a nonlinear dose-response curve, 6) cross reactivity with non-target antigens, 7) transmission of infection/disease, and 8) enhanced toxicity/exacerbation(s) in animal models of disease (Mordenti et al. 1996).

I. Study Design

The design of preclinical safety studies should focus on the selection of the appropriate model, including the animal species and the physiological state, i.e., age, size, normal vs disease, etc., route of administration, dose selection and treatment regimen, as well as selection of the endpoint with consideration of both efficacy and safety.

For over 20 years, conventional pharmaceuticals have generally been screened for toxicity in one rodent species and one non-rodent species, often without prior determination of species relevance to the particular agent being evaluated. However, recently there has been an increased emphasis in the use of pharmacokinetic data for defining species relevance and better extrapolation of exposure.

The selection of a relevant animal species should take into consideration the biology and mechanism of action (if known) of the specific product, as well as the proposed route and method of drug delivery. Selection of appropriate animal model(s) for preclinical investigation should consist of the identification of a relevant species for the product being tested. Relevancy can be based on metabolic profile, pharmacodynamic effects, presence of the receptor of interest and/or affinity for the receptor of interest.

In some cases, the selection of an animal model of disease, e.g., spontaneous or induced, that is generally used to assess activity or potency may be a more relevant model to assess safety. While non-human primates have not been required for assessing the safety of novel therapies – especially those involving human-derived cells or proteins – their use as an animal model has been recommended when a relevant specific safety issue can only be assessed in this species. Alternative systems for preclinical evaluation of safety and toxicity may also be needed to answer specific questions. These include in vitro systems or in vivo animal models, e.g., transgenic animals.

Ideally, the dosing regimen should mimic the proposed clinical study design. In practice, toxicity studies are conducted for a least the same duration as the intended clinical duration. Toxicology studies up to 6 months in duration may serve to support chronic administration in humans. Accelerating a treatment schedule to obtain a greater than anticipated human dose may be appropriate in certain cases. However, real time scheduling may also be important, for example, in determination of immunologically mediated responses, where time is required for generating the desired or unwanted immune response, e.g., vaccines.

J. Defining Exposure

Pharmacokinetic analysis is essential to the proper design, interpretation, and extrapolation of data from pharmacological and toxicological studies. Both single- and multiple-dose pharmacokinetic studies should be conducted in the same species that is used for toxicity testing.

Preclinical studies should not only be designed to identify a safe dose (the no-observable-adverse-effect-level, or NOAEL), but also a toxic dose in order to define a therapeutic index in humans. The cumulative experimental dose should also be higher than the expected total human dose. If the study is designed to assess toxicity, a toxic effect should be achieved whenever possible.

For biopharmaceuticals, in which activity is mediated through a direct cytotoxic effect, e.g., monoclonal antibodies or proteins conjugated to toxins, drugs or radionuclides, the optimal effect is generally related to the maximum tolerated dose (MTD). For other biopharmaceuticals, e.g., immunomodulators, an optimal biological dose (OBD), the lowest tolerated dose that maximizes desired activity, may be more important to determine.

K. Product-Specific Concerns

As previously discussed, the design of a preclinical safety assessment program for biological therapeutics is reflected by a flexible, case-by-case approach in order to identify and address product-specific issues and/or concerns. For example, the safety evaluation of growth factors should consider species-specific effects; immunogenicity, and the tumorigenic potential, based on the known biological activity of these agents. The proposed clinical use of the product also impacts on the type of preclinical safety studies that are performed. For example, growth factors used as wound-healing agents should be assessed for systemic exposure (immunoassays and bioassays) as well as for their effect on the local wound environment, to include stability of the agent in the wound, penetration into the soft tissue/bone layers, receptor binding/downregulation, the wound type, and the wound location. For neurotropic growth factors, additional concerns may include neurotoxicity or the unwanted expression/sprouting of various neuronal cell types and neurotransmitters.

The safety evaluation of monoclonal antibodies should consider the potential cross-reactivity with non-target tissues, immunogenicity, prolonged effects, i.e., long half-lives, conjugate and/or linker toxicity and "bystander" effect (such as is observed with use of a radiolabeled monoclonal antibody).

The primary safety concerns for tumor vaccines have been local reactogenicity, cross-reactivity with non-target tissues, a bystander effect to normal tissue and induction of an aberrant immune response (as the expression of surface antigens on target cells may be altered, resulting in an autoimmune response).

Product-specific concerns, regarding antisense therapies, include the preclinical assessment of sequence-specific effects, cellular toxicity from interaction of antisense with normal nucleic acid synthesis and degradation (Black et al. 1993, 1994).

There are many safety concerns regarding the use of cellular and gene-therapy products, which are dependent, in part, on the type of vector being used. Many different vectors that are directly administered to humans are currently in development. The primary vectors consist of replication-defective, recombinant retroviruses, adenoviral vectors, adeno-associated vectors and plasmid DNA. Safety concerns associated with these vectors include:

1. Vector dose
2. Route of administration
3. Selection of an appropriate animal model
4. The potential for viral activity
5. The potential for viral vector replication
6. Localization of the vector; persistence of the transduced cells and their expressed products
7. Overexpression of the therapeutic gene
8. Inappropriate immune activation
9. The potential for transfer of the vector gene to germ cell lines, with subsequent alterations in the genome
10. Generation of replication-competent virus, with the potential for rescue by subsequent infection with wild-type virus
11. Viral shedding and environmental spread
12. Overexpression of the transduced gene
13. The potential to transduce surrounding normal tissue.

Additional safety issues for adenoviral and adeno-associated vectors include the effects of expression of the inserted gene in the liver, as well as determination of the total number of viral particles and the ability of the particles to infect target cells and deliver the vector (due to the potential cytotoxicity of the viral particle itself).

The aforementioned safety concerns for retroviral vectors can also be applied to plasmid-vector products. In addition, plasmid vectors that are administered with lipid preparations or other facilitating agents should be assessed for the amount and identity of these agents in the final product formulation.

In addition to product-specific concerns, there may additional concerns for potential toxicities, which are dependent on the intended clinical indication and on the patient population. These include developmental toxicity (fertility, teratogenicity, behavioral), immunotoxicity (immunostimulatory/immunosuppressive assessment, as an intended or unintended effect), carcinogenicity and safety pharmacology (evaluating specific systems such as the central nervous system, CNS, and cardiovascular system).

L. Accessibility of Preclinical Safety Data

Enhanced communication through information technology initiatives should enhance the ability to file, store, search, retrieve and merge information from different studies. Electronic submissions should not only improve the timely access of information, but should also facilitate the quality of regulatory submissions and decision-making and, thus, reduce the time and cost for developing new therapies.

An overall willingness to share and/or publish information from preclinical studies, especially in the area of novel therapies, has aided in facilitating not only their clinical development, but also the development of regulatory guidelines. Annual scientific workshops and conferences have also provided a useful means for scientific discussion and debate between the regulators and the IND sponsors in emerging fields.

M. Clinical Studies

I. Early Development

The initial goal in human trials is to determine safety and dosage, as well as assess the initial evaluation of "development-limiting" side effects. Unlike conventional therapies, the first study in humans for novel biopharmaceutical therapies has generally involved patients rather than normal volunteers. Phase-1 studies to assess safety include the first introduction of a product into humans, use of an experimental product in a new indication or use of an approved product in a new indication. Clinical pharmacology studies are also referred to as phase-1 studies, but are designed throughout clinical development to assess exposure.

There is often some distinction in the design of studies, dependent upon the sponsoring body, which can be a company or a single investigator, especially in relation to the numbers of patients enrolled and the number of study sites. More flexibility in protocol design is encouraged early in development, including changes in primary endpoint(s), e.g., during phase-2 studies, which can lead to useful strategies to demonstrate clinical safety and effectiveness.

II. Late Development

Later phase clinical studies, designed as well-controlled trials, provide the primary basis for determining whether there is "substantial evidence" to support claims of effectiveness. In some cases, novel biopharmaceutical therapies have been approved with only a single clinical trial.

One (phase 3) trial may be appropriate when there is a limited number of patients, evidence for clinical benefit is overwhelming or the nature of the disease is so serious that only a single trial is warranted. However, for non-life-threatening diseases, when the clinical effect is minimal, more than one clinical trial is generally needed. The trial may not have to be a large double-blind placebo-controlled trial. Other phase-2 trials can support a well-executed pivotal study. Unfortunately, often neither dose nor endpoints have been adequately determined from the phase-2 studies. Additionally, small numbers of patients may have been treated in phase-1 studies. The ability to monitor meaningful clinical response is crucial. Products with marginal to moderate benefit may not be determined in small samples of patients.The climate may not be conducive for a well-controlled trial when there are early warning signs in preclinical or phase-1 early/early phase-2 studies or there is the potential for major therapeutic advances in a serious illness and widespread unbalanced publicity of preliminary favorable results.

N. Summary

Government, industry, academia and the public are key components influencing the regulatory environment. Regulatory issues have evolved with novel technologies in the past and can be expected to do so as well in the future. As novel therapies are developed and existing therapies are improved, communication, management and transparency of clinical development strategies, and the regulatory review process should ensure the availability of safe and effective therapies and provide the framework for the next generation of emerging technologies.

To ensure the continued success in introducing novel therapies into the clinic and their subsequent demonstration of clinical safety and effectiveness, sponsors and regulators need to work together throughout all stages of development. More specifically, regulatory authorities need to anticipate safety concerns inherent in novel technologies, create a regulatory environment that encourages innovation, respond to early dialogue with developers, make the review process transparent and manage the review process. Developers need to continue to seek innovative strategies for assessing safety and efficacy, design studies to identify problems, initiate an early dialogue with regulatory authorities, understand the regulatory review process, and prepare and manage a product development plan complimentary to the managed review process.

References

Black LE, DeGeorge JJ, Cavagnaro JA, Jordan A, Ahn CH (1993) Regulatory considerations for evaluation the pharmacology and toxicology of antisense drugs. Antisense Research and Development 3:339–404

Black LE, Ahn CH, Farrelly JG, DeGeorge JJ, Taylor AS, Cavagnaro JA, Jordan A (1994) Regulatory considerations for oligonucleotide drugs: updated recommendations for conducting pharmacology and toxicology studies prior to clinical trials. Antisense Research and Development 4:299–301

Cavagnaro JA (1993) Preclinical safety assessment of biological products. In: Mathieu M (ed) Biological development: a regulatory overview, Paraxel International Corp., MA, pp 23–40

Food and Drug Administration, Washington, D.C., Office of the Federal Register National Archives and Records Administration (1990) Code of Federal Regulation. Title 21, 1990. Chapter 1

International Conference on Harmonisation and related documents. Internet address: http://www.pharmweb.met/pwmirror/pw9/ifpma/ich5.html
Mordenti J, Cavagnaro JA, Green JD (1996) Design of biological equivalence programs for therapeutic biotechnology products in clinical development: a perspective, pharmaceutical research, vol. 13, 1427–1437
Office of Communication, Training and Manufacturers Assistance. Points to Consider and related documents for evaluation and review of biological products. CBER, HFM-40, 1401 Rockville Pike, Rockville, MD, 20852–1448, through voice mail at 800–835–4707 or 301–827–1800; fax 888–223–7329 or internet address CBER_INFO@a1.cber.gov
Siegel JP, Gerrard T, Cavagnaro JA, Keegan P, Cohen RB, Zoon K (1995) Development of biological therapies for oncologic use. In: VT DeVita, S, Hellman SA, Rosenberg JB (eds) Biologic therapy of cancer. Lippincott, Philadelphia, pp 879–890
Stromberg K, Chapekar MS, Goldman BA, Chambers WA, Cavagnaro JA (1994) Regulatory concerns in the development of topical recombinant ophthalmic and cutaneous wound healing biologics. Wound Rep Reg S155–164
U.S. Government Printing Office, HHS Publication No (FDA) 82, 1051.Public Health Service Act. Biological Products. Section 262 and 263

CHAPTER 2

Preparation of Clinical Trial Supplies of Biopharmaceuticals

A.S. LUBINIECKI, J.C. ERICKSON, C.L. NOLAN, R.G. SCOTT, P.J. SHADLE, T.M. SMITH, and D.W. ZABRISKIE

A. Introduction

I. Research Support Systems

For most monoclonal antibodies (MAbs), preparation of research supplies is relatively straightforward. Cell culture is done by brute force, since little time is available to optimize product gene expression or cell growth. Systems that are often employed include: 1) suspension culture in spinner vessels, roller bottles or other containers, 2) ascites production in mice, and 3) hollow-fiber bioreactors.

Purification of MAbs is typically accomplished by a series of chromatographic steps, e.g., by affinity chromatography on Protein A, and usually augmented by ion exchange chromatography (IEC). If serum is employed to grow the expression line, depletion of serum antibodies should be considered by passage of the serum through Protein A columns prior to use in cell culture medium; otherwise up to 50% of the antibody subsequently purified from the culture harvest may be unwanted bovine antibody.

The quality standards are generally imposed by the rigor and conscience of the individual investigators, the culture of the organization and good scientific principles. Economics is not usually a factor, however, speed is essential. The process may or may not need to be scaleable, and regulatory compliance is not at issue. If the material is to be used in animal studies, then care usually needs to be taken to minimize endotoxin contamination (this is discussed more fully in the next section).

B. Preclinical Studies

Initial assessment of safety in animals, formulation development, and evaluation of pharmacokinetics and/or pharmacodynamics in animal models requires materials that closely resemble the materials intended to be taken into the human clinical trials. The product should come from a representative manufacturing process; this typically means that materials are representative of those generated by a very similar method, e.g., same cell line, same medium and same chromatographic resins, as that used for the clinical supply. Minor differences in process parameters, such as cell growth, setpoints, buffers or resin load ratios, may occur as the process is optimized. Thus, the preclinical process is representative of, but not necessarily identical to, the clinical process in all major aspects, but some minor changes typically occur.

For a typical biopharmaceutical, the purity of the material need not be quite as high as the clinical materials. The preclinical materials should have a qualitatively similar impurity profile, but may contain more quantitatively, compared with the clinical materials, since this might lead to an overestimation of the safety of the new biopharmaceutical for subsequent human use. The preclinical materials need not be formulated in final excipients if the excipients are generally regarded as safe for the intended indication. Bulk supplies (without excipients, without final containers) may be employed for initial preclinical studies, although the more similar they are to the actual clinical drug product, the greater the confidence of applicability to the dosage form intended for use in man.

One further quality requirement is that the materials be substantially free of endotoxin/pyrogens to avoid provoking a pyrogenic response in the animals and activating their macrophages. Most endotoxins/pyrogens are derived from the cell walls of gram-negative bacteria or mycoplasma, although a few recombinant DNA (rDNA) biopharmaceuticals are inherently pyrogenic, such as alpha interferons (IFNs) and interleukin 1. The exact amount of endotoxin permitted will vary according to the age and species of animal. Meeting standards for human parenteral administration (<5 endotoxin units/kg body mass/8 h of infusion) is typically adequate for most animal uses (Food and Drug Administration 1987).

Another criterion for supplies destined for preclinical safety studies is that they be adequately characterized in terms of biophysical, biochemical and biological properties. For example, one should understand the biophysical and biochemical properties and be able to demonstrate that the materials are representative of those that will eventually be used in humans. The biological activity should be comparable enough to the preparations destined for use in man to allow the animal data to be deemed relevant in establishing the safety of the materials for human use.

Supplies intended for such studies need not be made under good manufacturing practice regulations (GMPs); however, it is common practice in many organizations to have the materials made under somewhat controlled conditions. For example, the materials, process and methods are traceable and documented, frequently with draft GMP documents. Alternatively, preparation can be fully documented in notebooks if the details are captured in sufficient detail to enable the preparation of preclinical supplies to be reproduced, and to support future claims for these processes and materials being representative, equivalent or comparable to clinical processes and materials. Typically, research materials are not used for preclinical purposes, although some pilot studies in animals (intended for dose ranging, not for establishing preclinical safety under good laboratory practices) may use research materials. Similarly, initial studies of formulation development may use research materials, with the risk that the results may not be indicative of the properties of subsequent clinical materials, especially if process changes and purity changes occur.

C. Clinical Supplies

I. Fermentation

Clinical production systems ideally reflect what will ultimately be used for commercial production; therefore, they will be discussed together and differences pointed out where relevant. A variety of fermentation systems have been successfully employed for the commercial preparation of MAbs (Table 1). In addition to these systems, several others have been used to prepare clinical supplies (Table 2). Clearly, this diversity shows that there is no requirement to use one system or another – only that the system consistently provides a product that meets predetermined quality attributes.

Small commercial markets for MAbs, such as those for in vitro diagnostic uses and for purification of other biological products, can be generally met by small-scale systems, such as ascites and hollow-fiber bioreactors. However, making substantial quantities (hundreds of grams to a few kilograms) is tedious in such systems, although still physically possible to do. Due to an inherent lack of scalability, producing more material using these systems requires more physical plant and labor, with relatively little economy of scale. Both ascites and hollow-fiber systems are also associated with quality limitations. Ascites production utilizes animals, which is increasingly less acceptable for manufacturing use for a variety of ethical and regulatory reasons, and also carries risk of animal-associated adventitious agents. Hollow-fiber technology was practiced by several firms, which are now defunct, including BioResponse and Endotronics (although the hollow-fiber technology of the latter survived the disappearance of its inventor). Hollow-fiber systems are non-homogeneous, which complicates scale-up, robustness and consistency from harvest to harvest. Several proteins produced in this way have shown batch-to-batch and

Table 1. Fermentation systems used for the commercial preparation of monoclonal antibodies (MAbs)

Product	Use	System
OKT3	Graft vs Host	Ascites
Panorex	Colon cancer	Perfused suspension culture
ReoPro	Restenosis prevention	Perfused suspension culture
Oncoscint O/V	Imaging ovarian cancer	Airlift suspension culture
Myoscint	Imaging	Hollow-fiber bioreactors
Anti-IFNalpha	Purification of IFNalpha	Microencapsulation
Verluma	Imaging	Batch suspension culture

IFN, interferon

Table 2. Other fermentation systems used to prepare clinical supplies for monoclonal antibodies (MAbs)

MAb	Use	System
Anti-hepatitis B_S Ag	Chronic hepatitis	Fluidized-bed reactor
Anti-CD20	B cell lymphoma	(Fed) batch suspension culture
Anti-RSV F protein	Prevention of RS	Batch suspension culture

within-batch consistency problems, including variable glycosylation and proteolysis. Ascites technology and hollow-fiber technologies are favored by some, especially university-based clinical investigators and small startup companies, because they are simple and require very little development work.

Several other systems were employed in the past for clinical and marketed MAbs, but are no longer available. These include the microencapsulation technology of Damon Biotech, the perfused ceramic cartridge of Corning, the fluidized bed technology of Verax and a variety of perfusion technologies originating at Monsanto, which were licensed to Invitron. Because their technologies were not cost-effective, all of these business units or organizations became defunct and their technologies disappeared along with them. In addition to lacking cost effectiveness, these methods also had major process-development impediments in common, such as non-homogeneity, lack of faithful scale-down/scale-up systems or poor robustness. These systems were typically difficult to validate. They remain interesting footnotes to biotechnology history, but are not considered further because of their inaccessibility.

Today, most clinical development and commercial production done by industry utilizes some form of suspension culture. Whether batch, fed-batch or perfused, or whether done in airlift or mechanically agitated vessels, suspension culture is what most developers and contract manufacturers have ultimately chosen. This method has proven to be the most cost-effective method for development and production. Simpler versions of this technology also are amenable to science-based process-development studies, as they have excellent scale-up/scale-down properties and well-understood homogeneity. These features are also usually associated with consistent batch-to-batch results and relatively simple validation efforts. More complex processes can be developed, but they take longer to reduce to practice. In general, the simple suspension culture systems are very robust and reliable, and the complex ones are a bit less so. Validation of processes for perfused reactors intended for long-term culture (months) can also be quite complicated, and product quality may be variable in some cases.

The period 1970–1990 was characterized by extensive investigation of fermentor design by industrial and academic organizations that attempted to improve and perfect various parameters for cell-culture optimization. This work provided a relatively modest gain over simple working models of suspension culture. Hydrodynamic shear was considered to be a major engineering problem for mammalian cell fermentations, but where it existed, it was rather easily solved by use of bulking agents to reduce cell damage. Better mechanical agitator seals eliminated the old problem of mechanical damage to cells by rotating bottom-drive agitators. Better seals and cleanable sanitary design provided improved robustness, lower contamination rates and lower endotoxin burden. Major gains in productivity came from medium development (just as for microbial fermentations), by increasing biomass and extending cell viability in time. Removal or reduction of serum, plus increased productivity, led to improved cost effectiveness.

II. Harvest

MAbs are generally relatively resistant to proteolysis, therefore, fermentation cycles in general can take quite long to maximize the amount of antibody produced. MAb titer typically continues to rise for a while after cell growth has ceased and viability begins to drop. Care must be taken to understand the structure–function relationships of the MAb to assure that the fermentation conditions and harvest time are chosen to maximize the content of recoverable molecules of acceptable quality. For most batch/fed-batch fermentations, this is generally a matter of 1 week or more; because perfused reactors run for extended periods, conditioned medium is collected (usually daily) and harvested repeatedly every few days for purification for a period of many weeks or even months. Glycovariants, proteolyzed product or deamidation products may accumulate to varying levels if the culture is continued for an excessive period.

Once the appropriate timing is determined for a given process, harvest can be accomplished by a variety of methods, usually centrifugation or filtration. Small volumes of research materials can be batch centrifuged in large-volume bucket rotors; larger scale volumes can utilize continuous-flow centrifuges. Filters of various types can be employed, including dead-end or depth-type cartridge filters or tangential-flow microporous filters. Filtration aids like diatomaceous earth are unpopular, due to reproducibility difficulties and regulatory concerns. Sedimentation at unit gravity is simple, effective and cost effective (just turn off the agitation system), but usually requires about 1 day to reach completion. Most organizations need to clean and resterilize production vessels quickly, and prefer filtration methods that require only a few hours to complete. Filtration methods tend to be simpler to develop than centrifugation methods and scale up very well. Filtration methods may also tolerate inconsistency in the feedstock (due to variable viability) better than centrifugation methods. Alternatively, centrifugation may be preferred for cultures containing high levels of debris, which sometimes can foul even the best-designed tangential-flow filtration system.

Harvest of perfused fermentors is slightly different, and a variety of approaches are used to separate the cells from the conditioned medium, either inside or outside the fermentor. These include the use of tangential-flow filters, the use of fluidized-bed bioreactors (where the cells growing attached to particles are retained in the reactor by gravity), inclined settling planes inside contained vessels and other ingenious pieces of engineering. Conditioned medium may be recycled into the fermentor after reoxygenation or, more typically, simply collected after one pass and stored for a few days (often refrigerated) before purification.

III. Purification

The goal of the purification process is to remove contaminants and impurities without compromising the structural or functional integrity of the product. In clinical development, the process must also meet regulatory criteria for the quality of its raw materials, the degree of consistency and control over the purification method, and the quality control (QC) analysis of in-process and bulk-product fractions. These concerns impose new limits on the selection of separations methods, compared with research methods. Furthermore, since the manufacturing process will need to be economically feasible, optimizing around yield is a new target that is in direct conflict with the needs to quickly move a product through development and into clinical trials.

Chromatography has been, and remains, the powerhouse means by which highly purified biopharmaceuticals are produced. Classical chromatographic methods, as in the chemical industry, exploit differences in physical properties, such as charge, size, hydrophobicity, and the like to purify product away from contaminants and impurities. Affinity methods use an immobilized ligand for which the product protein has an unusually high relative affinity, and may be related to a functional group on the product or class of products. Typically, a purification process consists of two or more chromatography steps, with buffer exchange, precipitation or concentration steps in between to link them together.

Classical methods, for example, IEC, are widely used and, in general, considered robust, relatively inexpensive and somewhat limited in selectivity. Improvements in manufacturing technology that are important for use in the GMP arena include new and improved selectivities, fast flowing bead properties, enhanced capacity, pH resistance and improved cleanability, which improves column life. The availability of different resin backbones has also led to observations that some proteins fractionate differently and that these properties, as well as the functional ligand, can be exploited.

Affinity-based methods are highly selective, often give a high yield and purification factor, and are generally high-capacity columns. Balanced against these apparent advantages may be disadvantages stemming from leaching of the affinity ligand from the resin, regulatory concerns that may be associated with the ligand, depending on its source, possible degradation of product on the affinity column [for example, immobilized metal ion affinity chromatography (IMAC) may oxidize products] and high costs of the resin. Improvements in manufacturing technology that have lifted affinity resins from the research lab into the GMP arena include lowered costs, fast flowing bead properties, enhanced capacity and improved cleanability, which improves column life. The ability to use many affinity columns for hundreds of batches has significantly lowered the cost barrier that once existed for these resins.

Size-exclusion chromatography (SEC) has become an infrequently used purification method, although still useful for buffer exchanges. SEC has several factors making it difficult to scale up beyond a critical point, among which are the need for a long and well-packed column bed (which may not be obtainable at large scale), a dependence upon wall effects that can impact on scalability and relatively slow flow rates. Together, these make SEC a method of last choice, as it is costly and does not offer as much purification power as other methods.

Improvements in cell culture and fermentation have directly impacted purification processes. Increases in expression levels, concomitant with the development of protein-free media, have permitted the use of affinity columns in capture steps, which was once not practical or feasible. Modern cell-culture methods can deliver MAb titers approaching 1 g/l, which is 10–100 times the concentration achieved in the past. The elimination of many of the animal-derived raw materials also reduces the burden of the purification process to remove pathogenic agents, which may have been introduced by those animal-derived materials. These improvements in cell culture have made the purification job easier.

D. Purification of rDNA-Derived Anti-RSV MAb

Purification of MAbs is relatively routine in the current era. Many organizations utilize Protein A-based affinity resins. Modern resins are much improved over initial versions in terms of cleanability, useful life, compression resistance and Protein A leakage rate. These may be operated as stationary packed bed resins or as fluidized bed resins in the bind-elute mode. Ion-exchange methods can be used for initial capture of MAbs or to further purify MAbs captured initially on affinity resins. Protein A-based affinity resins are advantageous, in that a single protocol can be utilized to capture many different MAbs with little modification, but ion exchange methods must be tailored to each MAb. Furthermore, ion-exchange protocols must be developed empirically, and frequently bear little relationship to the ionic properties predicted solely by examination of the amino-acid composition of the heavy and light chains, e.g., isoelectric point. This probably reflects the molecular complexity of MAbs, including factors such as local-charge distribution on the molecular surface (Table 3).

A fundamental principle in developing a purification process is to affect separation of the desired product-related molecules from the undesirable molecules, whether product related (impurities) or unrelated (contaminants).

Table 3. Purification summary for anti-RSV monoclonal antibody (Mab)

Step	Vol (l)	Mab (g)	Protein (g)	Yield (%)	Prot A[a]	IgG multime r[b]	Purity[c]
Harvest	416	392	477	100	0	ND[d]	ND
ProSep A	49	375	384	96	11.7	0.4	98.5
CM	21	116	117	93	ND	0.4	98.4
HIC	29	98	100	85	ND	<0.05	98.3
BBS	9	90	91	92	1.7	0.06	99.7
Overall	70						

IgG, immunoglobulin G; SDS-PAGE, sodium dodecyl sulfate polyacrylamide gel electrophoresis; HIC, hydrophobic interaction chromatography; BBS, bulk biological substance; ND, not done
[a] Protein A concentration in ng/mg MAb
[b] IgG multimer concentration as a percentage of total area scanned by reduced SDS-PAGE
[c] Purity as percentage of total area scanned in heavy and light chains by reduced SDS-PAGE
[d] Not determined

Among the undesirable molecules are medium components, host cell components, pharmacologically active or toxic molecules introduced during purification, and MAb-related variants, which are inactive or might lead to adverse reactions in vivo. Low molecular-weight molecules from medium or purification procedures usually can be effectively removed by ultrafiltration/diafiltration methods. Carefully designed purification steps based on orthogonal separation principles will usually remove proteins that are unrelated to the desired MAb. However, it is often difficult to resolve desired MAbs from their undesirable relatives, such as Mab–MAb oligomers or aggregates and protein A–MAb complexes, due to the similarity of their biochemical properties. Separations of such large molecules from MAbs would be difficult to achieve by size, charge or solubility-based methods. When baseline separation cannot be obtained, the collection of fractions and pooling of those meeting specified criteria is sometimes used to compensate for a lack of complete control or predictability of the process. Such a process may entail high losses in terms of yield, and suffer from lack of robustness and consistency. These often translate to regulatory concern over pooling criteria and result in higher QC costs.

To approach this complex problem, we have designed a three-step purification procedure to specifically reduce the levels of two product-related impurities of concern: MAb–MAb oligomers as well as Mab–protein A complexes. This process features hydrophobic interaction chromatography (HIC) following protein A chromatography and IEC to separate intact monomeric MAbs from MAb–MAb oligomers and Mab–protein A complexes, resulting in a pharmaceutically acceptable drug substance of at least 95% protein purity on sodium dodecyl sulfate polyacrylamide gel electrophoresis (SDS-PAGE). We have successfully applied HIC as a final step to purification of multiple rDNA-derived MAbs encompassing a variety of heavy- and light-chain combinations as described below, which were produced in serum-free cell culture. It is likely that this method would perform well on MAbs sourced from ascites or from serum-containing cell culture. Among the HIC ligands that we have employed are alkyl groups of various lengths and aryl groups, which can be immobilized to a number of inert carbohydrate or synthetic polymer supports. The method and its performance is outlined below; additional details can be found in SHADLE et al. (1995).

An expression system for rDNA-derived humanized MAb to respiratory syncytial virus (RSV) F protein was established in Chinese Hamster Ovary (CHO) cells as described in Chap. 6 by PORTER et al. MAb was harvested from cells grown in suspension culture by passing the culture through a tangential-flow filter of nominal 0.65-μm porosity at scales up to 416 l/batch. At small scale, batch centrifugation can also be employed to remove cells.

MAb was captured on an immobilized protein-A column, washed with phosphate-buffered saline (PBS) containing 0.1 M glycine, and eluted with 25 mM sodium citrate, pH 3.5. Clarified conditioned medium was applied to Prosep A (BioProcessing Ltd) at a flow rate of up to 1000 cm/h at a load ratio of up to 15 g immunoglobulin G (IgG)/l column volume. Multiple capture cycles were employed at large scale to minimize the investment in expensive affinity resin, while maintaining operating costs of multiple cycles at a practical level. Protein-A chromatography removed substantial amounts of DNA and cellular proteins and also concentrated the MAb by about tenfold. The eluates from all the protein-A cycles were adjusted to pH 3.5 with 2.5 M hydrochloric acid (HCl), held for approximately 30 min, the pH was adjusted to 5.5 with 1 M Tris base, and then all the eluates were pooled. The pH 3.5 hold step provided conditions that inactivated retrovirus activity, to mitigate concerns regarding putative retrovirus-like particles associated with CHO cells. The pH 5.5 adjustment prepared the MAb for IEC. This material was passed through 0.2-μm filters into a sterilized receiving vessel and stored at 4°C or frozen at 70°C until further processing resumed.

IEC was then employed to further remove protein and non-protein impurities. MAb was applied to CM Sepharose (Pharmacia) at a flow rate of 150 cm/h at a load ratio of $<$20 g IgG/l column volume. The column was washed with 10 mM sodium citrate, pH 5, and eluted with 40 mM sodium citrate, pH 6, containing 100 mM NaCl. This eluate was adjusted to 2 M guanidine hydrochloride (GuHCl) by adding 6 M stock solution to provide further opportunity for retrovirus inactivation. The GuHCl also helped solubilize the MAb for subsequent addition of ammonium sulfate. This solution was then prepared for HIC by mixture with an equal volume of 2.6 M ammonium sulfate.

The resulting solution was applied to an HIC column consisting of Toyopearl Phenyl-650 M (TosoHaas) at a flow rate of 150 cm/h and at a load ratio of $<$20 g/l resin. The column was washed with 3–5 volumes of equilibration buffer (1.3 M ammonium sulfate, 50 mM sodium phosphate at pH 7). A 20-column volume linear gradient of decreasing ammonium-sulfate concentration was then applied at a flow rate of 100–150 cm/h to elute the MAb as a major peak. Impurities eluted later in the gradient. This HIC procedure removed additional protein and non-protein impurities and contaminants, notably residual protein A, IgG multimeric forms, and host-cell DNA. The appropriate gradient fractions were pooled and concentrated by tangential-flow ultrafiltration to the desired concentration, diafiltered with a suitable purification buffer, and passed through a 0.2-μm filter into sterilized receiving vessels to become the bulk biological substance (BBS).

This purification process was easily scaled, and has provided equivalent results at 1 g, 40 g, and 125 g scales of operation. The process provided about sevenfold clearance of IgG multimeric forms and Protein A, and about one million-fold clearance of DNA, while still permitting 70% overall recovery yield. The process also provided a substantial amount of virus inactivation and removal capacity, which meets or exceeds the current standards of US and international regulatory bodies (Center for Biologics Evaluation and Review 1997; Committee on Proprietary Medical Products 1991; International Conference on Harmonization 1997). The process also worked well with other unrelated rDNA-derived MAbs with minor changes.

E. Product Quality Issues

I. Protein Purity

Protein purity is usually measured by SDS-PAGE, followed by densitometry of the gel. Purity by this method should exceed 95%. Product-related proteins are detected by Western blotting, isoelectric focusing (IEF), and high performance liquid chromatography (HPLC) methods, and are of less concern than any residual host cell and medium proteins. Host cell and medium proteins are also detected using Western blotting and enzyme-linked immunosorbent assay (ELISA) techniques and must be minimized due to concerns over their possible immunogenicity. Proteins used for affinity chromatography, e.g., protein A, or a monoclonal antibody, must also be removed from the product.

II. Protein Integrity

In addition to making sure that the biopharmaceutical product is substantially free of contaminating proteins, it must possess the correct primary, secondary and tertiary structure, so that activity and pharmacokinetics are consistent. Intermediates in the purification process must be monitored for degradation early in development so that conditions that harm the protein are avoided. This problem is most evident in steps where it is necessary to choose conditions that inactivate the most spiked model virus while leaving the product intact. Assays that are typically employed to verify integrity include N- and C-termini sequencing, peptide mapping following controlled proteolysis, isoelectric focusing, carbohydrate analysis, antigenic binding and potency. Special assays of Fc region function may be important for assuring the integrity and potency of some MAbs.

III. Microbial and Viral Safety

Some of the most difficult potential contaminants of biopharmaceuticals are living organisms. Since proteins are relatively labile and made from living cells, there is a possibility that viruses growing in the cells, or introduced from the media or workers may end up in the final product. To complicate matters, many production cell lines produce particles that look like viruses, so the product may look as though it is contaminated, when it is not. Previous generations of biological products, e.g., viral vaccines and plasma derivatives, have been demonstrated to be contaminated with microbes or viruses in some cases, which sometimes caused infections in product recipients (Horaud 1991). So far, this has not occurred with biopharmaceuticals, such as MAbs and rDNA proteins, probably because of better understanding of this history and better safety assurance (Lubiniecki 1998).

Safety is assured by a combination of appropriate assays to detect microbes and viruses, validation of well-designed processes to remove or inactivate them if they were present, and procedural and engineering controls on facilities and equipment. Cells and raw materials are tested extensively for possible viruses and contaminated cells or raw materials are discarded. Workers wear protective clothing. One or two special virus-inactivation steps are put in the process. Finally, each purification step is validated for removal of spiked model viruses.

Similar precautions are taken to ensure that bacteria do not contaminate biopharmaceuticals. Filters capable of removing bacteria and fungi are used throughout the purification process, and a testing program is established to confirm that adventitious agents have not contaminated the process or product.

IV. Other Contaminants

All cells contain DNA; therefore, there is a possibility that DNA may be present in a biopharmaceutical at low levels after purification. DNA is not known to be toxic, but some people have hypothesized that DNA could become functional in a product recipient, perhaps encoding for virus or an oncogene. Therefore, DNA must be reduced typically to less than 100 picograms per dose, based on a World Health Organization risk assessment study (WHO Study Group 1987).

Process chemicals that are added at various points during the process must also be removed. Particularly important are additives that are known to have pharmacological activity or toxicity. Examples include insulin (often used in cell culture medium), methotrexate (often used to amplify transfected plasmids), and chemicals used in purification (often used to inactivate viruses).

F. Process Design and Validation

Validation is a critical part of ensuring product quality. Process design is frequently driven by validation needs. This is true for MAbs and all rDNA-derived proteins. These concepts have also evolved considerably over the past decade. Our first cell culture process for soluble T4 (sT4) was under designed in several areas and validation was one of these. An improved process was created a few years later with an affinity step that significantly increased clearance of DNA and model retrovirus. Processes developed later for a therapeutic rDNA-derived protein and MAb were also designed to provide higher clearance levels. These are exemplified in Table 4.

I. Validation of Endotoxin Removal

Endotoxin in a biological product must be controlled to less than 5 EU/kg body weight per dose of product per 8 h of treatment (FOOD AND DRUG ADMINISTRATION 1987). Many therapeutic proteins are given at relatively high protein doses, so endotoxin must be controlled to very low levels. Assuming an average person's body weight of 70 kg and a protein dose of 100 mg, the endotoxin limit would be 0.35 EU/mg protein. For a 500 mg dose, the limit would be 0.07 EU/mg.

Endotoxin control for proteins prepared in cultured animal cells is usually a matter of excluding endotoxin from raw materials and maintaining effective cleaning procedures. However, for proteins expressed in E. coli, endotoxin is part of the host organism, and typically constitutes about 4% by weight of the harvest. Aggressive endotoxin removal is an important part of purification processing for proteins expressed in E. coli.

Endotoxin can be difficult to remove from a protein because it is not a single homogeneous species. Endotoxins are heterogeneous in their biophysical, biochemical and affinity properties. In addition, endotoxins are fairly large and may aggregate, so that they are usually concentrated along with the

Table 4. $\log_{10}$-fold removal of potential contaminants

Process	DNA	Endotoxin	Retrovirus	Poliovirus
sT4 (Process I)	6.3	ND	>11.6	>11.3
sT4 (Process II)	8.6	ND	>19.2	ND
Therapeutic	9.2	2.2	>28.8	8.9
Mab	9.5	1.0	>15.8	4.7

ND, not determined

protein product during ultrafiltration. Diafiltration is a step where endotoxin at very low levels in the diafiltration buffer can actually be concentrated to high levels in the product. In contrast, many chromatographic steps, e.g., IEC, HIC, RP-HPLC, can reduce endotoxin concentration. Because the presence of endotoxin may suggest a loss of process control or a reduction of quality level, it is very important to make sure that endotoxin is not added inadvertently at any step in the purification process. The first line of defense is to make sure that raw materials (including water) are controlled and tested to minimize the introduction of endotoxin. The use of water for injection (WFI) and ultrafiltration of buffers helps greatly in this area.

The second line of defense is housekeeping, which can minimize both the number and growth of microbes that make endotoxin. Procedures that control the flow of personnel, equipment and materials in and out of the area can reduce the importation of endotoxin. General cleanliness can reduce the opportunity for microbes to grow and spread in the area. Regular cleaning with a series of effective agents can reduce the number of microbes brought in, even with the strictest of procedures. Finally, storing equipment in a clean and dry state limits opportunities for microbial growth.

The third line of defense is to have robust column and equipment cleaning. Good cleaning comes from having equipment and chromatography media which are completely accessible to the cleaning agent, cleaning agents which effectively kill microbes and remove endotoxin and, finally, cleaning procedures that allow sufficient contact time and temperatures to allow the cleaning agents to do their jobs. Cleaning validation studies can be very helpful in identifying the source of recurrent endotoxin contamination, bioburden or other problems.

Modern chromatography media are capable of consistently providing proteins with negligible endotoxin when used in combination with other control procedures mentioned earlier. Table 5 shows some practical experience with endotoxin levels in recombinant proteins at Smith Kline Beecham Pharmaceuticals. In the early purification process for a vaccine, the product had 2.4 EU/

Table 5. Endotoxin levels in various bulk substances

Process	EU/mg[a]	*n*
Vaccine		
Process I	2.4	4
Process II	<0.18	1
sT4		
Process I	0.31	5
Process II	0.002	1
Thrombolytic	<0.02	7
Therapeutic	0.015	6
Mab	<0.01	

[a] Geometric mean of *N* runs

mg. An improved process provided product with less than 0.18 EU/mg, an improvement of more than an order of magnitude. Similarly, the early process for making sT4 produced a product with 0.31 EU/mg, but an improved process provided 0.002 EU/mg, which was more than two orders of magnitude better. Processes developed since that time have less than 0.01–0.02 EU/mg.

II. Process Validation of Model Virus Clearance

As with endotoxin, high protein doses drive the necessity to demonstrate large clearance factors of model viruses for recombinant proteins made in mammalian cell culture. CHO and other cell lines used to manufacture recombinant proteins contain virus-like particles (VLPs), which do not contain reverse transcriptase and are not infectious, but look like retrovirus particles in transmission electron microscopy (LUBINIECKI and MAY 1985). Although no biological activity has been ascribed to these VLPs, the validation exercise treats them as though they were infectious as a worst-case assumption. There is no generally accepted rule regarding how much virus clearance validation is enough; we have been uniformly successful to date employing the concept that validation must assure that there is less than 10^{-6} particles in a dose. Others have employed or suggested lower standards, ranging between 10^{-3} particles to 10^{-5} particles per dose, without obvious regulatory problems or problems in the field (CENTER FOR BIOLOGICS EVALUATION AND RESEARCH 1994). The concept of 10^{-6} particles per dose as a standard for acceptance is borrowed from validation of microbial sterility by terminal sterilization practices (LUBINIECKI et al. 1996).

Clearance (removal or inactivation) of retrovirus is a difficult task because the actual VLPs cannot be practically measured in the product. This has led to the use of model viruses to validate clearance. Live viruses are spiked into process feed streams and put through scaled-down columns. The eluates are analyzed for viral infectivity and clearance is calculated as the ratio of final level in the eluate to initial level in the feedstock, corrected for volume changes (INTERNATIONAL CONFERENCE ON HARMONIZATION 1997). It would be impractical to do these experiments at full scale for safety reasons and because of the unavailability of virus stocks in large quantities. Therefore, the clearance studies must be done on a scaled-down model of the chromatography column, outside the facility.

These validation experiments are very expensive and time consuming, therefore, process-development scientists cannot screen chromatography media and conditions to optimize virus removal. Generally, there is only enough time and money to test the final conditions and determine virus removal. Therefore, process-development scientists have to make use of historical data to assure that their processes will remove viruses. Unfortunately, virus-removal results vary with the details of a particular operation. Nevertheless, we are able to remove retroviruses to acceptable levels.

Table 6. Model retrovirus clearance

Chromatography mode	Range of $\log_{10}$ removal[a]
Cation Exchange	1–2
Anion Exchange	2–6
Hydrophobic Interaction	1–5
Affinity	3–4
SEC	1–3

SEC, size exclusion chromatography
[a] Murine leukemia virus was spiked into process intermediates; $\log_{10}$ removal refers to the ratio of detectable infectivity of spiked fluids before and after chromatography, corrected for volume changes due to chromatography

Table 6 shows the range of virus removal achieved in our experience with several different recombinant proteins. All the steps removed at least one $\log_{10}$ of virus, but different processes using the same mode of chromatography had results differing by up to four orders of magnitude. This shows that relatively minor differences in process conditions can result in significantly different clearance levels. However, repeated evaluations of the same chromatographic protocol have shown very consistent viral-clearance results (McAllister et al. 1996).

High levels of retroviral clearance are assured by making sure that any virus that might be left on a column is killed by the cleaning procedure and by having adequate controls to make sure that in-process material late in the process does not come into contact with material from earlier in the process.

G. Process Economics and the Future of Chromatography

Our view of biopharmaceutical development, particularly for chromatography, includes process automation and generic purification methods for selected classes of proteins. Automation of chromatography processes has the potential for increased monitoring and control of a process. Generic purification methods would be useful to speed up process development and reduce QC costs. As the topic of automation is complex and requires many years of activity to complete, it is often best to identify this possibility during development in order to have everything in place when it is truly needed. Accordingly, a brief description of the approach is included below.

I. Process Automation and Control

Automated systems are typically more complex and require more time to develop than manual systems. Since most biopharmaceuticals are developed as rapidly as possible, automation is not practical for products in early development. There is usually barely enough time to develop a manual process. In addition, information needed to automate a process is often not obtained until it has been run at full scale.

The development and capital costs of automation are other reasons why processes are not automated. We will present an analysis of total costs and show when it makes sense to automate a process, as the perceived cost is usually different from the actual cost. Total capital investment is increased by the computer, instrumentation and added validation, but is reduced by the need for fewer hold vessels, less space and increased capacity due to faster processing. Operating costs are lower with automation because fewer operators are required, cycle times are shorter and the success rate is higher. Cost to develop, validate and troubleshoot an automated process must also be considered. We estimate that it costs about $15 MM to develop an automated process.

The final analysis must take into consideration the projected cost of goods for the protein and the company's expectations for return on the investment to develop an automated process.

Table 7 lists recombinant proteins marketed in 1993. All of them, with the exception of tissue plasminogen activator (tPA), are given at low doses. By making some assumptions about cost of goods (COGS) as a fraction of sales, assuming that a certain fraction of the COGS comes from purification, and assuming a certain lifetime for products, the present value of all future purification costs can be calculated. This is shown in Table 8. If the present value of the purification costs is less than $15 MM (the cost to develop the automation), then automation would never be economical, even if the automated process

Table 7. Estimated sales for recombinant proteins

Product	Dose (mg)	Doses per course	Estimated 1992 sales ($MM)
Cytokines			
IL-2	2.6	28	25
INF-α	0.01	50–100	1200
GM-CSF	0.5	1442	53
G-CSF	0.28	40–160	544
IFN-γ	0.1	156	3
Hormones			
hGH	1–2[a]	156	800
EPO	0.02–0.04	156	1000
Insulin	0.1–0.5	1095	600
Mabs			
OKT3	5	10–14	50
Therapeutics			
tPA	100	1	278

IL, interleukin; INF, interferon; tPA, tissue plasminogen activator
[a] For 20 kg child

could purify the protein for free. If the present value of purification costs is higher than \$15 MM, automation might be economical, depending on how much was saved in cost of goods.

Figure 1 shows how much money could be saved (present value of purification savings) as a function of the percentage reduction in COGS brought about by automation. The horizontal line at \$15 MM is the break-even point. As was noted before, OKT3 MAb would never be economical to automate. On the other extreme, human growth hormone (hGH), has a very high cost of

Table 8. Purification costs for various biopharmaceuticals

Drug	Annual production (kg)	Annual sales (\$MM)	Estimated bulk COGS (\$/g)	Present value of purification costs (\$MM)
EPO	1	1000	16000	40
OKT3	2.5	50	1280	8
hGH	5	800	12480	156
tPA	30	278	798	60
Insulin	500	600	58	73

tPA, tissue plasminogen activator; hGF, human growth factor; COGS, cost of goods

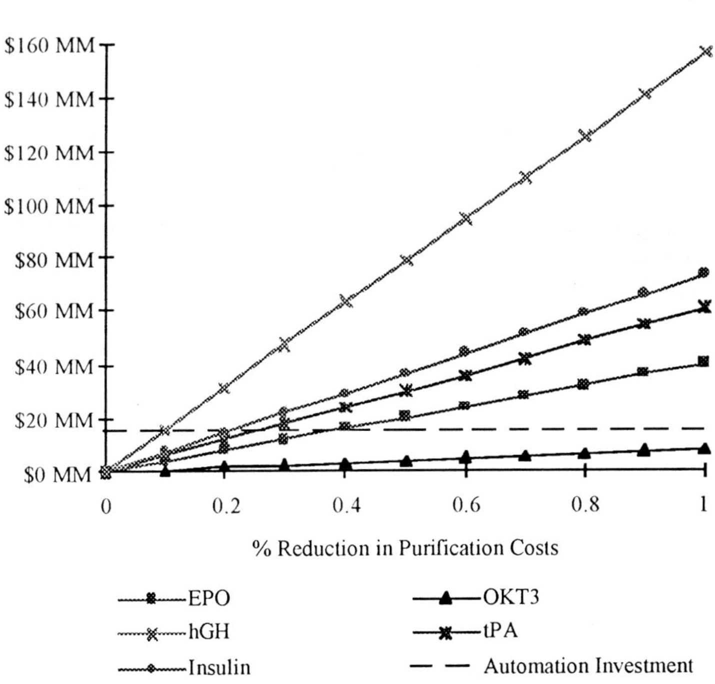

Fig. 1. Present value of purification savings as a function of reduction in purification costs

purification and so a modest 10% decrease in COGS would be economical. Insulin and tPA, which have intermediate purification costs, would require a 20–25% reduction in COGS in this model.

This analysis shows that products with low annual production costs (low sales volumes or protein production amounts) are usually not worth automating to achieve cost benefit. On the other hand, products with very high production costs (high sales volumes or high protein production) could benefit from being automated. For those products where there is incentive to automate, it would be more practical to do so if a second plant needed to be built to expand production.

II. Generic Purification Methods

Development of a biopharmaceutical is costly to a company, not only in terms of the money spent in development, but also in terms of the lost sales of product when product introduction is delayed by lengthy process development efforts. The patent expiration date is insensitive to how long it takes to develop a drug, so shorter development times lead to longer times that drugs can be sold under patent. When patents expire, generic competition usually reduces profits considerably. Therefore, there is an incentive to shorten development times for drugs.

About half of all new biological entities in clinical development are MAbs. About two-thirds of new biological entity INDs submitted to the Food and Drug Administration in 1992 were for monoclonal antibodies. Given these statistics, it is likely that many companies will be developing several MAbs.

If there were a generic purification technology for MAbs, it would greatly reduce process development time, so that clinical trials and approval could start earlier. Generic technology would also make more efficient use of multi-product pilot and manufacturing plants, and enable prospective automation of the process.

Biopharmaceutical chromatography has advanced significantly in the past 10 years, but there are still many areas for future work. Some of the most fruitful areas are:

- Faster flowing resins that would reduce capital and operating costs by allowing separations to be done on smaller equipment.
- Generic affinity methods like hexa-histidine tails that can be fused to proteins enabling purification on IMAC. Another useful generic affinity method would be a base-stable ligand for antibodies.
- Better methods for hydrophobic proteins, including nonflammable and non-toxic solvents that minimize aggregation and robust systems that resist clogging.
- On-line precipitate monitoring.
- Storage systems for process intermediates in 10–20 l frozen aliquots and reliable freezing and thawing methods for bulk proteins.

- Models that could predict behavior of columns with scale-up.
- Generic assays that would be ready as soon as process development started and predict product quality.
- Rapid LAL assay in 5 min accurate to ±tenfold rather than 60–90 min with accuracy of ±twofold.
- Better economic models that could be used by non-specialists and are more accurate.

H. Conclusions

In summary, biopharmaceutical development, in general, and chromatography, in particular, has significantly matured in the past 10 years. This seems quite possible for most applications under current development. The development of MAb products especially benefits from the maturation of biopharmaceutical technology. There are still challenges ahead for process-development scientists and it is up to them to turn the wish list into reality.

References

Center for Biologics Evaluation and Research (1994) Draft points to consider in the manufacturer and testing of monoclonal antibody products for human use. Rockville, MD

Center for Biologics Evaluation and Research (1997) Points to consider in the manufacture and testing of monoclonal antibody products for human use. Rockville, MD

Committee on Proprietary Medical Products (1991) Note for guidance: Validation of virus removal and inactivation procedures. Brussels

Food and Drug Administration (1987) Guideline on validation of the Limulus amebocyte lysate test as an end-product endotoxin test for human and animal parenteral drugs, biological products, and medical devices. Rockville, MD

Horaud F (1991) Introductory remarks: Viral safety of biologicals. Develop Biol Standard 75:3–7

International Conferences on Harmonization (1997) Viral safety evaluation of biotechnology products derived from cell lines of human or animal origin. Geneva

Lubiniecki AS (1998) Process validation. Develop Biol Standard in press

Lubiniecki AS, May LH (1985) Cell bank characterization for recombinant DNA mammalian cell lines. Develop Biol Standard 60:141–146

Lubiniecki AS, McAllister PR, Smith TM, Shadle PJ (1996) Process evaluation for biopharmaceuticals: What is appropriate in process evaluation? Develop Biol Standard 88:309–315

McAllister PR, Shadle PJ, Smith TM, Scott RG, Lubiniecki AS (1996) Use of a statistical strategy to evaluate sources of variability in viral safety experiments for a recombinant biopharmaceutical. Develop Biol Standard 88:111–121

Shadle PJ, Erickson JC, Scott RG, Smith TM (1995) Antibody Purification. US Patent # 5,429,756

World Health Organization Study Group (1987) Biologicals. Develop Biol Standard 68:69–72

CHAPTER 3

Proteins as Drugs: Analysis, Formulation and Delivery

C.R. MIDDAUGH and R. PEARLMAN

A. Introduction

The use of proteins as drugs is by no means new. Insulin, gamma-globulin and protein-containing vaccines have been routinely employed for decades. However, the advent of recombinant DNA technology has resulted in a dramatic expansion of interest in their pharmaceutical applications. It now appears that we can make virtually any desired protein in sufficient quantities for therapeutic use, although often with significant difficulty. It is considerably more problematic, however, to take the appropriate macromolecule and prepare it as a clinically acceptable drug substance. This problem arises from several sources. First, proteins are intrinsically less stable than their lower molecular weight (MW) pharmaceutical counterparts. Although this has turned out to be less of a problem than first anticipated, it remains a continuing challenge to formulate proteins that can be handled without damage throughout their entire lives; from their initial preparation, through distribution within the complex biomedical system, into their ultimate clinical use in a hospital or doctor's office or perhaps even in the home. Second, to make optimum use of a protein as a pharmaceutical agent, it is necessary to get it to the relevant in vivo site of action with maximum efficiency. A major potential power of proteins as therapeutic agents resides in their intrinsic compatibility with living systems. As critical components of virtually all biochemical processes, the presentation of a natural protein or one with specifically altered functional properties offers the opportunity to intervene in a pathological process with a high degree of specificity and minimal perturbation of normal processes. Currently, most protein products are simply administered systemically, for example intravenously, with the intrinsic specificity of the drug expected to provide sufficient efficacy. If it was possible to actually deliver a protein to a particular site in a temporally defined manner (sustained release, pulsatile, etc.), the utility of protein-based pharmaceuticals and vaccines could potentially be significantly expanded. A major stumbling block in this regard is the usual lack of knowledge concerning exactly how, where and when it would be best to deliver a specific protein. In general, our knowledge of the pathology, as well as the normal physiology of the usually complex biology of the situation in which we wish to intervene, is almost always grossly inadequate. This deficiency is particularly disappointing with regard to protein pharmaceuticals, since their exquisite specificity, which is potentially their most unique therapeutic property, cannot be fully utilized. Finally, there still exist significant challenges in the high-resolution analysis of protein pharmaceuticals. This is dramatically illustrated by the fact that when a new process is devised to produce a protein pharmaceutical with an already well-established safety and efficacy profile, it will often be necessary to perform new safety and clinical-efficacy trials to ensure the identity and quality of the protein derived from the new process. This is clearly an undesirable situation which, in fact, may result in a significant impediment to the production of proteins of higher quality and decreased cost.

In this chapter, we will discuss each of the three issues described above with the goal of defining the major problems in each area as we view them. In some cases, potential solutions will also be briefly considered. It is not our intention to provide a comprehensive review in each of these subjects since excellent, extensive treatments of each topic already exist. Rather, we will refer the reader to several of the more comprehensive, recent reviews about each subject and will assume some familiarity with each area as described in these texts.

B. The Analysis of Protein Pharmaceuticals

The ability to accurately define the structure of any protein that is being contemplated as a drug substance is absolutely essential to the pharmaceutical development process. Currently, such a characterization is performed by a series of chromatographic, electrophoretic, spectroscopic, immunological and biological measurements (Jones 1993). Despite the intrinsically higher accuracy and precision of the former four types of assays, the use of a biological (cell or organism based) method often remains critical, since there is, as of yet, little confidence in the ability of chemical and physical methods to consistently detect potentially subtle structural alterations that could be manifested as decreases in either the safety or efficacy of the product. The major exception to the need for a biological assay are proteins in which a well-defined enzymatic activity comprises the critical pharmaceutical activity. Although such macromolecules still require the same complex, multifaceted characterization as other proteins, a biological assay can usually be replaced by an in vitro enzymatic assay of superior accuracy and precision. The problem we will discuss here is the possibility of providing structural analyses of proteins with sufficient resolution to ensure their functional identity.

A critical question that needs to initially be addressed is the type and amount of information necessary to unequivocally establish the identity of any particular protein. A simplistic answer is easily provided: a data array (a matrix) specifying the three spatial coordinates and identity (C, H, O, N, S, etc.) of each atom in a protein uniquely defines any particular molecule. We will ignore, for the moment, the dynamic, fluctuating nature of proteins and problems of microheterogeneity. Only two methods, X-ray crystallography and nuclear magnetic resonance (NMR), are currently available, which provide the type of three-dimensional structural information desired. Both are usually considered primarily research tools and for a variety of reasons are not applicable to the routine pharmaceutical analysis of proteins. We will argue that, at least in principle, this need not necessarily be the case, and discuss each approach below.

The existence of a hierarchy of structure in proteins, however, suggests it may not be necessary to determine the actual location of each atom to ensure identity. A combination of various types of information acquired at different levels of structure can potentially provide a very detailed picture of a macromolecule. Could this less-direct approach provide an adequately high-resolution description sufficient to establish pharmaceutical identity? The answer to this question is uncertain, but it may ultimately be possible to employ this approach if certain carefully delineated requirements are fulfilled. Unfortunately, this has yet to be attempted. We view any such attempts to be at least partially a formal problem in information theory, in the sense that the type and amount of information necessary to define a protein's structure needs to be carefully defined. It is now clear that the chemical (covalent) identity of a protein can be well defined by modern methods of mass spectroscopy and/or peptide mapping and sequencing as discussed below. A corollary is that small amounts of impurities or variants, when present in a mixture, are exceedingly difficult to detect by such techniques. Thus, some form of separation (chromatography) coupled to a powerful structural-analysis method is often required. For example, the structure of the carbohydrate residues present in tissue-type plasminogen activator (tPA) and variants have been characterized by liquid chromatography coupled to mass spectrometry (GUZZETTA et al. 1993) to this end.

Secondary structure can be measured to an accuracy of several percent by circular dichroism (CD), Fourier transform infrared spectroscopy (FTIR) and, in some cases, one-dimensional NMR. The integrity of tertiary structure can be probed to only a limited extent by techniques such as ultraviolet absorption and intrinsic fluorescence, as well as chemical and immunological reactivity. Furthermore, knowledge that a protein is the proper chemical entity (correct unmodified sequence) with apparently native secondary structure does not guarantee the existence of correct tertiary structure. Nevertheless, it may be the case that less than complete direct knowledge of tertiary structure is actually necessary to ensure identity if the propriety of lower levels of structure are established. Proper quaternary structure is probably the least difficult problem, since a combination of permeation chromatography, static and dynamic light scattering, as well as equilibrium and velocity analytical ultracentrifugation can fairly rigorously establish the number of subunits, as well as higher aggregation states under most solution conditions. We will not attempt to answer the question of how much information is ultimately necessary for structural definition, but rather outline below how one might simultaneously measure a number of different parameters that reflect a variety of aspects of protein structure. Availability of a convenient experimental approach would, hopefully, drive the development of the appropriate theoretical framework.

It should also be remembered that proteins often exist in multiple forms, that is, they can exhibit microheterogeneity. Whether this is due to different levels or types of glycosylation, or other post-translational modifications, such as oxidation, deamidation etc., it is important to discern what effects, if any, such modifications may have on the activity and safety of the protein. This also produces potential major problems in structural analysis, as will be indicated later. In the case of human growth hormone (hGH), the readily oxidized forms are the mono- and di-methionine sulfoxides at Met14 and Met125, and these forms retain full bioactivity (BECKER et al. 1988). The methionine sulfoxide at Met170 produces a loss in potency, but is very difficult to form. Deamidated hGH at position Asn149 and Asn 152 (forming the corresponding aspartate variants) can be prepared by incubation of the protein in basic solutions. These deamidated species, again, retain full bioactivity, whereas a dimeric form of hGH manifests a loss of potency (BECKER et al. 1987). Each of these forms can be chromatographically resolved, at least in principle, permitting structural analysis of the individual variants.

I. X-Ray Crystallography

X-ray crystallography currently offers the accepted definitive method for determining the three-dimensional structure of molecules, including proteins. The only major concern with this approach is that the structure elucidated is that of the molecule in the crystalline, rather than solution state. A minor caveat is that highly flexible regions of a protein (e.g. the termini) may not be well resolved. The general consensus, however, based on a plethora of evidence, is that the crystalline state is predictive of, if not entirely identical to, the solution state. Nevertheless, for a variety of reasons, this method is assumed not to be applicable to the pharmaceutical analysis of proteins. First, crystallization of proteins is usually considered to be more art than science. Many proteins are thought to be difficult, if not impossible, to crystallize with sufficient order and crystal size, to permit acquisition of high-resolution data. Furthermore, the quantities of protein necessary for successful crystallization (tens of milligrams) are often not readily available. Second, to solve a structure using this method, isomorphous heavy-metal derivatives of the proteins must be obtained to permit the phase of the diffracted X-rays to be determined. In some cases, however, knowledge of a protein's polypeptide backbone structure (perhaps by homology to a related molecule of known structure) may obviate this need. Third, the entire process is both very labor- and time intensive. Complete determination of the structure of a protein by this method can take many months or even years. Such a situation is clearly unacceptable to scientists analyzing pharmaceutical macromolecules where data usually needs to be obtained in days to weeks.

Nevertheless, it seems possible that, in the near future, crystallography has the potential to play a significant role in the pharmaceutical analysis of proteins. Several factors encourage this view. Most importantly, it should not be necessary to perform actual structural determinations (generation of electron-density maps). Rather, one need merely compare a series of X-ray-diffraction patterns of a test protein with a standard preparation of known integrity. Thus, the problem becomes one of comparing a multi-dimensional data matrix with the elements consisting of crystallographic reflections defined by spatial coordinates and intensities. The similarity of such matrices could then constitute evidence of structural identity at the resolution of the diffraction pattern. Methods such as discriminant analysis could be used for comparative pattern recognition and appropriate statistical analysis. In principle, information about dynamic aspects of the protein's structure are also available through the temperature dependence of the pattern and "patches" present in the diffraction pattern arising from diffuse X-ray scattering due to correlated intramolecular motions (FAURE et al. 1994).

The critical question would seem to be whether such data could be routinely obtained. In many (and perhaps the majority of) cases, this may be possible. Protein concentrations of several to tens of milligrams per milliliter are necessary for crystallization. Ultrafiltration procedures offer rapid (within hours) methods to achieve such concentrations. This task is also facilitated by the quantities of protein that are typically available at the pharmaceutical stages of development. Thus, a second major concern becomes our ability to crystallize a particular protein. Recent advances suggest, in fact, that a surprisingly large number of proteins can be crystallized under one or more sets of conditions. While they may be small, it appears that crystals of sufficient quality can often be reproducibly obtained with persistent effort. This is reflected in the eventual appearance (usually within a few years) of crystallographically determined structures of many soluble proteins of current interest, especially during drug-discovery processes. Thus, although obtaining diffractable crystals is not automatic, it may rarely prove to be the limiting problem. Rather, the critical factor in obtaining the necessary diffraction data should be the availability of a protein solution of sufficient concentration and purity.

As described above, processes such as glycosylation and deamidation often introduce significant microheterogeneity into protein preparations. Although chemical or enzymatic conversion of a polydisperse protein mixture to a more homogeneous state prior to crystallization offers one potential solution to this problem, this is obviously unsatisfactory from a pharmaceutical perspective. It would seem better to attempt to prepare more homogeneous proteins through improved production and isolation methods. Without a solution to the microheterogeneity problem, however, this will remain a significant barrier to the use of this approach with many proteins.

Recent advances in both radiation sources and detectors have also greatly reduced the time necessary to obtain high-resolution diffraction data. Approximately a dozen synchrotron beam lines are now available worldwide, with another dozen or so projected to be available in the near future. In fact, over 20% of the protein crystallography papers published between 1992 and 1993 employed synchrotron radiation (Ealick and Walter 1993). The high photon flux available from this radiation leads to very rapid data collection, as well as better spot-to-spot resolution and signal-to-noise ratios. Simultaneously, film-detection methods have been extensively replaced by image plates/storage-phosphor systems (Pflugrath 1992). In addition, charged-coupled device (CCD)-based X-ray detectors are beginning to appear, which produce a rapid, directly computer-processable readout of data. Both of these integrating X-ray position-sensitive detectors feature high spatial resolution and dynamic ranges, as well as low noise, the ability to count at high rates, and large active areas. The combination of synchrotron sources and these new detectors means that complete data sets can now be obtained in many cases within a few hours.

In summary, recent advances in X-ray crystallography, combined with the need for only comparative data, suggest that this method may soon be ready to play a tentative role in the pharmaceutical analysis of proteins. This potentially provides the ability to argue that the three-dimensional structure of a test protein is unchanged at the relatively high 2–3 Å resolution of this approach.

II. Nuclear Magnetic Resonance

Like X-ray crystallography, NMR has generally been considered to be of little help in analyzing proteins under pharmaceutical conditions, although its general power as a structural tool for proteins is well recognized (Bax 1991). Many of the potential problems associated with this method, such as the need for high protein concentrations and the complexity of the data analysis, are similar to those of crystallography. As discussed above, this need not completely discourage the pharmaceutical use of NMR. In such applications, a two-dimensional nuclear Overhauser effect (NOE) proton spectrum of a test protein could be compared with a standard spectrum. A comparative data analysis reduces to that described above for crystallography with crosspeaks playing an analogous role to X-ray reflections. An advantage of NMR is that the solution state of the protein is examined. Unfortunately, this approach offers a major disadvantage that significantly reduces its potential utility. As the size of a protein increases, the line width of the resonances increases, limiting application, in most cases, to proteins of less than 20 kDa MW. Although this problem can be overcome, to some extent, by the combined use of isotopically labeled proteins and three- and four-dimensional techniques, an upper limit of approximately 35–40 kDa is still apparent (Clore and Gronenborn 1994). Unfortunately, isotopic labeling will be unacceptable for most of the applications of concern here. It remains, however, that in the case of smaller proteins of pharmaceutical utility, proton NMR does have the potential to provide the type of high-resolution structural resolution desired.

For larger proteins, some type of breakthrough seems necessary for NMR to become of direct applicability to the pharmaceutical analysis of protein structure. The nature of such an improvement is not obvious. The use of higher magnetic fields seems unlikely to have much of an impact in this case. Problems connected with protein microheterogeneity remain with this technique as well.

III. Mass Spectroscopy

Recent advances in mass spectroscopy (MS) have led to the application of this high-resolution technique to proteins and other large molecules (Carr et al. 1991; Chait and Kent 1992). In particular, the introduction of electrospray (ES) and matrix-assisted laser desorption (MALD) methods permits the measurement of protein MWs with better than single-dalton accuracy in molecules of moderate size or less. These techniques, therefore, permit the covalent structure of a protein to be rigorously defined. In general, they provide little conformational information, although recent work suggests that limited three-dimensional structural characterization might be obtainable by the ES method (Feng and Konishi 1993; Mirza et al. 1993). The two methods are complementary, with each possessing distinctive strengths and weaknesses. In ES–MS, molecules from a flowing liquid are ionized at atmospheric pressure by injection through an electrostatically charged needle and, consequently, passed directly into the mass spectrometer. A characteristic of this process is that a series of multiply charged molecules are formed, due to the many potential sites of cation association on proteins. While each peak permits an independent measurement of MW, the spectra are complicated and, thus, less amenable to the detection of contaminants or degradates. The sample requirements for ES–MS are also rather stringent, since salts significantly interfere with analysis. This problem is somewhat offset, however, by the ease of coupling of ES instruments to chromatographs, thereby permitting the analysis of complex mixtures.

In contrast, MALD–MS usually produces only a few peaks and can be performed on samples under a wide variety of solution conditions, including in the presence of many solutes, such as salts. In this procedure, a high concentration of laser-absorbing matrix material is dissolved with the sample and this mixture is dried. Illumination of this solid by a laser pulse causes volatilization of the protein–matrix mixture, and the subsequently ionized protein is accelerated down a time-of-flight analyzer. Most of the protein molecules produced by this method are singly ionized, thus, simplifying analysis and often permitting contaminants and degradates to be identified. Overall, MALD–MS is somewhat more sensitive than ES methods (<1 pmol of protein can often be detected), but of somewhat lower accuracy and resolution.

Both methods permit changes in MW arising from protein modifications, such as glycosylation, phosphorylation, acetylation, proteolysis and oxidation of Met and Trp residues to be easily detected. More subtle changes, such as those produced by disulfide formation or breakage (Δ MW ~2 Da), deamidation of Asn or Gln residues (Δ MW ~1 Da), or the presence of a C-terminal amide (Δ MW ~1 Da), are more difficult to detect, but can, in fact, be reliably determined with the use of proper calibration standards, especially with proteins of MW <20,000. Thus, in many cases, the chemical identity of a protein can be established by these methods. Furthermore, the high resolving power and MW accuracy of MS can be employed in peptide-mapping analysis (Arnott et al. 1993): ES–MS can be combined with high-performance liquid chromatography (HPLC) to identify individual peaks, while MALD–MS alone can be used to directly generate an interpretable peptide fragmentation pattern (Billeci and Stults 1993). As previously mentioned, it is also possible that MS alone, or in combination with time-dependent proteolysis or other chemical modifications, could be used to provide conformational information, but the potential resolution of such approaches are uncertain. We emphasize again, however, that MS procedures are generally insensitive to protein-conformational features and are, therefore, unlikely to be employed to establish the unique, three-dimensional structure of a macromolecule. Nevertheless, the high information content of MS measurements could potentially be utilized in combination with lower resolution methods to achieve a comprehensive, molecular characterization, as we will discuss next.

IV. Multiple Parametric Approaches

If the information necessary to define the structure of a protein cannot be obtained by a single, information-rich method, then the use of some combination of procedures can be employed. This is, of course, the currently used approach in which chromatographic, electrophoretic, spectroscopic, immunological and biological methods are sequentially implemented (Jones 1993). As described above, major problems with this approach include the time, effort and complexity involved, as well as ultimate reliance upon some type of biological assay to ensure structural propriety. One possible step forward would be a single device that simultaneously performs a carefully selected set of the analyses desired (Middaugh 1990). A large number of different types of such machines can be envisioned. Here, we describe only one such system, based on a selection of some of the more critical structural parameters necessary for the definition of a typical pharmaceutical protein. In general, such a device is envisioned to simply consist of a chromatographic system with multiple detectors. Several options exist for the front-end chromatographic component, based on the property of the protein that we wish the separation process to be sensitive to. It would seem best to also avoid nonaqueous (organic) solvent systems due to their potential for protein-structure disruption. Therefore, either an ion-exchange or molecular-sieve column (separating on the basis of charge and size, respectively) are reasonable choices by these criteria. Alternatively, if the problem of small sample volume can be overcome, capillary electrophoresis could be used. The choice of detectors is currently primarily limited by availability and overall system compatibility.

A logical first choice for a detector is a simple ultraviolet (UV)/visible absorption device. This should be either of the diode array or fast-scan grating type to permit entire spectra to be rapidly obtained. Analysis of protein UV spectra in the form of second derivatives can potentially provide simultaneous evaluation of the average local environments of Trp, Tyr and Phe residues in a protein and, thus, serve as a sensitive monitor of tertiary structure (MACH et al. 1993). In principle, the spectral properties of the peptide-bond absorption peak near 190 nm could provide information about the protein's secondary structure content. Because of high extraneous absorptive interference in this region, however, this would almost certainly be better accomplished by a second detection component based on far-UV CD (JOHNSON 1990). Such a measurement also presents several difficulties (low sensitivity, interference by many solutes, etc.), but the construction of an appropriate specialized CD detector is technically feasible. The other attractive secondary-structure-sensitive method, which involves measurement of infrared absorption by the protein's amide bonds, is overly complicated by even lower sensitivity in the solution state and the intense infrared absorption of water (SUREWICZ et al. 1993).

A third attractive detection step would involve measurement of the intrinsic UV-fluorescence emission of a protein. Far-UV excitation (260–300 nm) produces primarily emission from a protein's Trp residues (BURSTEIN et al. 1973). Although such measurements provide only a probe of the immediate environment of the limited number of indole side chains in most proteins, these few residues are frequently very sensitive to the three-dimensional structure of many proteins. Another advantage of this technique is that Raleigh scattered light is also produced by the protein, and this can be simply detected (usually at right angles) during the same measurement (DOLLINGER et al. 1992). The intensity of this scattered light (see below) is related to the dimensions of the scatterer (protein) and could thus serve as a convenient monitor of protein size. A major requirement for obtaining useable information from fluorescence-emission data is that the entire spectral region of interest needs to be examined. Again, this can now be simply accomplished by either fast scanning or diode-array technology.

The size or state of oligomerization and aggregation of a protein is another crucial molecular parameter of interest. Although elution position in a front-end molecular-sieve separation provides information in this regard, size data is probably better obtained by measurement of the angular dependence of the steady-state intensity of scattered light and/or a dynamic light-scattering determination, in which the autocorrelation function of the scattered light is analyzed. Both techniques have the required sensitivity and are available commercially in a chromatographic-detector format (CLAES et al. 1990; WYATT and PAPAZIAN 1993). In theory, a combination of these two measurements can provide the radius of gyration, hydrodynamic radius and MW of a protein. This level of analysis is probably unnecessary, however, since as argued above, comparative measurements employing a protein standard of the same identity as the test molecule should be all that is necessary to establish aggregation state.

Finally, it would seem optimal to pass the analyzed material to a mass spectrometer for definitive chemical identification. At this point, ES–MS would seem the better approach, but MALD–MS methods employing deposition of samples and auxiliary matrix addition onto some type of receptive device for subsequent analysis may also soon be possible.

Numerous variations of the type of instrument outlined can be imagined, in which rigorous chemical identification (MS) is combined with analysis of secondary (CD), tertiary (UV absorption, fluorescence), and quatemary (light scattering) structure in a chromatographic format (electrostatic or size information) to provide a comprehensive description of a macromolecule. Then, by comparison with a standard protein of the same type, the identity of the molecule can arguably be established. One can even imagine the optical portion housed in a single package, perhaps employing CCD technology. Such a "protein machine" should require only submilligram quantities of protein and could also be used to analyze stressed samples (employing elevated temperature or extremes of pH) providing both real-time and accelerated-stability information. We conclude this section on a tentative note, however, in the form of a question: how could one unequivocally demonstrate the ability of such a system to detect any and all potential changes in a target protein? As indicated above, the intrinsic microheterogeneity of most proteins presents a significant challenge to such a system. At this point, the less than satisfactory answer must be that the detection of all potential structural alterations would not be possible, rather, the currently employed procedure would need to be used; all possible relevant forms of molecular change must be generated and demonstrated to be resolvable by the combined approach.

V. Miscellaneous Comments

An area not directly addressed by the methods described previously is the analysis of contaminants and degradates. It seems, at least in the immediate future, highly specific individual assays would still need to be developed for each protein. It is possible, however, that MS or some type of multiple-parameter protein-analysis system could be an advance in this regard as well. Neither crystallography nor NMR seems likely to prove helpful, despite their potential power in structural verification of homogeneous macromolecules.

Another area in which advances might be anticipated, however, concerns more biologically based assays. A strong trend already exists toward the replacement of animal studies with cell culture-based measurements. For example, the biological activity of growth factors is now often measured by their stimulation of mitogenesis of cultured cells. Similarly, many proteins that initially exert their effects by binding to cell-surface receptors can be analyzed in terms of the binding process itself. A simple extrapolation of the success of these approaches suggests that analysis of critical intracellular responses induced by a particular pharmaceutical protein might be of significant utility in monitoring the protein's biological activity. Advances in this area could take several forms. For example, the development of pH-sensitive microsensors has permitted the biological responses of several proteins to be quantitated in terms of protein-induced changes in pH in the medium of cultured cells (McConnell et al. 1992). Commercial instrumentation has recently become available based on this work, which should permit data of adequate precision to be obtained. Similarly, the development of commercial plasmon-resonance technology, which detects changes in the concentration of agents near an optical surface as an immediate consequence of binding phenomena, could also prove of pharmaceutical utility (Granzow and Reed 1992). A variety of other physically and biologically based sensors, such as quartz-crystal microbalances (Lasky and Buttry 1990), which have the potential to be applied to new bioassays, should become available in the next several years. One focus could be on the events that take place during the complex molecular-signaling pathways that underlie many intracellular information-transduction mechanisms. For example, changes in intracellular calcium levels and alterations in membrane potential are easily detectable by fluorescence methods. It will, of course, be incumbent upon the pharmaceutical community to demonstrate that such approaches both serve as accurate and biologically realistic monitors of protein bioactivity, as well as possess the ruggedness and precision required of pharmaceutical assays, but this seems quite plausible.

Finally, it seems probable that the analysis of vaccines is on the verge of significant progress. Vaccines usually consist of a relatively complex mixture of components. Major forms of vaccines include attenuated and inactivated viral particles, high MW polysaccharides, polysaccharide conjugated to various proteins and protein complexes, peptides and peptide conjugates, protein aggregates (natural and recombinant), naked DNA, etc. These materials are usually analyzed by either their induction of antibody titers in test animals or some type of cell-based plaque assay. Unfortunately, such techniques are of low throughput and high variability, often resulting in major analytical problems. Structural characterization, if any, is usually limited to simple electrophoretic or chromatographic methods. As simpler peptide and recombinant protein-based vaccines become available, the more comprehensive type of analyses currently employed for pure protein pharmaceuticals can be expected to be employed (Volkin et al. 1994). Such vaccines are usually of reduced immunogenicity, however, and higher resolution methods are also needed to analyze the more complex types of vaccines. In fact, it would seem that the first steps can immediately be taken. It is clear, for example, that detailed information about the size, shape and aggregation state of even complex entities (viral or protein particles, polysaccharide–protein conjugates) can be obtained by a combination of molecular-sieve chromatography, static and dynamic light scattering, and analytical equilibrium and velocity sedimentation. Similarly CD and FTIR can be employed to extract secondary structure information, intrinsic fluorescence applied to tertiary structure characterization and mass spectroscopy used to obtain more detailed chemical data. A variety of fluorescent extrinsic optical probes sensitive to the polarity of their binding sites can also be used to explore both local and global aspects of various structural features of vaccine molecular complexes. These methods are equally applicable to polysaccharides and nucleic acids, either separately or in combination with protein components. It is not our purpose to discuss this issue in any depth, rather to simply point out that the analysis of vaccines is hopefully moving toward the situation that currently exists with purified proteins, namely, employment of a combination of methods that characterize an extensive variety of structural features of the major vaccine components to achieve functional description. To what extent such methods can increase our confidence in the quality of vaccine entities remains to be seen, although total elimination of more biologically based assays seems only a remote possibility in the immediate future.

C. Formulation

In the early days of biotechnology, there was a good deal of concern about our potential ability to formulate proteins as drugs. This was directly due to the relatively low stability of proteins compared with their lower MW pharmaceutical counterparts. While the formulation of proteins remains an ongoing challenge, it has become clear that most, if not all, proteins can, in fact, be prepared in a pharmaceutically acceptable form. The most probable mechanisms of degradation of proteins have become increasingly better defined (CLELAND et al. 1993; VOLKIN and MIDDAUGH 1992). This has, in turn, led to the identification of both methods of protein preparation and storage, as well as physiologically acceptable excipients that serve as starting points in most formulation efforts.

Briefly, the major covalent changes seen in proteins under moderate conditions include: (1) deamidation of Asn (as well as isoAsp formation) and to a lesser extent Gln residues, (2) cystine destruction, Cys oxidation and thiol-disulfide exchange, (3) oxidation of Met residues, (4) photooxidation of aromatic side chains, (5) peptide-bond cleavage, and (6) glycation and carbamylation of amino groups. The most common noncovalent changes seen in proteins involve partial or complete unfolding, and various types of aggregation, including precipitation. In the case of oligomeric proteins, subunit dissociation can also occur. Another problem, which sometimes appears, is adsorption of proteins to container surfaces (BURKE et al. 1992). This can be either a consequence of one or more of the aforementioned processes or may simply be a reflection of the intrinsic affinity of the amphipathic surface of proteins for a variety of interfaces. This is a phenomenon that is usually most evident at low protein concentrations, where a significant portion of the protein may be surface bound. Currently employed analytical methods are generally able to detect the type of problems indicated, although novel structural alterations and improved analytical methods can be expected to maintain active interest in this area. The importance of accelerated stability studies to these efforts cannot be overemphasized. Although the use of elevated temperature, extremes of pH and addition of chemically active modifying agents (oxidants, UV light, etc.) do not necessarily lead to results that can be quantitatively extrapolated to pharmaceutical storage conditions (typically −70°C to 4°C for protein solutions or solids), they do provide critical identification of susceptible regions in proteins. This, in turn, allows assays to be developed that have an established ability to detect hypothetical alterations in proteins at defined detection limits. If changes in a pharmaceutical are detected and deemed unacceptable (not all alterations need require prevention), then several classes of excipients have been identified to minimize particular problems.

The substances most commonly used to stabilize solution formulations of proteins include carbohydrates, amino acids, detergents and buffering salts at optimized pHs (Wang and Hanson 1988; Wang and Pearlman 1993). Detergents are usually employed to minimize aggregation and surface adsorption, presumably through their direct interactions with proteins (Thurow and Geisen 1984). Several physiologically acceptable examples of such amphiphiles exist and their use in moderate concentration seems to present few problems, although we see no reason to employ them unless absolutely necessary. Carbohydrates and amino acids at relatively high concentrations can nonspecifically increase the conformational stability of proteins by causing preferential hydration of macromolecules (Arakawa et al. 1991; Arakawa et al. 1993). In addition, these, as well as other classes of compounds, can often bind to the native forms of proteins, thus, enhancing stability. It is the use of specific excipients that we believe to be an approach that will become increasingly prominent. Several examples of this method of formulating protein pharmaceuticals already exist.

tPA is a 527-amino acid glycoprotein used to dissolve blood clots as a means of intervention in acute myocardial infarction. It presented challenging formulation and stabilization problems because of its poor aqueous solubility and lability in solution (Bennett et al. 1988; Nguyen and Ward 1993). The limiting reactions of tPA in solution included a clip occurring between Arg 275 and Ile 276 (converting single-chain tPA to 2-chain tPA), and aggregation, resulting in loss of enzymatic activity and potency as determined by a cot-lysis assay. The presence of these reactions meant that tPA had to be stabilized by formulation as a freeze-dried product.

The poor solubility of tPA at neutral pH (~100 μg/ml), however, precluded lyophilization and reconstitution of the clinical 100-mg dose, unless a means of increasing the solubility was found. The issue of solubility is further complicated because tPA exists as an approximately 50/50 mixture of two glycosylated species; namely type I, which possesses three glycosylation sites, and type II, which has two sites glycosylated (SPELLMAN et al. 1989). These two species have slightly different, but very low solubilities. In investigating the solubility of tPA as a function of pH and various buffer species, it was found that arginine, amongst several other agents, enhanced the solubility of tPA by several orders of magnitude. This increase in solubility was attributed to the binding of positively charged amino acids to the kringle-2 region of the molecule (CLEARY et al. 1989). It was also observed that by binding arginine to this region of the protein, the conversion of 1-chain tPA to 2-chain tPA was remarkably reduced. Studies were performed to optimize the solution pH and arginine concentration to balance the solubility and stability of the protein.

Arginine also serves the function of acting as a bulking agent for the freeze-dried product, since the protein requires a matrix to support it in a stable freeze-dried cake. In addition to formulation components, however, one has to also consider the processing conditions. It has recently been reported (HSU et al. 1994) that the freeze-drying cycle can greatly affect the stability of tPA upon storage. It was found that rapid freezing of the solution to be lyophilized resulted in a larger surface area in the freeze-dried cake, compared with use of a slower freezing rate. Upon accelerated storage conditions of the freeze-dried material (at 50°C), samples prepared with the fast-freeze process exhibited greater macroscopic aggregation than those produced by the slower freezing process. This finding is contrary to the commonly held view that freezing protein solutions as fast as possible is a preferred technique, as it minimizes potential degradation processes.

Another example of specific excipient stabilization involves the use of polyanions in the formulation of the mitogenic protein, acidic fibroblast growth factor (aFGF). This protein is relatively unstable to the extent that it is partially unfolded at physiological temperatures (DABORA et al. 1991). In addition, it contains a particularly reactive thiol group, whose oxidation leads to aggregation (and inactivation) of the protein. The structural instability of aFGF appears to be, at least in part, due to a large crevice in the protein, which is also involved in binding polyanions, an interaction of apparently great functional significance to the protein (VOLKIN et al. 1993). In vivo, the FGFs often appear to be primarily stored extracellularly, bound to polyanionic substances, such as the heparan sulfates, where they can be enzymatically released in active form in response to various physiological stimuli. Extensive studies have demonstrated the polyanion binding site of aFGF to be remarkably nonspecific, with the protein able to strongly complex to a wide variety of polyanionic polymers (sulfated polysaccharides, nucleic acids, etc.) and small molecules (inositol poly-phosphates and -sulfates, sulfated cyclodextrins, polyphosphorylated nucleotides, etc.), with a resultant effective stabilization of the protein (VOLKIN et al. 1993). The binding of these polyanions also results in significant protection of the major oxidizable thiol group in aFGF, as well as a reduction in the rate of deamidation of the protein. On this basis, it has been possible to prepare a number of quite stable formulations of this relatively labile protein by including either molar excesses of various multiply sulfated or phosphorylated small molecules or lesser quantities of polyanionic polymers (Tsai et al. 1993). In addition, inclusion of small quantities of a chelating agent further reduce oxidation by metals. Various nonionic polymers can also be included as viscosity-enhancing agents to more easily permit topical use of aFGF as a wound-healing agent.

Recently, Gombotz et al. (1994) reported on the use of polyvinylpyrrolidone (PVP), with MWs ranging from 10 kDa to 40 kDa, to stabilize an immunoglobulin M (IgM) antibody against thermally induced aggregation. The authors found that while concentrations of PVP in the range 0.1–1.0% stabilized the protein, higher concentrations led to increased aggregation. In addition, the higher the MW of the PVP, the greater the aggregation induced. The authors suggest that specific interactions occur between the polymer and antibody, wherein a small amount of PVP binds preferentially to the native antibody, resulting in stabilization. At higher concentrations, however, scanning calorimetry demonstrates a lowering of the protein's unfolding temperature, which could result in a less-stable system.

Another example of the influence of formulation on the storage stability of a lyophilized protein is seen with hGH (Pikal et al. 1991; Pearlman and Oeswein 1992). Freeze drying, again, prevents many degradation reactions from occurring in this case, but small amounts of dimer form upon storage of the lyophilized material at 40°C. It was found that by adding glycine in the preferred molar ratio of 100:1 (glycine to hGH), aggregation was minimized. In this study, varying amounts of mannitol and glycine were employed as a matrix for the lyophilized cake.

DNase is used as a therapeutic agent to reduce the viscoelastic properties of purulent sputum in patients suffering from cystic fibrosis (CF). The protein is isolated as a mixture of native protein and material deamidated at Asn 74. Separation of the deamidated from the native protein was made by tentacle ion-exchange chromatography, with subsequent analysis by tryptic peptide mapping and MS (Frenz et al. 1994). The deamidated DNase has a lower activity as measured by an in vitro enzymatic assay.

DNase is administered to the lungs as an aerosol, by means of an air-jet nebulizer. Because of the hyperreactive nature of the airways of CF patients, very few excipients were available for formulation of the protein. Since many commonly used buffering agents have the potential to cause pulmonary sensitivity, DNase was formulated as an isotonic saline solution, stabilized with calcium, but without a buffering agent (Cipolla et al. 1994). Such an approach posed a potential problem for long-term storage of the protein, since small amounts of hydroxide ion may leach out of glass, causing a rise in pH, effecting an increase in the rate of deamidation. A solution to this problem was found by storing the formulated DNase in plastic ampoules, made of high-density polypropylene, which did not exhibit the pH fluctuations observed in glass vials.

As indicated above for tPA and hGH, despite advances in our ability to formulate proteins in solution in a stable manner, many preparations are freeze dried (lyophilized) to ensure an adequate shelf life. Until recently, the preparation of a lyophilized protein formulation was primarily an empirical process. This situation has changed significantly, however, in the last several years (Franks 1990; Hsu et al. 1991; Pikal et al. 1991). The processes of freezing, as well as primary and secondary drying, are now much better understood, especially in terms of the glass-transition temperature (Tg′). For example, it is now clear that the rate of freezing needs to be carefully controlled and that primary drying should be performed at a temperature below Tg′. Initial differential scanning calorimetry (DSC) studies appear to be particularly useful in optimizing freeze-drying variables. It has also recently been argued that it is possible to separately address problems that may occur in freezing and drying processes through the use of distinct excipients (Carpenter et al. 1993). In this regard, polyethylene glycol (PEG) has been shown to protect several proteins from damage during freezing (and thawing), but provides little stabilization during drying. In contrast, several simple sugars protected the same proteins during dehydration. Thus, independent analysis of these two quite different processes may offer a much more effective approach to the rational design of lyophilized formulations. The use of FTIR to examine the conformational properties of proteins in the solid state has proven particularly useful in such studies (Prestrelski et al. 1993). Thus, much more detailed preformulation analyses of proteins employing especially DSC and FTIR to probe bulk solution and protein-conformational properties, respectively, in terms of both system-phase changes and individual freezing and drying processes, may lead to generally applicable rules for lyophilization processes.

Finally, we need to consider the very real possibility that, in the near future, formulation problems may often be addressed by direct engineering of the protein pharmaceutical itself. This is certainly the case when instability constitutes the major obstacle. Although the production of new epitopes and, therefore, potential enhancement of immunogenicity is a concern with such altered proteins, this is unlikely to be an insurmountable problem. The origin of conformational stability has now been dealt with in extensive studies of the effect of site-specific mutations on this property in numerous proteins (Fontana 1991; Fersht et al. 1993). It is clear that the stability of a protein can often be dramatically increased by small improvements in the packing of its apolar interior or the creation of a novel intramolecular interaction (salt bridge, hydrogen bond), as well as other subtle changes. Similarly, the presence of a chemically labile side chain, such as an aspartic residue, which tends to deamidate, or an oxidizable thiol group, can be obviated by an acceptable mutation at that site. Obviously, the introduction of such alterations produces a new pharmaceutical entity. Therefore, it will be imperative that formulation considerations be undertaken at the earliest possible time (at the basic research stage) if the engineering approach is to be successfully employed.

There also exists another, rather intermediate approach in which one can chemically derivatize a protein to produce a product that presents reduced formulation difficulties. By far, the most common of these modifications is the covalent addition of PEG to proteins (SMITH et al. 1993). The presence of PEG can have a variety of potentially positive effects on a protein, including enhancing its stability and lipid solubility, reducing its immunogenicity and, perhaps most importantly, prolonging its rate of plasma clearance. Despite the dramatic nature of this modification, it is often possible to retain the biological activity of a protein if proper reaction conditions can be identified. Consequently, a significant number of proteins have been successfully modified. In fact, one PEG-modified protein has already been approved for human use (bovine adenosine deaminase), and a second has been recommended for approval (L-asparaginase) at the time of writing this paper. These results argue that covalent modification of proteins may prove more extensively useful when applied to formulation problems, although the intrinsic heterogeneity often introduced by this approach remains problematic.

D. Delivery

The majority of protein pharmaceuticals are administered by injection, which, for obvious reasons, is very unpopular with patients and those administering the medication. This uncomfortable and stigmatized route of administration ultimately limits the more widespread use of protein drugs. However, when no alternative route of administration exists and substantial benefit is derived from use of the drug, there is little alternative to this approach. For a number of chronic applications, patients have been successfully instructed in self administration of intramuscular (IM) or subcutaneous (SC) injections (of insulin, growth hormone, and interferon, for example). While progress in needle technology and administration, including needleless jet-injection devices, represent potentially significant improvements in therapy, there remains a strong desire for alternative approaches. Many and varied approaches to the non-injectable delivery of protein drugs exist in the literature, but very few have successfully survived the lengthy clinical-trial approval process.

The parenteral route for proteins almost seems inevitable, since their large size and relatively fragile nature does not generally permit them to survive the rigors of the gastrointestinal (GI) tract. In contrast, the oral route is by far the most preferred path of administration for traditional (low MW) drugs. The GI tract has as its primary function the digestion of a variety of foodstuffs and is replete with numerous enzymes (including proteases), bile salts and other agents to accomplish this task. Such an environment is clearly not compatible with the aim of delivering an intact protein to the mesenteric blood supply. Even with successful passage across the GI mucosa, a protein would have to survive first-pass metabolism through the liver. Furthermore, the large molecular dimensions of proteins, relative to traditional drugs, generally impede ready transport across various biological membranes – particularly epithelial membranes of the GI tract, mucosa or skin.

In a particularly prescient paper, DAVIS (1990) describes the barriers to oral absorption of peptides and offers a critical assessment of several recent claims to successful oral delivery of protein drugs. The author goes on to show the naive nature of many approaches, especially in the case of microemulsions and particulate carriers, and the necessity for a rigorous understanding of the mechanisms involved in protein absorption via the oral route. The need for long-term safety studies is also cogently argued. Several additional recent papers have reviewed the challenges and possibilities for traversing the GI tract as a delivery port for proteins (PUSZTAI 1989; Nellans 1991; SMITH et al. 1992). A common theme, again, is the enormous difficulty posed by the oral route for such metabolically labile species as peptides and proteins. The problem of obtaining reproducible and consistent absorption is also seen as an enormous challenge. Nevertheless, a number of groups continue to explore the use of particulate carriers to protect the protein in the GI tract, or as a means of effecting absorption via a putative endocytotic process. Based upon the extremely low number of particles available for absorption by endocytosis (EBEL 1990), the delivery of therapeutic amounts of protein by this method should prove very difficult. There exists, however, a promising application of this idea to the GI lymphatic tissue (the Peyer's patch) for use in orally administered vaccines (ELDRIDGE et al. 1990). In such an approach, an antigen is encapsulated in a polymer and delivered orally to the GI tract. Small ($<5\,\mu$m) particles are taken up with low efficiency into the Peyer's patches, and detectable levels of antigen-specific secretory IgA can sometimes be detected. Unfortunately, the aim of delivery of most protein drugs is exactly the opposite, since absence of an immune response is a desired outcome.

Even before the advent of biotechnology-based drugs, there was considerable interest in improving the delivery of proteins by non-oral, non-parenteral means. This interest has sharply increased, now that recombinant proteins have become available. A number of recent papers examine the potential for delivery of proteins and peptides by such means (LEE 1990; WEARLY 1991; TALMADGE 1993). While there are many excellent papers and research programs investigating these novel delivery approaches, it is again sobering to realize that very few such systems have been approved for actual clinical use, despite intense effort. We believe the reasons for this are manifold, but that the major problem is one of complexity. Alternative drug-delivery systems, even for lower MW drugs, are technologically challenging, and proteins add additional dimensions of difficulty – both in the formulation and processing stages as well as in their biological availability. Despite this, a number of interesting approaches have been described in both preclinical and clinical studies, including the nasal delivery of proteins, using either absorption enhancers or mucoadhesives, which produce a longer retention time at absorption sites, thereby permitting longer periods of uptake (EDMAN and BJORK 1992; MACHIDA et al. 1993). Absorption enhancers generally act by breaking down the barrier properties of biological membranes and, thus, a significant potential exists for damage to the nasal epithelia after chronic exposure. Another promising method involves delivery of aerosol forms of proteins to the lungs (PATTON and PLATZ 1992; NIVEN et al. 1993). A major advantage of the pulmonary delivery route is that the lung presents a large absorptive area with minimal degradative enzymatic activity. A problem that needs to be addressed with dosing to the lungs is the large variability and generally inefficient delivery possible with currently available aerosol devices. The recent licensing of DNase, which is administered as an aerosol to the lungs by an air-jet nebulizer, encourages further exploration of this area.

I. Controlled-Release Dosage Forms

Because of the discomfort of frequent injections, one strategy is to develop controlled-release technology to reduce injection frequency for chronically administered drugs (OKADA et al. 1994). While the patient is still confronted with an injection, reducing the frequency to once a month, compared with daily administration, is a decided benefit. This approach has been successfully applied to the luteinizing hormone releasing hormone (LHRH) antagonist leuprolide (OKADA et al. 1988), a small (10-amino acid residue) peptide. It involved microencapsulation of the peptide in a biocompatible, biodegradable co-polymer of polylactic and polyglycolic acids (PLGA). The microencapsulated product need only be injected once monthly, compared with the requirement of daily injections of the unencapsulated peptide. SALTZMAN (1993) has reviewed the potential use of antibodies in controlled-release systems designed to provide a more powerful approach to immunotherapy. He also describes a method by which antibodies can be linked to degradable polymers. Upon administration of these complexes, the antibody is slowly released in a predictable fashion as the polymer backbone degrades. A particularly exciting extension of this technology is that vaccines with a single dose of microencapsulated antigen might replace the current multiple series of injections (MORRIS et al. 1994).

A number of groups are applying various microencapsulation strategies to protein drugs, and the search for degradable materials for such applications is also the subject of much effort (PEPPAS and LANGER 1994). Such an approach is difficult to apply to proteins, however, because the typical conditions for the microencapsulation process (organic solvents, high shear rates, etc.) are very destructive for many labile macromolecules (JONES et al. 1994). In this regard, some novel microencapsulation techniques have been developed, which have been specifically oriented toward the encapsulation of proteins (EPPSTEIN et al. 1990; GOMBOTZ et al. 1991). In these cases, encapsulation conditions have been specifically tailored to minimize protein degradation and, thus, seem of some promise.

One such protein-microencapsulation technique (GOMBOTZ et al. 1991), employs PLGA as the polymer. In this method, a protein is stabilized and suspended as a fine powder in a methylene chloride/polymer solution, which is sprayed as fine droplets into a tank of liquid nitrogen. The spherical microspheres fall through the liquid nitrogen and settle onto a bed of frozen ethanol. As this mixture is gradually warmed, the ethanol solubilizes the methylene chloride, resulting in a fine suspension of microencapsulated protein. The low temperature and absence of an aqueous phase help to stabilize the protein during the process. The authors report a number of apparently stable microcapsules prepared using this technique.

A problem that confronts a microencapsulated protein is that once administered, the protein is present in an essentially aqueous environment at 37°C. In order to achieve prolonged activity, the protein must then be stable for an extended period of time in this potentially difficult environment. One attempt to address this problem uses hydrophobic biodegradable polymers to protect labile proteins upon administration (TABATA et al. 1993). Polyanhydrides consisting of poly(fatty acid dimer), poly(secbacic acid) and various copolymers have been used to prepare small ($\sim$100 μm) microspheres, into which a number of proteins are encapsulated. Such preparations show significantly less degradation of the protein when incubated at 37°C, compared with the non-encapsulated protein.

Another approach to improving the delivery of therapeutic proteins is by specifically targeting the agent to its site of action. In theory, targeting should lead to improvements in therapy, since it should result in a more efficient utilization of the drug, as well as a reduction in the amount of drug delivered to unwanted sites (a reduction in side effects). For example, preferential delivery of colloids to the bone marrow by coating the particles with polymers has been reported (PORTER et al. 1992). Liposomes, with various antibodies or other ligands attached, have also been shown to enhance the delivery of drugs to target organs (MORI et al. 1993). While a considerable body of literature exists concerning various methods of targeting, very few of these approaches have been successfully applied in the clinic. Nonspecific uptake of the targeting moiety is often cited as the reason for failure, although a significantly higher dose of payload may be delivered to the desired site. By far, the most successful application of targeting has been in the diagnostic area, where a number of products have been approved. This success engenders hope that pharmaceutical applications will eventually be possible.

II. A Practical Delivery Challenge: Insulin

It is instructive to use the widely prescribed protein drug insulin as an example of current and potential delivery attempts. Insulin is employed to treat both insulin-dependent and -independent diabetes. Because it requires chronic (often several injections per day) parenteral therapy, numerous and varied attempts to improve its delivery have been reported. These range from improvements in injection devices to attempts at oral and nasal delivery and development of an artificial pancreas.

The need to improve the control of insulin administration (and blood glucose levels) was heightened last year with the publication of a large cooperative study of insulin use in the management of diabetes (THE DIABETES CONTROL AND COMPLICATIONS TRIAL RESEARCH GROUP 1993). This extensive, multi-center study involved over 1400 patients and concluded that better insulin and blood-glucose control (achieved by intensive blood-glucose monitoring and either multiple daily insulin injections or insulin pumps) resulted in drastically reduced complications (diabetic retinopathy) in the patient population. Thus, the need for superior delivery of insulin is clear, and a means of improving compliance and feedback for dosing would be highly desirable.

In a recent review of the state of the art in insulin delivery (SAUDEK 1993), many of the current and future approaches to improving delivery, compliance and overall insulin therapy are described. Interestingly, at the present time, actual significant improvements for the patient are generally limited to improved injection technology and glucose monitoring. Improved injection technology involves such seemingly straightforward approaches as finer needles, needleless injectors, pen injectors, external and internal pumps and so forth. Oral delivery and other nonparenteral approaches, even after decades of research, are still not available to the patient. This observation is not meant to trivialize improved injector technology. Such enhancements are vital to the comfort and compliance of the patient. Rather, it illustrates in an all too sobering fashion just how difficult the challenge of improving delivery of a protein drug can be.

Many laboratories have pursued the oral route for the delivery of insulin. These approaches often include a means of stabilizing insulin in the GI tract (HOVGAARD et al. 1992) and agents to enhance absorption as well as inhibit protease action (PATEL et al. 1991; MORISHITA et al. 1992a,b). The lack of success of these efforts is very discouraging and needs to be viewed in the context of just how formidable a barrier the GI tract presents to proteins. Again, we recommend that one should be very wary of claims of high bioavailability of orally delivered proteins. In our experience, such results have not lived up to careful scrutiny. Other researchers have examined the nasal route for insulin delivery, with reports of modest bioavailability (RYDÉN and EDMAN 1992). Phosphatidylcholine and other surfactants have been used as permeation enhancers in the nasal cavity (BECHGAARD et al. 1993), but the tissue-damaging effects of chronic administration of such formulations may limit their usefulness (JIAN and LI WAN PO 1993). The lung has also been investigated as a port of entry for insulin (OKUMURA et al. 1992; LAUBE et al. 1993), and these studies present, perhaps, the most encouraging results. As noted above, however, dose reproducibility needs to be addressed particularly, since insulin possesses a relatively narrow therapeutic window. In summary, the delivery of protein drugs is a daunting endeavor, but one that presents a real need for the patient and a significant challenge to the researcher. We feel that the ultimate utility of proteins as therapeutic agents is limited by our ability to effectively and efficiently deliver these products of biotechnology, and this justifies continued efforts despite the current lack of success.

References

Arakawa T, Kita Y, Carpenter JF (1991) Protein-solvent interactions in pharmaceutical formulations. Pharm Res 8:285–291
Arakawa T, Prestrelski SJ, Kenney WC, Carpenter JF (1993) Factors affecting the short-term and long-term stabilities of proteins. Adv Drug Del Rev 10:1–28
Amott D, Shabanowitz J, Hunt DF (1993) Mass spectrometry of proteins and peptides: sensitive and accurate mass measurement and sequence analysis. Clin Chem 39: 2005–2010
Bax A (1991) Experimental NMR techniques for studies of biopolymers. Curr Opin Struct Biol 1:1030–1035
Bechgaard E, J¿rgensen L, Larsen R, Gizurarson S, Carstensen J, Hvass (1993) Insulin and didecanoyl-L-a-phosphatidylcholine: in vitro study of the transport through rabbit nasal mucosal tissue. Int J Pharm 89:147–154
Becker GW, Tackitt PM, Bromer WW, Lefeber DS, Riggin RM (1988) Isolation and characterization of a sulfoxide and a desamido derivative of biosynthetic human growth hormone. Biotechnol Appl Biochem 10:326–337
Becker GW, Bowsher, RR, Mackellar WC, Poor ML, Tackitt PM, Riggin RM (1987) Chemical, physical, and biological characterization of a dimeric form of biosynthetic human growth hormone. Biotechnol Appl Biochem 9:478–487
Bennett WF, Builder SE, Gatlin LA (1988) Stabilized human tissue plasminogen activator compositions. United States Patent Number 4,777,043
Billeci TM, Stults JT (1993) Tryptic mapping of recombinant proteins by matrix-assisted laser desorption/ionization mass spectrometry. Anal Chem 65:1709–1716
Burke CJ, Steadman BL, Volkin DB, Tsai PK, Bruner MW, Middaugh CR (1992) The adsorption of proteins to container surfaces. Int J Pharm 86:89–93
Burstein EA, Vedenkina NS, Ivkova MN (1973) Fluorescence and the location of tryptophan residues in protein molecules. Photochem Photobiol 18:263–279
Carpenter JF, Prestrelski SJ, Arakawa T (1993) Separation of freezing- and drying-induced denaturation of Iyophilized proteins using stress-specific stabilization. I. Enzyme activity and calorimetric studies. Arch Biochem Biophys 303:456–464
Carr SA, Hemling ME, Bean MF, Roberts GD (1991) Integration of mass spectrometry in analytical biochemistry. Anal Chem 63:2802–2824
Chait BT, Kent SBH (1992) Weighing naked proteins: practical, high-accuracy mass measurement of peptides and proteins. Science 257:1885–1894
Cipolla D, Gonda I, Shire SJ (1994) Characterization of aerosols of human recombinant deoxyribonuclease I (rhDNase) generated by jet nebulizers. Pharm Res 11: 491–498
Claes P, Fowell S, Woollin C, Kenney A (1990) On-line molecular size determination for protein chromatography. Am Lab 22:58–62
Cleary S, Mulkerrin M, Kelley R (1989) Purification and characterization of tissue plasminogen activator kringle-2 domain expressed in Escherichia coli. Biochemistry 28:1844–1890
Cleland JL, Powell MF, Shire SJ (1993) The development of stable protein formulations: a close look at protein aggregation, deamidation, and oxidation. Crit Rev Therapeutic Drug Carrier Sys 10:307–377
Clore GM, Gronenborn AM (1994) Structures of larger proteins, protein-ligand and protein-DNA complexes by multidimensional heteronuclear NMR. Protein Sci 3:372–390
Dabora JM, Sanyal G, Middaugh CR (1991) Effects of polyanions on the refolding of human acidic fibroblast growth factor. J Biol Chem 266:23637–23640

Davis SS (1990) Overcoming barriers to the oral administration of peptide drugs. Trends Pharmacol Sci 11:353–355

The Diabetes Control and Complications Trial Research Group (1993) The effect of intensive treatment of diabetes on the development and progression of long-term complications in insulin-dependent diabetes mellitus. New Engl J Med 329:977–986

Dollinger G, Cunico B, Kunitani M, Johnson D, Jones R (1992) Practical on-line determination of biopolymer molecular weights by high-performance liquid chromatography with classical light-scattering detection. J Chromatogr 592:215–228

Ealick SE, Walter RL (1993) Synchroton beamlines for macromolecular crystallography. Curr Opin Struct Biol 3:725–736

Ebel JP (1990) A method for quantifying particle absorption from the small intestine. Pharm Res 7:848–851

Edman P, Bjsrk (1992) Nasal delivery of peptide drugs. Adv Drug Delivery Rev 8:165–178

Eppstein DA, Schryver BB (1991) Controlled release of macromolecular peptides. United States Patent Number 4,962,091

Faure P, Micu A, Perahia D, Doucet J, Smith JC, Benoit JP (1994) Correlated intramolecular motions and diffuse X-ray scattering in Iysozyme. Struct Biol 1:124–128

Feng R, Konishi Y (1993) Stepwise refolding of acid-denatured myoglobin: Evidence from electrospray mass spectrometry. J Am Soc Mass Spectrom 4:638–645

Fersht AR, Jackson SE, Serrano L (1993) Protein stability: experimental data from protein engineering. Philos Trans R Soc London, Ser A 345:141–151

Fontana A (1991) Analysis and modulation of protein stability. Curr Opin Biotechnol 2:551–560

Franks F (1990) Freeze drying: from empiricism to predictability. Cryo-Letters 11:93–110

Frenz J, Shire SJ, Slikowski MB (1994) Purified forms of DNase. United States Patent Number 5,279,823

Gombotz WR, Healey MS, Brown LR (1991) Very low temperature casting of controlled release microspheres. United States Patent Number 5,019,400

Gombotz WR, Pankey SC, Phan D, Drager R, Donaldson K, Antonsen KP, Hoffman AS, Raff HV (1994) The stabilization of a human IgM monoclonal antibody with poly(vinylpyrrolidone). Pharm Res 11:624–632

Granzow R, Reed R (1992) Interactions in the fourth dimension. Bio/Technology 10: 390–393

Guzzetta AW, Basa LJ, Hancock WS, Keyt, Bennett WF (1993) Identification of carbohydrate structures in glycoprotein peptide maps by the use of LC/MS with selected ion extraction and special reference to tissue plasminogen activator and glycosylation variant produced by site directed mutagenesis. Anal Chem 65:2953–2962

Harris RJ, van Halbeek H, Glushka J, Basa LJ, Ling VT, Smith KJ, Spellman MW (1993) Identification and structural analysis of the tetrasaccharide NeuAcα (2→6)Galβ (1→4)GlcNacβ (1→3)Fucα1O-linked to Serine 61 of human factor IX. Biochemistry 32:6539–6547

Hovgaard L, Mack EJ, Kim SW (1992) Insulin stabilization and GI absorption. J Controlled Release 19:99–108

Hsu CC, Ward CA, Pearlman R, Nguyen HM, Yeung DA, Curley JA (1991) Determining the optimum residual moisture in Iyophilized protein pharmaceuticals in: Developments in Biological Standardization: Intemational Symposium on Biological Product Freeze- Drying and Formulation, Vol. 74, Karger, Basel, pp 255–271

Hsu CC, Pearlman R, Bewley TA, Yeung DA, Koe GS, Brooks DA, Nguyen HM (1995) The effect of freezing rate during lyophilization on the stability of tissue-type plasminogen activator. Pharm Res 12:69–77

Jian L, Li Wan Po A (1993) Effects of insulin and nasal absorption enhancers on ciliary activity. Int J Pharm 95:101–104
Johnson WC Jr (1990) Protein secondary structure and circular dichroism: a practical guide. Proteins: Struct, Func, Genet 7:205–214
Jones AJS (1993) Analysis of polypeptides and proteins. Adv Drug Delivery Rev 10:29–90
Jones AJS, Nguyen T, Cleland JL, Pearlman R (1994) New delivery systems for recombinant proteins – practical issues from proof of concept to clinic. In: Lee VHL, Hashida M, Mizushima M (eds) Trends and future perspectives in peptide and protein drug delivery. Drug targeting and delivery, Harwood Academic Publishers, Gmbh, Amsterdam, The Netherlands
Laube BL, Georgopoulos A, Adams GK (1993) Preliminary study of the efficacy of aerosolized insulin delivered by oral inhalation in diabetic patients. JAMA 269: 2106–2109
Lasky SJ, Buttry DA (1990) Development of a real-time glucose biosensor by enzyme immobilization on the quartz crystal microbalance. Am Biotech Lab 2:8–16
Lee VHL (ed) (1990) Peptide and protein drug delivery, Marcel Dekker, Inc., New York
Mach H, Volkin DB, Middaugh CR (1995) Spectroscopic techniques to study protein folding: ultraviolet absorbance spectroscopy. In: Shirley BA (ed) Methods in Molecular Biology: Protein Stability and Folding Protocols 40:91–114
Mashida M, Sano K, Arakawa M, Hayashi M, Awazu S (1993) Absorption of recombinant human granulocyte colony-stimulating factor (rhG-CSF) from rat nasal mucosa. Pharm Res 10:1372–1377
McConnell HM, Owicki JC, Parce JW, Miller DL, Baxter GT, Wada HG, Pitchford S (1992) The cytosensor microphysiometer: biological applications of silicon technology. Science 257:1906–1912
Middaugh CR (1990) Biophysical approaches to the pharmaceutical development of proteins. Drug Develop Ind Pharm 16:2635–2654
Mirza UA, Cohen SL, Chait BT (1993) Heat-induced conformational changes in proteins studied by electrospray ionization mass spectrometry. Anal Chem 65: 1–6
Mori A, Kennel SJ, Huang L (1993) Immunotargeting of liposomes containing lipophilic antitumor prodrugs. Pharm Res 10:507–514
Morishita M, Morishita I, Takayama K, Machida Y, Nagai T (1992a) Novel oral microspheres of insulin with protease inhibitor protecting from enzymatic degradation. Int J Pharm 78:1–8
Morishita I, Morishita M, Takayama K, Machida Y, Nagai T (1992b) Hypoglycemic effect of novel oral microspheres of insulin with protease inhibitor in normal and diabetic rats Int J Pharm 78:9–16
Morris W, Steinhoff MC, Russell PK (1994) Potential of Polymer microencapsulation technology for vaccine innovation. Vaccine 12:5–11
Nellans HN (1991) Mechanisms of peptide and protein absorption. Paracellular intestinal transport: modulation of absorption. Adv Drug Delivery Rev 7:339–364
Niven RW, Lott FD, Cribbs JM (1993) Pulmonary absorption of recombinant methionyl human granulocyte colony stimulating factor r-hu-G-CSF after intratracheal instillation to the hamster. Pharm Res 10:1604–1610
Nguyen TH, Ward C (1993) Stability characterization and formulation development of Alteplase, a recombinant tissue type plasminogen activator. In: Wang YJ, Pearlman R (eds) Stability and characterization of protein and peptide drugs: Case histories, Plenum Press, New York, pp 91–134
Okada H, Ogawa Y, Yashiki T (1987) Prolonged release microcapsule and its production. United States Patent Number 4,652,441
Okada H, Heya T, Igari Y, Yamamoto M, Ogawa Y, Toguchi H, Shimamato T (1989) One- month-release injectable microspheres of a superactive agonist of LHRH,

leuprolide acetate. In: Therapeutic Peptides and Proteins. Formulation, Delivery and Targeting, Marshak D, Liu D (eds) Cold Spring Harbor Laboratory, NY, pp 107–112
Okada H, Yamamoto M, Heya T, Inoue Y, Kamei S, Ogawa Y, Toguchi H (1994) Drug delivery using degradable microspheres. J Controlled Release 28:121–130
Okumura K, Iwakawa S, Yoshida T, Seki T, Komada F (1992) Intratracheal delivery of insulin. Absorption from solution and aerosol by rat lung. Int J Pharm 88:63–74
Patel DG, Ritschel WA, Chalasani P, Rao S (1991) Biological activity of insulin in microemulsion in mice. J Pharm Sci 80:613–614
Patton JS, Platz RM (1992) Pulmonary delivery of peptides and proteins for systemic action. Adv Drug Delivery Rev 8:179–196
Pearlman R, Bewley TA (1993) Stability and characterization of human growth hormone. In: Wang YJ, Pearlman R (eds) Stability and characterization of protein and peptide drugs: case histories, Plenum Press, NY, pp 1–58
Pearlman R, Nguyen T (1992) Pharmaceutics of protein drugs. J Pharm Pharmacol 44: 178–186
Pearlman R, Oeswein JQ (1992) Human growth hormone formulation. United States Patent Number 5,096,885
Peppas NA, Langer R (1994) New challenges in biomaterials. Science 263:1715–1720
Pflugrath JW (1992) Developments in X-ray detectors. Curr Opin Struct Biol 2:811–815
Pikal MJ, Dellerman KM, Roy ML, Riggin RM (1991) The effects of formulation variables on the stability of freeze-dried human growth hormone. Pharm Res 8:427–436
Porter CJH, Moghimi SM, Davies MC, Davis SS, Ilum L (1992) Differences in the molecular weight profile of poloxamer 407 affect its ability to redirect intravenously administered colloids. Int J Pharm 83:273–276
Prestrelski SJ, Arakawa T, Carpenter JF (1993) Separation of freezing- and drying-induced denaturation of lyophilized proteins using stress specific stabilization. II. Structural studies using infrared spectroscopy. Arch Biochem Biophys 303:465–473
Pusztai A (1989) Transport of proteins through the membranes of the adult gastro-intestinal tract – a potential for drug delivery? Adv Drug Delivery Rev 3:215–228
Saltzman WM (1993) Antibodies for treating and preventing disease: the potential role of polymeric controlled release. Crit Rev Therapeutic Drug Carrier Syst 10:142
Saudek CD (1993) Future developments in insulin delivery systems. Diabetes Care 16 (Suppl 3):122–132
Smith PL, Wall DA, Gochoco CH, Wilson G (1992) Oral absorption of peptides and proteins. Adv Drug Delivery Rev 8:253–290
Smith PAG, Dewdney JM, Fears R, Poste G (1993) Chemical derivatization of therapeutic proteins. Trends Biotech 11:397–403
Spellman M, Basa L, Leonard C, Chakel J, O'Connor J (1989) Carbohydrate structures of human tissue plasminogen activator expressed in Chinese hamster ovary cells. J Biol Chem 264:14100–14111
Surewicz WK, Mantsch HH, Chapman D (1993) Determination of protein secondary structure by Fourier transform infrared spectroscopy: a critical assessment. Biochemistry 32:389–394
Tabata Y, Gutta S, Langer R (1993) Controlled delivery systems for proteins using polyanhydride microspheres. Pharm Res 10:487–496
Talmadge JE (1993) The pharmaceutics and delivery of therapeutic polypeptides and proteins. Adv Drug Delivery Rev 10:247–300
Thurow H, Geisen (1984) Stabilisation of dissolved proteins against denaturation at hydrophobic surfaces. Diabetologia 27:212–218
Tsai PK, Volkin DB, Dabora, JM, Thompson, KC, Bruner MW, Gress JO, Matuszewska B, Keogan M, Bondi JV, Middaugh CR (1993) Formulation design of acidic fibroblast growth factor. Pharm Res 10:649–659

Volkin DB, Middaugh CR (1992) The effect of temperature on protein structure. In: Ahern TJ, Manning MC (eds) Stability of protein pharmaceuticals, Part A: Chemical and physical pathways of protein degradation, Plenum Press, NY, pp 215–247

Volkin DB, Tsai PK, Dabora JM, Gress JO, Burke CJ, Linhardt RJ, Middaugh CR (1993) The stabilization of acidic fibroblast growth factor by polyanions. Arch Biochem Biophys 300:30–41

Volkin DB, Burke C, Marfia K, Middaugh R, Oswald B, Hennessey J, Orella C, Hagan A, Sitrin R, Oliver C (1997) Biophysical characterization of hepatitis A virus (HAV). J Pharm Sci 86:666–673

Wyatt PJ, Papazian LA (1993) The interdetector volume in modern light scattering and high performance size-exclusion chromatography. LC-GC 11:862–872

Wang Y-CJ, Hanson MA (1988) Parenteral formulations of proteins and peptides: stability and stabilizers. J Parent, Sci Tech 42 Suppl:53–526

Wang YJ, Pearlman R (eds) (1993) Stability and characterization of protein and peptide drugs: Case histories, Plenum Press, NY

Wearley L (1991) Recent progress in protein and peptide delivery by non invasive routes. Crit Rev Therapeutic Drug Carrier Systems 8:331–394

CHAPTER 4

Strategies for Dealing With the Immunogenicity of Therapeutic Proteins

M.L. Nucci, R.G.L. Shorr, and A. Abuchowski

A. Introduction

As proteins and peptides (hereinafter referred to collectively as "proteins") are major components of biological processes, it is logical to assume that their potential for use as therapeutics would be extremely high. However, the recognizable need for a defense system has implications for the administration of protein-based pharmaceuticals (Table 1). Minor differences in the primary structure of proteins – even as little as one amino acid from the human form of the protein – have been shown to be sufficient for immune system recognition.

During the past 15 years, techniques to manufacture large quantities of proteins through recombinant technology have become routine. Prior to recombinant technology, the majority of proteins and peptide pharmaceuticals were derived from non-human sources. Recombinant technology was seen as the ultimate solution to the problem of unwanted immunological responses to heterologous proteins through its ability to produce large quantities of human proteins.

It soon became evident that human protein therapeutics could still initiate an immune response. Clinical experience has shown that the development of antibodies to therapeutic proteins and peptides is common, even for recombinant human proteins. Antigen-specific antibodies have been shown to develop following the administration of: recombinant insulin (Fineberg et al. 1983), recombinant tissue-type plasminogen activator (Reed et al. 1990), recombinant human growth hormone (Kaplan et al. 1986), recombinant interleukin-2 (Allegretta et al. 1986), recombinant interferon (Itri et al. 1987), recombinant granulocyte-macrophage colony stimulating factor (Gribben et al. 1990), chorionic gonadotropin (Musch et al. 1981), leutinizing hormone releasing factor (Lindner et al. 1981), and blood-derived factor VIII (Roberts and Cromartie 1984). The development of antibodies may be associated with serious immunological consequences, e.g., anaphylaxis and/or death, neutralization of protein efficacy due to either clearance and/or protein inactivation, or may have no clinical effect at all (antibody is produced, but has no effect on protein activity or circulating life).

The "immunogenicity of proteins and peptides is considered to be the most serious issue of concern in the use of peptide and proteins as parenterals"

Table 1. Factors affecting the immunogenicity of proteins and peptides

Structure: primary and tertiary
Dosage size and frequency
Route of administration
Source of protein
Disease state
Genetics

(Martis 1986), yet, the value of protein pharmaceuticals is high enough to stimulate the development of strategies to allow for the safe and efficacious use of proteins as drugs. For example, protein drugs can always be administered in a clinical setting so as to have available the necessary resources to treat possible hypersensitivity or anaphylaxis, though this restriction is associated with an increase in drug cost and decrease in the patient's quality of life.

Immunological responses associated with neutralization or decreased circulatory life can be dealt with by increasing the drug dose over the duration of the dosing period, such as with blood-clotting factor (Roberts and Cromartie 1984), in which 15% of patients treated with plasma-derived human factor VIII developed neutralizing immunoglobulin G (IgG) (predominantly IgG_4) antibodies. This also increases the drug cost as well as the potential for serious side effects.

In certain cases, a tolerance of the offending antigen may be used to overcome the neutralization effect of the antibodies, and this procedure has been used with success in patients treated with factor VIII (Brackman and Gormsen 1977; Brackman and Egli 1981; White et al. 1983; Ewing et al. 1988), insulin (Gossain et al. 1985) and polyethylene glycol (PEG)-modified adenosine deaminase (ADA; Chun et al. 1993).

When no clinical effect is noted following antibody production, there is obviously no effect on therapeutic value or potential, but one must keep in mind the long-term effects of antibody production in such a situation. There are no hard and fast rules to govern whether such antibody production might lead to serious problems later on. Serum sickness, which is usually associated with large doses of protein therapeutics, has not been seen with either insulin or GH treatment, but has developed following the administration of streptokinase (Alexopoulos et al. 1984; McGrath et al. 1985; McGrath and Patterson 1985). However, streptokinase continues to be used, as its risk–benefit ratio is considered to be acceptable (McPherson and Livingston 1989).

Despite these problems, many programs continue to evaluate protein compounds for the clinic because of their direct connection to many disease conditions and their potential as successful therapeutics. Research is focused on the development of techniques to deliver proteins while avoiding possible anaphylactic or hypersensitivity reactions. Delivery systems under study include encapsulation, route of administration and conjugation with carriers. Technologies, such as protein and genetic engineering, are also under consideration, sometimes in combination with a delivery system.

B. Case Histories of Protein Therapeutic Development

Based on the variability of proteins, it is highly unlikely that any single strategy will prove to be effective for every protein. The strategies chosen to reduce immunogenicity will vary, based on the protein's inherent characteristics and its intended use. During development, much work focuses on techniques and methodologies to deliver proteins as efficiently as possible, while avoiding or decreasing any possible anaphylactic or hypersensitivity reactions.

A computer search of the *Description* category of the *Physicians Desk Reference* (PDR; MEHTA 1993) for proteins showed that there are at least 25 approved pharmaceuticals either based on proteins or containing proteins as excipients. Of the approved protein-based products on the market, five of the most commonly recognized are insulin, GH, asparaginase, glucocerebrosidase and OKT3, representing the protein classes of hormones, enzymes and monoclonal antibodies (mAb). Examination of the historical development of each illustrates some of the strategies available for producing an effective protein-based pharmaceutical, as well as the diverse nature of successful management of protein immunological responses.

I. Insulin

Of all protein-based pharmaceuticals approved for use, the one most frequently associated with the value of proteins as therapeutics is insulin, used in the treatment of insulin-dependent diabetes. Insulin has been used in the United States since 1922, and the history of its development is well reviewed elsewhere (BLISS 1982; MAYER 1982; MACLEOD 1978).

Initial clinical use of insulin resulted in the development of unwanted immunological side effects (WILLIAMS 1922; TUFT 1928), which led to further purification of the extract, resulting in a crystalline insulin (recrystallized insulin), which was first introduced into clinical practice in the late 1940s (OWENS 1980).

In spite of purification, the treatment of human patients with either a bovine or porcine source of insulin generally induced an immune response in which the presence of antibodies in the blood could be detected (BERSON and YALOW 1959, 1964, 1966; BERSON et al. 1956; CHANCE et al. 1976). Impurities in commercially available insulins noted by partition chromatography (CARPENTER 1958), disc electrophoresis (MIRSKY and KAWAMURA 1966) and gel filtration (STEINER 1967; STEINER et al. 1968) were identified as being primarily responsible for the development of circulating antibodies. Studies with homologous insulin in pigs (Lockwood and Prout 1965; Brunfeldt and Deckert 1966), cows (Renold et al. 1966) and humans (Deckert et al. 1972) suggested that the impurities in the insulin preparations acted as adjuvants for antibody development (Schlichtkrull et al. 1974). An additional chromatographic step was introduced into the manufacturing process of several commercially available insulins, resulting in the manufacture of monocompetent insulin preparations.

The changeover to highly purified insulin was not without immunological consequences (Christy et al. 1977; Leslie 1977; Reisner et al. 1978). Local allergic reactions were noted to occur in approximately 10% of patients (Galloway et al. 1982), though switching the patient from recrystallized insulin to monocompetent insulin was accompanied by a decrease in insulin-specific antibody titers in most patients (Williams 1922; Andreani 1973; Kurzt and Nabarro 1980; Fineberg et al. 1982).

The development of recombinant technology, which allowed for the production of commercial quantities of human proteins, was harnessed to produce a human-derived insulin, which was assumed would be free from all immunological responses associated with the heterologous insulins on the market. Human-derived insulin became available commercially in the United States in 1982. Even though bovine insulin differed from human insulin by only three amino acids, and porcine differed by only one amino acid (Grammar et al. 1984), it was believed that even this minor difference was enough to result in the elimination of allergic responses following administration of the human form of the protein.

The majority of patients who switched to human insulin demonstrated a decrease in antibody titers to insulin (di Mario et al. 1986; Brodgen and Heel 1987) and a decrease in allergic reactions (Fineberg et al. 1982; Grammar et al. 1984; Fireman et al. 1982; Wiles et al. 1983). Decreases in pre-existing anti-insulin IgG antibodies of up to 58.6% were noted in 15 patients treated with recombinant human insulin for 11–12 months. Reduction in the IgG_1, IgG_3 and IgG_4 subclasses were noted, with the greatest decrease in the IgG_4 subclass. Local allergy resolution, improvement of lipoatrophic lesions and some decrease in insulin requirements were also noted (Kumar 1993). Patients that began therapy with human insulin also had lower levels of insulin-specific IgE antibodies than did patients who previously received animal insulin (Fireman et al. 1982).

However, human insulin [both semisynthetic human insulin (Carveth-Jahnson et al. 1982; Altman 1983) and human insulin of recombinant-DNA origin (Wiles et al. 1983)] retained the ability to cause generalized allergic reactions in human patients. IgE antibody to insulin-antigenic determinants was shown to be responsible for most immediate-type reactions to insulin preparations (Lieberman et al. 1984; Grammer 1986), while most cases of systemic insulin allergy were associated with an interruption and resumption of treatment (Brodgen and Heel 1987; Dykiewicz et al. 1994; Feinglos and Jegasothy 1979). In certain cases, a previously existing hypersensitivity to animal insulins was associated with hypersensitivity to human insulin (LIEBERMAN et al. 1984; GARCIA-ORTEGA 1984; BLANDFORD et al. 1984). One patient demonstrated cross reactivity to both porcine- and bovine-insulin antibodies, and it was not until that patient was withdrawn from all insulin sources for a period of 18 months, then reintroduced to human insulin, that the allergic response was eliminated. This problem, and others associated with changeover in form of insulin, led to the development of procedures for the clinical use of the compound in order to control the development of an immune response (AALBERSE et al. 1983a,b).

Insulin is an example of a successful protein therapeutic, yet its use has been (and continues to be) associated with the need for techniques and strategies to deal with unwanted immunogenic responses. Recent papers dealing with the development of antibodies following insulin administration have detailed protamine-specific antibodies mimicking insulin hypersensitivity (DYKIEWICZ et al. 1994), or have implicated zinc (zinc and protamine are excipients in various insulin preparations) as being responsible for altering the three-dimensional structure of insulin and altering its immunogenicity (FEINGLOS and JEGASOTHY 1979). Research continues to examine ways to improve the administration of the drug in order to reduce side effects and better mimic the body's precise delivery of insulin.

II. Growth Hormone

GH, used to treat children with GH insufficiency, was originally prepared from cadaveric human pituitaries. On average, 45% of children treated developed GH-specific antibodies (PREECE 1986). As with insulin, the high incidence of unwanted immunological side effects was the stimulus to develop new processing methods to improve the purity of the material.

Subsequent to the introduction of improved purification methodologies, antibody formation decreased, such that only 10% of all children being treated developed antibodies specific for GH (LUNDIN et al. 1991). The purest preparation available at that time (Crescormon, which is no longer available) was associated with antibody formation in only 2% of all patients (LUNDIN et al. 1991).

Cadaveric GH was pulled from the United States market in 1985 due to deaths from Creutzfeld-Jacob syndrome, in which the drug was implicated as a carrier of the agent that caused the neurological disease (NORMAN 1985). The inability to purify the cadaveric GH, as well as restrictions on its clinical use due to a lack of cadaveric human pituitaries, catalyzed the development and manufacture of a recombinant form of the hormone. Recombinant-produced GH became available in the United States in 1985. The first clinical experiences resulted in most patients developing anti-GH antibodies within a few months of treatment (BIERICH 1986). As with insulin, antibody production was shown to be associated with an impurity in the preparation. In the case of GH, impurities from the *E. coli* used to manufacture the GH were demonstrated as the antigens. Further purification resulted in the development of a GH with low immunogenicity (Lundin et al. 1991).

Fully human GH was first approved for human use in the United States in 1985. The first report of antibodies specific for this form of GH was noted after 12 months of treatment in 1.3% of patients (Wilton and Gunnarson 1988). Clinical trial experience between 1985 and 1989 demonstrated an incidence of GH-specific antibodies of 1.1%, with no incidence of allergic reaction. There was no antibody-related growth attenuation and no effect on clinical treatment (Ludin et al. 1991), indicating that GH was of low immunogenicity.

Unlike other proteins, antibodies produced by GH may actually potentiate its clinical effect. It has been shown that the biological activity of GH can be potentiated by complexing with mAb specific for GH. In hypophysectomized rats, injections of bovine GH complexed with mAb induced a greater increase in body weight and serum insulin-like growth-factor 1 (Walles et al. 1987) than when injected alone. These results have been confirmed by other investigators (Holder et al. 1985; Aston et al. 1987; Wang et al. 1990). In some patients, the half-life of the drug was actually extended after the development of antibodies (Underwood et al. 1974).

III. Asparaginase

In 1961, asparaginase was isolated from guinea-pig serum and identified as the agent responsible for anti-lymphoma activity (Broome 1961). When *E. coli* asparaginase was also shown to have anti-tumor activity (Mashburn and Wriston 1964), it became possible to produce material in sufficient quantities for clinical trials, with subsequent approval in the United States in the late 1960s. Dose-limiting hypersensitivity reactions developed in 3–78% (Kurtzberg et al. 1993) of patients treated, with serious anaphylactic reactions occurring in less than 10% (Evans et al. 1982). Antibody neutralization diminished therapeutic efficacy, either by decreasing activity and/or increasing clearance (Kurtzberg et al. 1993). The potential for adverse immunological responses was shown to increase with the number of doses, becoming more common upon repeated exposure.

The potential for asparaginase in the treatment of leukemia provided the motivation to develop techniques to overcome its immunological side effects. Unlike insulin and GH, recombinant technology played no role in the therapeutic development of asparaginase. Development was, instead, associated with refinement of fermentation techniques, the comparison of several forms of asparaginase for immunological potential (Asselin et al. 1993) and dosing strategies to minimize hypersensitivity (Tallal et al. 1970; Jones et al. 1979; Ertel et al. 1979; Sallan et al. 1983; Clavell et al. 1986). Dosing at daily intervals (Ellison et al. 1991), rather than weekly intervals (Land et al. 1989), decreased hypersensitivity from 73% to less than 3%. Intramuscular injections, rather than intravenous administration, were chosen as the preferred route, due to the increased risk of hypersensitivity upon intravenous administration (Nesbit et al. 1979).

Alternate forms of the enzyme, with low cross reactivity, were made available to treat patients hypersensitive to the *E. coli* form. Recently, a new form of the *E. coli* asparaginase coupled with PEG became available. PEG-asparaginase (ONCASPAR[R]) is indicated for patients who are hypersensitive to the non-PEG-modified form of the enzyme. This protein drug was developed specifically to address the immunological side effects associated with the use of asparaginase. The final product incorporated the technology of PEGylation, which is associated with a decrease in immunogenicity and antigenicity of the base protein (Asselin et al. 1993). Modification of asparaginase with PEG extends the usable therapeutic window, by allowing for extended treatment regimens at lower doses (Keating et al. 1993).

IV. Glucocerebrosidase

If there are any diseases in which protein therapeutics demonstrate their highest potential, it is in the arena of inherited metabolic disorders. Over 5000 genetic disorders have been identified, 30% of which are recessive or X-linked and are often due to a lack of a specific protein product (Collins 1991; Belmont and Caskey 1986).

Gaucher's disease is an inherited metabolic disorder resulting from a deficiency of the enzyme glucocerebrosidase (β-glucosidase) (GC). Though certain forms of the disease are not associated with clinical symptoms, type-II Gaucher's is usually fatal, with death usually occurring before 2 years of age.

Purified human placental carbohydrate-modified GC (Ceredase) was approved for human use in 1991 in the United States for the treatment of Gaucher's disease. Unlike the other proteins previously discussed, GC was modified with carbohydrate to target the protein to the macrophages of the reticuloendothelial cells, where it was believed GC had to be delivered to provide efficacy. When administered intravenously to two patients over a 3-week period, no IgG or IgM antibodies were noted. Patients received either 11 mg or 28 mg total in nine doses (Britton et al. 1978).

Beneficial responses to clinical treatment (Barton et al. 1990) led to increased numbers of patients receiving carbohydrate-modified GC. When serum from those patients was tested by a sensitive enzyme-linked immunosorbent assay (ELISA), a significant number of positive responses for GC-specific antibodies were noted in both normal and Gaucher's patient's sera. The high number of positive responses were interpreted by the authors as false positives (Murray et al. 1991).

Although antibody response to repeated carbohydrate-modified GC administration has remained low, and the percentage of patients with clinical sequelae following antibody production is also low (two of five patients with demonstrable anti-GC antibodies had mild antibody-mediated reactions

(Pastores et al. 1993); 90% of all patients treated with carbohydrate-modified GC develop antibodies to the protein within the first 6 months (Richards et al. 1993). A recombinant form of glucocerebrosidase was developed and approved in the United States.

Like GH, recombinant carbohydrate-modified GC (Cerezyme) was developed to address the concern of possible viral contamination as well as answer the need for increased clinical quantities of material for patient use (Grabowski et al. 1993). It was approved for clinical use in the United States in 1994. Clinical effectiveness and antibody formation to the recombinant form was similar to that noted with the placental form of the enzyme (Grabowski et al. 1993).

The frequency of reactions to administered recombinant carbohydrate-modified GC was higher than the frequencies reported for patients treated with other recombinant human-tissue-derived therapeutic proteins. It was theorized that the lack of the hormone during fetal development resulted in its being recognized as foreign, even when the human form of the enzyme was used (Richards et al. 1993).

A new form of GC to treat Gaucher's disease recently began clinical trials. This form of the enzyme is not modified with carbohydrate for targeting purposes, rather it is modified with PEG. Development of this form of the protein is based on the concept that GC does not require mannose targeting to achieve therapeutic effect.

The enzyme is a recombinant human protein produced in a baculoviral vector. The attachment of PEG to the protein results in a compound with increased circulating life (Cho et al. 1994) and decreased immunogenicity. No results are currently available from the clinic, but in preclinical studies, PEG-GC was able to degrade both synthetic and natural substrates (Yang et al. 1994).

V. OKT3

The success rate of transplantation medicine greatly increased upon the development and approval in the United States in 1986 of the mouse mAb OKT3 to reverse organ-allograft rejection. This antibody reacts with the T3 complex on killer T cells and renders them inactive. T3 cells are the cells responsible for graft rejection.

However, the use of OKT3 is not without side effects. Chills, fever, headache, gastrointestinal (GI) upset and, occasionally, pulmonary edema, aseptic meningitis and intragraft thromboses (Thistletzwaithe et al. 1988; Abramowsicz et al. 1989) have all been associated with its use. More importantly, the use of OKT3 is limited by the development of specific antibodies (Vigeral et al. 1986; Thistlewaithe et al. 1984; Goldstein 1987; Jaffers et al. 1986; Goldstein et al. 1986), which are generally anti-idiotypic (Chatenoud et al. 1986) and oligoclonal (Chatenoud et al. 1986). These antibodies result in a loss in the therapeutic efficacy of OKT3 (Hirsch et al. 1989) and, in some cases, have resulted in anaphylactic shock after repeat treatment (ABRAMOWICZ et al. 1992; WERIER et al. 1991).

It has been postulated that either a humanized mAb specific for T3 or the development of different idiotypes of anti-CD3 mAbs would result in an OKT3 with decreased immunogenicity (HIRSCH et al. 1989). Research on humanized anti-CD3 monoclonals has demonstrated that the humanized version, even when presented as an Fc fragment, was able to suppress cytolytic T-cell activity at a level comparable to OKT3 (ALEGRE et al. 1992, 1994; WOODLE et al. 1992). However, such research has not, to date, been applied to testing in the clinic. Studies focusing on the development of techniques to help patients develop a tolerance to OKT3, using cyclosporine and azathioprine, have shown little success (CHATENOUD et al. 1986).

C. Strategies Under Development for Increasing the Therapeutic Value of Proteins and Peptides

Various strategies for increasing the therapeutic value of proteins are being researched. Though varied in their technology and method, each attempts to deliver proteins with a minimum of immunological side effects. The strategies discussed will represent only a few of those under development, but these are the ones leading the way to an expanded use of proteins in therapy.

I. Encapsulation

The enclosure or entrapment of protein pharmaceuticals is posited on the premise that a compound so enclosed is protected from immunological attack by the body. The protected protein would then be free from the immunological side effects associated with protein drugs.

There are several enclosing methods under development, which utilize either naturally occurring capsules (erythrocytes), naturally occurring materials (lipids, dextran, starch) or synthetic materials (usually polymers). The capsules can be membranous (as in liposomes and semipermeable microspheres) or rigid and impermeable (CHANG 1964, 1972; MACHY and LERMAN 1987; GREGORIADIS 1984). These materials are generally believed to be non-immunogenic, though in certain cases, they have been shown to elicit an immune response specific for the material. Polyacrylamide, which is in use as a non-immunogenic carrier for haptens, when coupled with bovine serum albumin initiated the production of antibodies specific for polyacrylamide (DREWES et al. 1978). Liposomes (ALLISON and GREGORIADIS 1974; KINSKY and NICOLETTI 1977), polyacrylamide capsules (EDMAN and SJOHOLM 1982), polyacrylstarch and precipitated dextran–starch capsules (ARTURRSON et al. 1985) have all been shown to elicit immune responses.

Synthetic capsules can be composed of a number of different substances, including lipids, polymethacrylate (KREUTER 1983a–c), polyacrylamide (Oppenheim 1981), polyalkylcyanoacrylate (Brasseur et al. 1980; Couvreur et al. 1979, 1980, 1982) or polyacrylstarch (Arturrson et al. 1984). If desired, these capsules can be modified by the surface attachment of a number of different markers for either targeting, labeling or sensing. These markers can be chemical, fluorescent, magnetic or radioactive, depending on the intended target of the capsule.

Encapsulation can allow for passive or active targeting, or sustained release of a protein pharmaceutical. Passive targeting is the naturally occurring distribution of the encapsulated protein after administration, usually by injection. The distribution can be affected by the size, charge density, fluidity, or hydrophobicity or lipophilicity of the carrier. Small colloidal particles used as protein carriers are generally taken up by the cells of the reticuloendothelial system (RES), especially those found in the liver and spleen (Storm et al. 1991). To profit from clearance by the RES, capsular technology is being used to treat diseases of the RES, such as leishmaniasis, cryptococcosis and histoplasmosis, as well as to increase the in vitro or in vivo (Storm et al. 1991) tumoricidal activity of macrophages through the encapsulation of interferon or macrophage-activating factor.

Active targeting involves the use of a targeting agent, which is linked to the protein carrier, such as a mAb specific for certain cells or receptors. Active targeting allows for the direct and specific interaction of the drug with the target tissue, and protection of the body and protein from interaction with each other (Tomlinson 1988).

Sustained protein-drug release can potentially eliminate the high peak levels associated with single-dose administration and increase the therapeutic efficiency of a protein drug. However, sustained release of a protein pharmaceutical is generally not possible with an unprotected protein, especially if the protein is strongly immunogenic. The encapsulation of protein drugs can allow for sustained release through the formation of a depot after injection, from which the drug is released over a period of time.

In both visceral and cutaneous leishmaniasis, liposome-encapsulated antimicrobial drugs were effective in suppressing the disease in hamsters (Alving et al. 1978), mice (New et al. 1978) and dogs (Chapman et al. 1984). A prophylactic effect of liposome-encapsulated drugs was noted when the liposomes were injected 1 week before leishmanial infection (Alving et al. 1980). Toxicity of the liposome-encapsulated drugs was much higher than the free drug, but when compared with the therapeutic efficacy, the therapeutic index for the liposome-encapsulated drug was approximately 40 times that of the free drug (Alving et al. 1978).

In cryptococcosis, liposome-encapsulated amphotericin B was more effective than the free drug in mice (Graybill et al. 1982). Increased efficacy was mainly due to the ability to deliver large doses of drug without increased toxicity, as noted when the equivalent doses of free and liposomally encapsulated drug were administered. There was no increased therapeutic efficacy of the encapsulated drug when given at the same dose as the free drug (Graybill et al. 1982). Liposome-encapsulated amphotericin B was shown to be successful in the treatment of histoplasmosis in humans (LOPEZ-BERESTEIN et al. 1985).

When interferon-γ and muramyl tripeptide phosphatidylethanolamine were incorporated into liposomes, resistance of mice against Listeria monocytogenes (L. monocytogenes) infection increased 33- to 66-fold, compared with free drug. The RES clearance of the liposomes resulted in more rapid clearance of L. monocytogenes infection from the liver and spleen (MELISSEN et al. 1993).

Unfortunately, to date, all capsules developed for non-RES disease indications have shown problems restricting the activity to the intended target site (WIDDER et al. 1979, 1982), as well as achieving a circulating life of sufficient duration to achieve a significant therapeutic effect. However, new formulations of long-circulating liposomes have resulted in a resurgence of interest in liposomes for non-RES-disease therapy. When added to the liposome structure, ganglioside GM_1 increased tumor uptake and circulating life (GABIZON 1988). These stabilized liposomes, to which PEG was attached to their surface (ALLEN 1994; BLUME and CEVC 1993; NEEDHAM et al. 1992; SENIOR et al. 1991), demonstrated a reduced affinity for the RES and circulating half-lives of several hours (ALLEN et al. 1991; BLUME and CEVC 1990). They were shown to localize in high concentrations in solid tumors (GABIZON 1992; HUANG et al. 1992) and at sites of infection (BAKKER-WOUDENBERG et al. 1993). In mice, PEG liposomes delivered doxorubicin more effectively than free doxorubicin, resulting in long-term survival of 50% of mice implanted with squamous lung carcinoma (AHMAD et al. 1993).

As passive targeting will not prove effective for all disease indications, especially those in which the target tissues are widely spread, e.g., leukemia or metastases, direct targeting of capsules is under evaluation. To achieve direct targeting, ligands specific for the target tissue can be attached to the surface of the capsule. Desialyted fetuin (GREGORIADIS and NEERUNJUN 1975), erythrocyte membrane glycoprotein (JULIANO 1976), antibodies (WEISSMAN et al. 1975), or purified receptors or antireceptor molecules (JULIANO and STAMP 1976) have all been used to target capsules. Magnetic microspheres have been used in conjunction with an external magnetic field over the specific target site (WIDDER et al. 1983) in order to direct the encapsulated drug.

Early formulations of non-stabilized liposomes, when targeted with antibody, were shown to have a reduced circulating time as the antibody increased uptake by the RES (ALLEN 1994). Though stabilized liposomes also demonstrated a decreased circulating time when targeted with antibody (TORCHILIN et al. 1992), their circulation time was sufficiently extended by stabilization to still provide a long circulating time. These targeted liposomes were shown to localize to murine lung tissue (HEATH 1987), infarcted rabbit myocardium (GABIZON et al. 1989) and lung endothelial cells (MAYER et al. 1986). Though therapeutic efficacy was not determined, it was shown that the targeted liposomes were able to bind to specific in vivo targets.

II. Non-parenteral Routes of Administration

It has been shown that the route of administration affects the immunological response to proteins, usually based on the rapidity with which the protein is exposed to the immune system. In those cases, where a single injection of a protein drug is required, parenteral delivery of protein drugs is generally not an issue. However, non-parenteral routes (oral, transmucosal, pulmonary, transdermal, etc.) are preferred routes for long-term administration of protein drugs. Unfortunately, the larger a molecule, generally, the greater the problems that must be overcome in terms of transport and stability across biological membranes. Proteins are among the largest of drugs.

Among the non-parenteral delivery systems, transmucosal delivery via the vaginal, buccal or rectal routes has been shown to be more successful at delivering protein drugs than oral administration. However, of all the possible non-parenteral routes, oral administration is probably the preferred in terms of patient compliance and ease of beginning and ending administration. Unfortunately, oral delivery of proteins has been problematic at best, as proteins are very susceptible to the acidic environment of the gut as well as to proteolytic digestion by GI enzymes. Oral delivery also subjects protein drugs to first-pass hepatic metabolism, decreasing their active residence time. Additionally, as most proteins are of a high molecular weight (MW) and are hydrophilic, they do not easily cross the intestinal mucosa.

Little success has been shown for oral protein-drug delivery. Bioavailabilities of less than 1% are common (Davis 1992). The larger the drug administered, the smaller the amount of material transported across the membranes. Enhancement of oral delivery has been attempted by the protection of the protein through liposome encapsulation (Smith et al. 1992), the use of enhancing agents (Muranishi 1990), or microparticulates and microemulsions (Ritschel et al. 1990), to little positive effect.

In general, transmucosal delivery has been more successful than oral delivery. Nasal delivery avoids first-pass degradation by the liver and is in use in the United States for several approved protein drugs, including luteinizing hormone-releasing hormone analogs, desmopressin, oxytocin and calcitonin. The bioavailability of nasally delivered proteins is correlated with their MW (Fisher et al. 1987), with molecules greater than 10 kDa requiring bioadhesives and/or enhancing agents to achieve a decent rate of absorption (Fisher et al. 1991). Bioavailability on the order of 40%, compared with intravenous, intramuscular or subcutaneous administration, is common (Davis 1992). The most serious disadvantages of nasal delivery are the negative effects of the drugs and/or excipients on nasal ciliary cells, of which proper functioning is essential, as well as the presence of proteases and peptidases in the mucosa, which may act as a barrier to absorption of the intact, active protein (Stratford and Lee 1985; Lee 1986).

Pulmonary delivery for localized therapy has been shown to be a viable and effective means of treatment for certain lung diseases. Pulmonary delivery for systemic effect has not been shown to be as effective, since delivery externally from the lungs is limited. The ability of the drug to interact with endogenous transport systems and its lipophilicity, controls the external diffusion of drug molecules (ENNA and SCHANKER 1972). With presently available administration methods, 1% deposition in the alveoli is common. Greater deposition is dependent on nebulization and/or increased time of administration (STRATFORD and LEE 1985).

Vaginal delivery is most commonly used for the treatment of sex-specific diseases, where absorption of the drug has been of interest only from the point of view of toxicity. Though it is known that certain compounds, such as insulin (FISHER 1923; ROBINSON 1927), TSH (COLCHER et al. 1990) and various antigens (BENZIGER and EDELSON 1983; STRAUS 1961; FRANKLIN and DUKES 1964; ROIG DE VARGAS-LANARES 1978), are absorbed from the vaginal membranes, bioavailability appears to be variable and insufficient with cyclic changes in the vaginal tissue, affecting the absorption of vaginally administered drugs. However, studies with various protein hormones have indicated absorption sufficient to produce a systemic response (PERACCHIA 1978; WINDSOR and CRONHEIM 1961). As with other forms of non-parenteral delivery, larger MW proteins require the use of enhancing agents to facilitate absorption.

Buccal delivery shows promise in that the oral mucosa is less prone to damage or irritation than nasal mucosa, it bypasses hepatic first-pass metabolism, and promises better patient compliance than rectal or vaginal delivery. Tissue-barrier characteristics are drug dependent, with the bioavailability of buccally delivered drugs being relatively small, indicating that absorption enhancers or enzyme inhibitors to decrease drug degradation during tissue transit are required (DE VRIES et al. 1991). Buccal delivery systems must be designed to avoid swallowing or salivary washout, and must remain in intimate contact with the mucosal membranes for a prolonged period of time (VEILLARD et al. 1987; MEYER et al. 1984; NAGAI 1985).

Transdermal delivery of proteins is limited by the structural nature of the skin and physiochemical properties of the drug in question. Large molecules, such as proteins, are not well absorbed in their intact form due to their low diffusability. However, transdermal delivery has several advantages over parenteral delivery, including the avoidance of first-pass metabolism, potentially decreased side effects and the relative ease in beginning and ending drug therapy (CHIEN 1983; GUY and HADGRAFT 1985). Several protein drugs are currently approved for transdermal delivery. Transdermal delivery has been associated with allergic contact sensitivity, usually at the site of application, though sites of previous application may show a reaction, indicating residual antigen in the skin (TROZAK 1985).

Transdermal delivery of most drugs is not possible by simple diffusion. Even though a drug may penetrate the stratum corneum, the intercellular pathway is the major barrier to drug permeation (ELIAS and FRIEND 1975). In addition, the skin in metabolically active and drug transformation may occur after application, rendering the protein inactive (FRIENKEL 1983; MARTIN et al.

1987). Methods to increase percutaneous absorption include iontophoresis (Pikal 1990), occlusion (Barry et al. 1984), vehicle manipulation (Kubota et al. 1990), phonophoresis (Murphy and Hadgraft 1990) and absorption enhancers (Williams and Barry 1992).

III. Targeting

Drug targeting is based on the principle that cells display specific markers on their surfaces, to which complementary compounds can be directed. By attaching a protein to these complementary compounds, direct drug delivery can be achieved, with a decrease in systemic drug toxicity and an increase in therapeutic efficacy. Many different types of compounds have been proposed for site-specific delivery, including antibodies or antibody fragments, naturally occurring or synthetic polymers, carbohydrates, or particulate carriers such as liposomes or synthetic particles.

Targeted delivery can be achieved by the recognition of physiological mechanisms, which result in selective uptake of administered materials. For example, since liposomes are selectively taken up by the RES, they can be used to deliver drugs to treat or activate RES macrophages. As macromolecular carriers are selectively directed to the lysosomal compartments of cells following their pinocytic capture (Duncan 1985), they can be used to deliver drugs into the cell. Carbohydrates, hormones, drugs, glycoproteins and lectins can be used to provide targeting specificity through specific cell-surface receptors. A large number of mammalian cells express cell-surface sugar-binding proteins, to which glycoconjugate-modified proteins can be targeted. Superoxide dismutase (SOD) was selectively delivered to macrophages by the attachment of mannose or galactose (Fujita et al. 1992a) and demonstrated therapeutic efficacy against hepatic injury following ischemia/reperfusion (Fujita et al. 1992b).

Due to their specificity, the most common targeting agents are antibodies, either whole or fragments. Antibodies can be directly attached to a protein or attached to the surface of a carrier. Specific tumor localization of radiolabeled antibodies for human melanoma antigens (Hellstrom et al. 1984; Larson et al. 1984), carcinoembryonic antigens (Gasparini et al. 1988; Haller 1988) and other cell-surface antigens (Jain 1990; Andrew et al. 1986; Starling et al. 1989) has been demonstrated. In animal-model studies, antibody–drug conjugates or antibody–toxin conjugates demonstrate superior effects as anti-tumor agents, when compared with free drug (Blakely 1992). When conjugated to antibody, tumor necrosis factor (TNF) significantly increased cytotoxic activity against cells resistant to TNF alone, with this increased activity appearing to be mediated by antibody binding to the cell surface (Zuckerman et al. 1987). When coupled to antibody, alpha-interferon was also more selective and effective than free alpha-interferon (Rosenblum et al. 1994; Alkan et al. 1988).

The size of the antibody molecule (50–150,000 Da) has been shown to limit the ability of an antibody–drug conjugate to penetrate all tissues, especially tumor tissues. When used for anti-tumor therapy, intact antibody tends to accumulate in the perivascular region of the tumor (KENNEL et al. 1991; BORMAN et al. 1991). The generation of antibody fragments by both enzymatic and recombinant techniques (COLCHER et al. 1990) results in a compound that retains specificity, but is able to penetrate tumors better than the intact antibody. To increase tumor penetrance, single-chain antigen-binding proteins (sFvs) (MW approximately 20,000 Da) were tested. When sFvs directed to the human transferrin receptor were bound to exotoxin, human cells bearing the targeted receptor were killed (BATRA et al. 1990). sFvs specific for the p55 subunit of the human interleukin-2 receptor were able to selectively kill cell lines with that receptor active in a mixed leukocyte reaction (BATRA et al. 1990; CHAUDHARY et al. 1990). When used in radioimmunoguided surgery, the sFv CC49 localized primary tumors 86% of the time, compared with 75% for the intact antibody B72.3. CC49 successfully targeted metastatic disease 97% of the time, compared with 63% of the time for the intact antibody (COHEN et al. 1991). sFvs can be made with novel effector functions or can be produced as bifunctional sFvs, thus, extending their delivery capability. While Fab and F(ab′)2 fragments accumulate in the kidney, sFvs do not (SANDHU 1992).

Antibodies can also be engineered to retain the variable (V) domain and antigen-binding specificity or the constant (C) domain and effector functions. When an intact V domain was coupled with *Staphylococcus aureaus* nuclease, the specificity of the binding site and nuclease activity were combined (NEUBERGER et al. 1984). When CD4 was fused to C domains, the resultant compound, in the presence of complement, lysed human immunodeficiency virus (HIV)-infected CD4+ cells (SMITH et al. 1987).

When translated to the clinic, to date, there has been little evidence of significant clinical response with the use of antibodies as targeting agents (GHOSE et al. 1977). In humans, antibodies coupled to chlorambucil, daunorubicin or neocarzinostatin showed some positive anti-tumor benefits (GHOSE et al. 1977; MELINO et al. 1985; TAKAHASKI et al. 1988), while normal tissue binding limited the use of antibody-ricin-A conjugates in clinical trials (GOULD et al. 1989).

As the majority of antibodies in clinical trials are murine based, their use is often limited, due to the development of antibodies to the non-human component of the conjugate. However, adverse effects following antibody development have been minimal, with an incidence of mild allergic phenomena of approximately 5% (KHAZAELI et al. 1994). The most serious consequence of antibody development appears to be a shortening of circulating life (KHAZAELI 1989; KHAZAELI et al. 1988; MEREDITH et al. 1992), with reaction to injection varying, based on the antibody injected and disease state (KHAZAELI et al. 1994). Patients that were immunocompromised showed less antibody development. Patients with cutaneous T-cell lymphoma had a higher anti-antibody response than patients with chronic lymphocytic leukemia (Khazaeli et al. 1994).

When humanized, immunogenicity has been shown to decrease, as shown with the mouse mAb 17–1A, which was found to be completely non-immunogenic in nine out of ten patients after multiple injections (Khazaeli et al. 1994; LoBuglio et al. 1989). Immunogenicity can also be decreased by the attachment of PEG (Goodman et al. 1990) or low-MW dextran (Fagnani et al. 1990), which reduces immunogenicity in xenogeneic hosts.

Immunogenicity may be more problematic with toxin conjugates, as toxins are immunogenic by themselves (Godal and Fodstad 1989). Clinical trials with ricin A were accompanied by an antibody response to both components within 10–12 days, following the initial injection (Byers et al. 1989). High titers of antibodies to ricin A and murine antibodies were found in all patients receiving the conjugates (Byers et al. 1989; Spitler et al. 1987; Weiner et al. 1989). In a number of animal models, toxicity due to non-specific binding was an issue in the administration of antibody coupled to radionuclides, in spite of a regression noted in some of the studies (Blakely 1992).

Techniques to limit non-specific binding involving the use of external magnetic sources coupled with the administration of a magnetic antibody-drug conjugate or bispecific antibodies, in which one domain recognizes the target antigen (the tissue surface) while the other recognizes a chelated radionuclide, are being examined and have been shown to result in a decrease in non-specific toxicity (Le Doussal et al. 1990; Kurtzman et al. 1988). Bispecific antibodies, where one domain recognizes the target and the other recognizes the drug, are also under development. These have been shown to increase the fibrin specificity and thrombolytic potency of tissue plasminogen activator (tPA), and single- and double-chain urokinase-type plasminogen activator (Bode et al. 1989; Charpie et al. 1990; Kurokawa et al. 1989; Branscomb et al. 1990). By choosing an antibody that binds to a specific region on the protein adjacent to activation sites for unwanted properties, modulation of these activities of plasminogen activator was noted (Bos et al. 1992). Bispecific antibodies specific for a target tissue and a drug compound can also be used to prelocalize antibody to the target cell, wait for systemic clearance of the antibody and then administer therapeutic compound, which binds to the target.

Combinations of delivery systems can also be designed to take advantage of properties of each. For the delivery of genes, the combination of antibody, polymer and poly-L-lysine was used to specifically deliver DNA, yielding an increased transfection level without an increase in non-specific transfection (Trubetskoy et al. 1992). The antibody OX36, coupled to avidin, increased the brain delivery of a vasoactive intestinal peptide (VIP) analog, resulting in a 65% increase in cerebral blood flow, compared with no increase for the same dose of VIP administered without the carrier (Bickel et al. 1993).

IV. Conjugation

The attachment of compounds to the surface of proteins to reduce immunogenicity is based on the fact that surface determinants of proteins are responsible for the resulting immunological response after administration. By attaching compounds to the surface, one can hide or disguise these determinants and, theoretically, be able to diminish the immune recognition of the protein with the result that the protein should circulate longer and have fewer adverse side effects. Additionally, should antibodies develop, their ability to react with the conjugated protein drug should also be diminished, due to an inability of antibody to react with the surface determinants. Early studies on the antigenicity of proteins demonstrated that modification of proteins with certain polymers resulted in a loss of antigenic potential (Sela 1969; Habeeb et al. 1958; Jones and Leskowitz 1965).

Many compounds are under investigation as modification agents, including dextran, *N*-(2-hydroxypropyl) methacrylamide copolymers (HPMA), divinyl ether-maleic acid (DIVEMA), styrene-maleic acid/anhydride (SMA), and PEG. Generally, the compound is covalently bound to the protein, only to be released during the normal degradation of the adduct. However, if required for efficacy, it is possible to provide for controlled degradation and specific targeting when designing the adduct, through the proper addition of targeting agents and/or degradable linkers. A variety of compounds can be linked together to perform specific functions, such as targeting, solubilization and transportation (Yoyokama 1992). The protein may be bound directly to the polymer or through a spacer. Spacer design is generally dependent on whether the protein needs to be released from the polymer to express activity.

As most of these compounds are macromolecular and the majority of synthetic polymers with MW greater than 80–100,000 Da cannot be excreted by the kidneys (Kopecek 1981), pinocytosis and transfer into lysosomes is the major intracellular degradation route (Trouet et al. 1982). Drug–polymer linkers can be designed to be stable in the circulation and extracellular space, while capable of hydrolysis once internalized in the cells.

Polymers can be used to increase drug efficacy by functioning as stabilizers and protectants, altering biodistribution and decreasing toxicity (Maeda et al. 1991). As with liposomes, some polymers are taken up preferentially by certain cells and, thus, may be used to primarily target a particular tissue. Macromolecular drug conjugates are generally transferred intracellularly to lysosomes. Based on the design of the conjugate, the drug can be delivered intracellularly post-degradation of the linkage by lysosomal enzymes.

Most polymers are chosen because of their low biological activity. DIVEMA is used because, by itself, it expresses anticancer activity through immune system stimulation (Fung et al. 1979). When conjugated to doxorubicin, the resulting drop in drug toxicity allowed for increased dosages (Hirano et al. 1986). Within 60 days of injection into mice with an intraperitoneal tumor, 50% of the mice receiving the doxorubicin conjugate were alive, while all the mice receiving only free doxorubicin were dead within 3 weeks (HIRANO et al. 1986).

Naturally occurring polymers, such as dextran, have been shown to alter the physiochemical properties of protein drugs, resulting in increased circulating life, decreased immunogenicity, increased thermal stability and resistance to proteolysis (MIYATA et al. 1988). SOD, with 34% of its free amino acids modified with dextran (MW 80,000 Da), retained 67% activity, but had a higher immunogenicity than SOD in vivo (MIYATA et al. 1988). Asparaginase modified with dextran displayed a decreased immunogenicity and a 50% reduction in activity (WILEMAN et al. 1983, 1986). Higher MW dextrans were shown to increase immunogenicity of attached proteins, haptens or peptides (LEES et al. 1994). Oxidized dextran coupled via lysine residues to soybean trypsin inhibitor retained significant activity, and demonstrated an increased circulating life and decreased immunogenicity by PCA (TAKAKURA et al. 1989a,b). Doxorubicin attached to oxidized dextran had an LD_{50} approximately 3- to 4-fold higher than free doxorubicin, but a much higher therapeutic efficacy and increased circulating life (FUJITA et al. 1991).

Dextran-modified horseradish peroxidase combined with liposomes demonstrated increased intracellular stability, such that 9% of the enzyme was present in the liver 24 h after administration, compared with no free enzyme detectable in the same time frame (MUMTAZ and BACHHAWAT 1992). The rate of degradation of the dextran conjugate was decreased, compared with free enzyme. Of the conjugate present in the lysosome, 70% was detectable at 24 h, compared with 5% of the free enzyme (MUMTAZ and BACHHAWAT 1992).

HPMA has also been used to modify doxorubicin, and the resultant compound demonstrates an increase in circulating life from 2 min to approximately 60 min in mice (SEYMOUR et al. 1990). Tumor drugs levels of 7.5 μg/g were achieved following administration of HPMA-doxorubicin conjugates, compared with 0.55 μg/g following administration of free doxorubicin. The modified doxorubicin was noted to have decreased cardiotoxicity (MAEDA et al. 1992).

When the anti-tumor protein neocarzinostatin was coupled to SMA, the antigenicity of the base protein decreased 8–16 times (MAEDA et al. 1984), and the compound demonstrated enhanced stability in aqueous, as well as the lipid medium (HIRAYAMA et al. 1986). In clinical trials in patients with advanced inoperable disease, survival time increased from a mean of 4 months to survival rates of 30–90% at year 3 and year 4 (MAEDA 1991; KONNO et al. 1983, 1984). More recent clinical trials have indicated that this adduct is weakly antigenic, but its clinical efficacy was not compromised (MAEDA et al. 1992), and the compound was approved in Japan for the treatment of hepatocellular carcinoma.

Of all the modification technologies, PEG conjugation appears to provide the most consistent pattern in changes of protein characteristics. Recent reviews discuss, in detail, the successful use of PEG in protein modification (DUNCAN and SPREAFICO 1994; KATRE 1993). PEG compounds generally display increased circulating lives (by 3- to 486-fold, dependent on the protein modified), reduced immunogenicity and antigenicity, and increased solubility (DELGADO et al. 1992). Decreased activity of the protein, common following modification, is offset by the increase in circulating time. The increased circulating life allows for decreases in the dosage size, thus, resulting in a decrease in toxicity as an indirect effect of PEG modification.

PEG is the only modification polymer conjugated to a protein drug that has achieved Food and Drug Administration approval. PEG conjugated to ADA (ADAGEN) was approved in 1990 for the treatment of ADA-deficient severe combined immunodeficiency disease. Patients currently undergoing gene therapy for this disease are treated concurrently with ADAGEN. Patients have been receiving ADAGEN for up to 8 years, with no diminishment in therapeutic efficacy. Though several patients did develop antibodies specific for the conjugate, they were able to develop a tolerance and continue therapy (CHUN et al. 1993).

PEG conjugated to asparaginase (ONCASPAR) was approved in the United States in 1994 for the treatment of patients hypersensitive to existing asparaginase therapy for acute lymphocytic leukemia. Both of these approved proteins demonstrate increased circulating lives, decreased immunogenicity and antigenicity, and increased stability (HO et al. 1986, 1988; DAVIS et al. 1981). Similar changes in the characteristics of other base proteins following modification have resulted in clinical trials to study the therapeutic use of PEG-SOD, PEG-interleukin-2, PEG-interferon, PEG-GC and PEG-hemoglobin.

PEG-conjugates have also been used to suppress the production of IgE response to specific allergens. When ovalbumin, pollen allergens and bacterial allergens were conjugated to PEG, they markedly depressed the ongoing IgE-antibody formation in sensitized animals through the activation of suppressor cells (SEHON and LEE 1981; LEE et al. 1981). However, in clinical trials on PEG-honeybee venom, allergen-specific tolerance could not be induced in patients with ongoing IgE responses (DREBORG and AKERBLOM 1990).

V. Protein Engineering

The biological activity of proteins is governed by surface structure and surface determinants. These determinants are based on the three-dimensional structure of the proteins, which, in turn, is based on the amino acid sequence, or primary structure, and the ways in which these amino acids interact with each other and their environment. The presence of non-protein components, such as carbohydrates, also affects the protein's biological activity. The distribution of carbohydrates is noted to be a "major determinant of the biological deposition and potential safety of therapeutic proteins" (TOMLINSON 1989). Therapeutic benefits as well as adverse side effects are ultimately due to the shape and activity of the protein. Improvement of the clinical efficacy can be accomplished by the engineering or remodeling of the protein itself.

By changing the amino acid sequence or other components of a protein through protein engineering, one can alter the circulating life, the ability of the protein to interact with receptor sites, and recognition by the immune system. Protein stability (Kaarsholm et al. 1993), specificity (Witkowski et al. 1994) or environmental adaptability (Arnold 1993) can be altered by protein engineering. New proteins can be designed through the addition of new functions (Higaki et al. 1992), the incorporation of non-peptidic residues (Baca et al. 1993; Tuchscherer et al. 1992) or by protein design (Johnsson et al. 1993; Ball 1994). Protein engineering can be used to humanize antibodies for therapy, resulting in molecules with improved pharmacokinetics and reduced immunogenicity (Winter and Harris 1993; Adair 1992), or it can be used to alter the thermal stability of food enzymes (Goodenough and Jenkins 1991).

Proteins can be altered by either site-directed mutagenesis or post-translational modification. Site-directed mutagenesis is a process whereby the DNA that codes for the amino acid structure is mutated to precisely alter the protein. Several recent reviews discuss the basics of the procedures (Fersht and Winter 1992; Petersen and Martel 1998). The process itself is straightforward, however, the resulting change(s) must be examined for each mutation. A single mutation can have significant effects on biological activity, as is noted for the single-point mutation that is responsible for sickle-cell anemia.

Site-directed mutagenesis has been shown to be successful in changing the biological activity of proteins. tPA was engineered at residue 120 from asparagine to glutamine. In mice, the mutated tPA circulated longer than the wild-type tPA, and was more active in promoting thrombolysis of a labeled clot (Lau et al. 1987). When mutated tPA variants were compared to commercially available recombinant tPA (Activase^{R}), the rates of clearance of the two variants were decreased 9-fold and 18-fold in rabbits. Plasma plasminogen-activator activity clearance in rabbits was also decreased 9-fold and 22-fold, relative to the recombinant version. A 50% clot thrombolysis was achieved with doses 8.6-fold and 3-fold lower than the wild type (Larsen et al. 1991).

Post-translational modification typically consists of side-chain modifications by phosphorylation, glycosylation or farnesylation. As the amino acids of terminal chains appear to be responsible for mediating intracellular metabolic stability of a protein (Bachmair et al. 1986), post-translational modification to alter terminal or internal protein sequences can result in therapeutic benefits. ADA, which is used for the treatment of ADA-deficient severe combined deficiency disease, is normally found only in the cytosol. The protein was modified through the fusion of a β-lactamase signal sequence with its amino terminus. The engineered protein displayed antigenic, catalytic, conformational and immunological properties similar to human ADA. When cloned into a mammalian expression vector and transfected into mouse fibroblasts and myoblasts, ADA activity was demonstrated in the media (Hughes et al. 1994).

Individual residues of proteins may also be engineered by phosphorylation, glycosylation, sulfation, methylation, farnesylation, prenylation, myristylation or hydroxylation. Changes in the distribution of carbohydrates on the surface of proteins can alter the blood circulating life, recognition by the immune system and the ability to access sites of action (Tomlinson 1989). Post-translational modification by the addition of mannose to take advantage of preferential uptake and endocytosis by macrophage receptors was used to improve the therapeutic efficacy of glucocerebrosidase for Gaucher's disease. Only a small increase in uptake was noted. Uptake was mediated by a receptor distinct from the classical mannose receptor that binds mannose-bovine serum albumin (Sato and Beutler 1993). When GC purified from human placental tissue or produced by recombinant techniques was modified by the addition of mannose to the terminus and administered randomly to 15 patients with Gaucher's disease, no difference in therapeutic outcome was noted (Grabowski et al. 1995). IgG-antibody formation of 40% was noted in patients receiving the tissue-derived protein, compared with 20% for patients receiving the recombinant protein (Grabowski et al. 1995).

SOD modified with mannose was preferentially targeted to mouse peritoneal macrophages, compared with non-modified or galactosylated SOD. Macrophages incubated with mannosylated SOD demonstrated a stronger inhibitory effect on the release of superoxide anion from phorbol myristate acetate-stimulated macrophages (Takakura et al. 1994). The in vivo disposition of human-recombinant SOD (Fujita et al. 1992), albumin (Nishikawa et al. 1992), soybean-trypsin inhibitor (Takakura et al. 1989a,b) and uricase (Fujita et al. 1990; Yasuda et al. 1990) were all shown to be controllable by modification with various sugars.

D. Choosing the Proper Strategy for a Protein Therapeutic

The choice of the strategy for a protein therapeutic is based ultimately on the desired characteristics of the final product. These, in turn, are based on the indications(s), the desired route of administration and, most importantly, the characteristics of the base protein.

Choice of a strategy to reduce immunogenicity should take into account what effect the strategy will have on the protein's characteristics. Will the strategy inactivate or decrease the desired protein activity? Will the strategy result in an extension of activity such that the protein becomes toxic? An extensive review of the effects of the methodology on other proteins, especially those that are similar in nature, will assist in predicting the effect the technology will have on the base protein.

Rigorous testing of the protein after it is incorporated with the desired methodology is essential. Proteins are extremely sensitive to change, therefore, an extensive determination of activity, pharmacokinetics and toxicity are required. Though this may seem like a significant amount of effort, it is no more than is normally required of the drug development process. What is important to keep in mind, when looking at alternate methodologies to reduce the immunogenicity of protein therapeutics, is that once the protein is linked to some technology, it is a new compound and cannot be defined as the base protein.

E. The Future of Protein Therapeutics

The clinical application of proteins is beginning to be realized. The choices of technologies to overcome the limitations associated with protein therapeutics and their refinement continues to produce products that are successful in negotiating the regulatory hurdles in the United States and other countries.

What of the future? As the intricacies of the body and its recognition and receptor systems are further defined, it may be possible to use some of the strategies defined above, or some combination, to produce a protein drug that displays a high level of efficacy and a low to non-existent level of unwanted side effects. Gene therapy offers the potential to cure metabolic disorders and cancer (Tepper and Mule 1994; Freeman et al. 1993; Moseley and Caskey 1993); selective immunomodulation offers the possibility that the immune system can be selectively manipulated during a protein drug's dosing schedule.

Until the time that we discover the final cure for the diseases that require protein therapy, the strategies described will make it possible for physicians and clinicians to provide patients with the best protein-based treatment. Even if we do not achieve the realm of the "final cure", we will continue to be able to use the strategies already developed, or further refine them, to allow the use of proteins as therapeutics.

References

Aalberse RC, van der Gaag R, van Leeuwen J (1983) J Immunol 130:722–726
Aalberse RC, Heber D, Liebermann J (1983) Clin Exp Immunol 52:164–172
Abramowicz D, Crusiaux A, Goldman M (1992) N Engl J Med 327:736
Abramowicz D, Schandend L, Goldman M, Crusiaux A, Veerstraeten P, De Pauw L, Wybran J, Kinnaert P (1989) Transplantation 47:606–608
Adair JR (1992) Immunol Rev 130:5–40
Ahmad I, Longenecker M, Samuel J, Allen TM (1993) Cancer Res 53:1484–1488
Alegre ML, Collins AM, Pulito VL, Borsius RA, Olson WC, Zivin RA, Knowles R, Thistlewaithe JR, Jolliffe LK, Bluestone JA (1992) J Immunol 148:3461–3468
Alegre ML, Peterson LJ, Xu D, Sattar HA, Jeyarajah DR, Kowalkowski K, Thistlewaithe JR, Zivin RA, Jolliffe L, Bluestone JA (1994) Transplantation 57: 1537–1543
Alexopoulos A, Raine AEG, Cobbe SM (1984) Eur Heart J 5:1010–1012
Alkan SS, Towbin H, Hochkeppel HK (1988) J Interferon Res 8:25–33
Allegretta MA, Atkins B, Dempsey RA, Bradley EC, Konrad MW, Childs A, Wolfe SN, Mier JW (1986) J Clin Immunol 6:481–490
Allen TM (1994) Adv Drug Del Rev 13:285–309

Allen TM (1994) Trends Pharmacol Sci 15:215–220
Allen TM, Hansen C, Martin F, Redemann C, Yau-Young A (1991) Biochim Biophys Acta 1066:77–82
Allison AC, Gregoriadis G (1974) Nature 252:252
Altman JJ (1983) Lancet ii:524
Alving CR, Steck EA, Chapman WL Jr, Waits VB, Hendricks LD, Swartz GM Jr, Hanson WL (1978) Proc Natl Acad Sci USA 75:2959–2963
Alving CR, Steck EA, Hanson WL, Loizeaux PS, Chapman WL Jr, Waits VB (1980) Ann Life Sci 26:2231–2238
Andreani D (1973) Excerpta Med Int Congr Ser 316:68–75
Andrew SM, Pimm MV, Perkins AC, Baldwin RW (1986) Eur J Med 12:168–175
Arnold FH (1993) FASEB J 744–749
Arturrson P, Edman P, Sjoholm I (1984) J Pharmacol Exp Ther 231:705–712
Arturrson P, Edman P, Sjoholm I (1985) J Pharmacol Exp Ther 234:255–260
Asselin BL, Whiten JC, Coppola DJ, Rupp IP, Sallan JE, Cohen HJ (1993) J Clin Oncol 9:1780–1786
Asselin BL, Whitin JC, Coppola DJ, Rupp IP, Sallan SE, Cohen HJ (1993) J Clin Oncol 11:1780
Aston R, Holder AT, Ivanyi J, Bomford R (1987) Mol Immunol 24:143–150
Baca M, Alewood PF, Kent SB (1993) Protein Sci 2:1085–1091
Bachmair A, Finley D, Varshavsky A (1986) Science 234:179–186
Bakker-Woudenberg IAJM, Lokerse AF, ten Kate MT, Mouton JW, Woodle MC, Storm G (1993) J Infect Dis 168:164–171
Ball P (1994) Nature 367:323–324
Barry BW, Southwell D, Woodford RJ (1984) Invest Dermatol 82:49–52
Barton NW, Brady RO, Doppelt SM, Hill SC, Bisceglia AM, Hill SC, Verderese CL, Graham OC, Argoff CE, Grewel RP, Yu K-T (1990) Clin Res 38:457a
Batra JK, Chaudhary VK, FitzGerald D, Pastan I (1990) Biochem Biophys Res Commun 171:1–6
Belmont J, Caskey CT (1986) Developments leading to gene therapy. In: Kucherlapati R (ed) Gene therapy. Plenum, New York, pp 411–440
Benziger DP, Edelson J (1983) Drug Metab Rev 14:137–168
Berson SA, Yalow RS (1959) J Clin Invest 38:2017–2025
Berson SA, Yalow RS (1964) Diabetes 13:247–259
Berson SA, Yalow RS (1966) Am J Med 40:676–690
Berson SA, Yalow RS, Bauman A, Rothschild MA, Newerly K (1956) J Clin Invest 35:170–190
Bickel U, Yoshikawa T, Landaw EM, Faull KF, Pardridge WM (1993) Proc Natl Acad Sci USA 90:2618–2622
Bierich JR (1986) Acta Paediatr Scand 325:13–18
Blakely DC (1992) Acta Oncol 31:91–97
Blandford et al (1984) JAMA 251(11):1459–1460
Bliss M (1982) The discovery of insulin. McClelland and Stewart, Toronto
Blume G, Cevc G (1990) Biochim Biophys Acta 1029:91–97
Blume G, Cevc G (1993) Biochim Biophys Acta 1146:157–268
Bode C, Runge MS, Branscomb EE, Newell JB, Matsueda GR, Haber E (1989) J Biol Chem 264:944–948
Boerman OC, Mijnheere EP, Broers JLV, Voojis GP, Ramaekers FCS (1991) Int J Cancer 48:457–462
Bos R, Siegel K, Otter M, Nieuwenhuizen W (1992) Fibrinolysis 6:173–182
Brackman HH, Egli H (1981) Treatment of hemophilia patients with inhibitors. In: Seligohy U, Rimon A, Horoszowski H (eds) Hemophilia. Castle House, London, p 113
Brackman HH, Gormsen J (1977) Lancet ii:933
Branscomb EE, Runge MS, Savard CE, Adams KM, Matsueda GR, Haber E (1990) Thromb Haemost 64:260–266

Brasseur F, Couvreur P, Kante B, Deckers-Passau L, Roland M, Deckers C, Speiser P (1980) Eur J Cancer 16:1441–1445
Britton DE, Leinikki PO, Barranger JA, Brady RO (1978) Life Sci 23(25):2517–2520
Brodgen RN, Heel RC (1987) Drugs 34:350–371
Broome JD (1961) Nature 191:1114–1115
Brunfeldt K, Deckert T (1966) Acta Endocrinol (Copenh) 47:367–370
Byers VS, Rodvien R, Grant K, Durrant LG, Hudson KH, Baldwin RW, Scannon PJ (1989) Cancer Res 49:6153–6160
Carpenter FH (1958) Arch Biochem Biophys 78:539–545
Carveth-Johnson AO, Mylvaganam K, Child DF (1982) Lancet ii:1287
Chance RE, Root MA, Galloway JA (1976) Acta Endocrinol (Copenh) 83[Suppl 205]: 185–196
Chang TMS (1964) Science 146:524–545
Chang TMS (1972) Artificial cells. Thomas, Springfield
Chapman WL, Hanson WL, Alving CR, Hendricks LD (1984) Am J Vet Res 45:1028–1030
Charpie JR, Runge MS, Matsueda GR, Haber E (1990) Biochemistry 29:6374–6378
Chatenoud L, Baudrihaye MF, Chkoff N, Kries H, Goldstein G, Bach JF (1986) J Immunol 137:830–838
Chatenoud L, Jonker M, Villemain F, Goldstein G, Bach J (1986) Science 232:1406–1408
Chaudhary VK, Gallo MG, FitzGerald DJ, Pastan I (1990) Proc Soc Natl Acad Sci USA 87:9491–9494
Chien YW (1983) Drug Dev Ind Pharm 9:1291–1330
Cho M, Gilbert CW, Campbell K, Pschenyczny V, Martin R, Ginns E, Shorr RGL (1994) FASEB J 8:A94
Christy M, Deckert T, Nerup J (1977) Clin Endocrinol Metab 6:305–322
Chun JD, Lee N, Kobayashi RH, Chaffee S, Hershfield MS, Stiehm ER (1993) Ann Allergy 70:462–466
Chun JD, Lee N, Kobayashi RH, Chaffee S, Hershfield MS, Stiehm ER (1993) Ann Allergy 70:462
Clavell LA, Gelber RD, Cohen HJ, Hitchcock-Bryan S, Cassady JR, Tarbell NJ, Blattner SR, Tantracahi K, Leavitt P, Sallan SE (1986) N Engl J Med 315:657–663
Cohen AM, Martin EW, Lavery I, Daly J, Sardi A, Aitken D, Bland K, Mojzisik C, Hinkle G (1991) Arch Surg 126:349–352
Colcher D, Bird R, Roselli M, Hardman KD, Johnson S, Pope S, Dodd SW, Pantoliano MW, Milenic DE, Schlom J (1990) J Natl Cancer Inst 82:1191–1197
Collins FS (1991) Hosp Pract 26:93–98
Couvreur, Kante P, Grislain L, Roland M, Speiser PJ (1982) Pharm Sci 71:790–792
Couvreur P, Kante B, Lenaerts V, Scaidteur V, Roland M, Speiser PJ (1980) Pharm Sci 69:199–202
Couvreur P, Kante B, Roland M, Guiot P, Bauduin P, Speiser PJ (1979) Pharm Pharmacol 31:331–332
Davis S, Abuchowski A, Park YK, Davis FF (1981) Clin Exp Immunol 46:649
Davis SS (1992) J Pharm Pharmacol 44:186–190
de Vries ME, Bodde HE, Verhoef JC, Junginger HE (1991) Crit Rev Ther Drug Carrier Sys 8:271–303
Deckert T, Andersen OO, Grundahl E, Kerp L (1972) Diabetologia. 8:358–361
Delgado C, Francis GE, Fisher D (1992) Crit Rev Ther Drug Carr Syst 9:249–304
di Mario UD, Arduini P, Tiberti C, Lombardi G, Pietravalle P, Andreani D (1986) Diabetes Res Clin Pract 2:317–324
Dreborg S, Akerblom EB (1990) Crit Rev Ther Drug Carrier Syst 6:315–365
Drewes PA, Kamp AO, Winkelman JW (1978) Experientia. 34:316–318
Duncan R, Spreafico F (1994) Clin Pharmacokinet 27:290–306
Duncan R (1985) Crit Rev Ther Drug Carrier Syst 1:281–310

Dykiewicz MS, Kim HW, Orfan N, Yoo TJ, Leiberman P (1994) J Allergy Clin Immunol 93:117–125
Edman P, Sjoholm I (1982) Life Sci 30:327–330
Elias PM, Friend DS (1975) J Cell Biol 65:180–191
Ellison RR, Mick R, Cuttner J, Schiffer CA, Silver RT, Henderson ES, Woliver T, Royston I, Davey FR, Glicksman AS, Bloomfield CD, Holland JF (1991) J Clin Oncol 9:2002–2015
Enna SJ, Schanker LS (1972) Am J Physiol 223:1227–1231
Ertel IJ, Nesbit ME, Hammond D, Weiner J, Sather H (1979) Cancer Res 39:3893–3896
Evans WE, Tsiatis A, Rivera G, Murphy SB, Dahl GV, Denison M, Crom WR, Barker LF, Mauer AM (1982) Cancer 49:1378–1383
Ewing NP, Sanders NL, Dietrich SL, Kasper CK (1988) JAMA 259:65–68
Fagnani R, Hagan MS, Bartholomew R (1990) Cancer Res 50:3638–3645
Feinglos MN, Jegasothy BV (1979) Lancet i:122–124
Fersht A, Winter G (1992) Trends Biochem Sci 17:292–295
Fineberg SE, Galloway JA, Fineberg NS, Rathbun MJ (1982) Diabetes Care 5:107–113
Fineberg SE, Galloway JA, Fineberg NS, Rathbun MJ, Hufferd S (1983) Diabetologia 25:465–469
Fireman P, Fineberg SE, Galloway JA (1982) Diabetes Care. 5:119–125
Fisher AN, Brown K, Davis SS, Parr GD, Smith DA (1987) J Pharm Pharmacol 39:357–362
Fisher AN, Farraj NF, O'Hagan DT, Jabbal-Gill I, Johansen BR, Davis SS, Illum L (1991) Int J Pharm 74:147–156
Fisher NF (1923) Am J Physiol 67:65–71
Franklin RR, Dukes CD (1964) J Am Med Assoc 190:682–683
Freeman SM, Whartenby KA, Abraham GN, Zweibel JA (1993) Adv Drug Del Rev 12:169–183
Frienkel RK (1983) In: Goldsmith LA (ed) Biochemistry and physiology of the skin, vol I. Oxford University Press, Oxford, p 328
Fujita H, Okamoto M, Takao A (1991) Drug Delivery Syst 6:133–138
Fujita T, Furitsu H, Nishikawa M, Takakura H, Sezaki H, Hashida M (1992) Biochem Biophys Res Commun 189:191–196
Fujita T, Nishikawa M, Tamake C, Takakura Y, Hashida M, Sezaki H (1992) J Pharmacol Exp Ther 263:971–978
Fujita T, Nishikawa M, Tamaki C, Takakura T, Hashida M, Sezaki H (1992) J Pharmacol Exp Ther 263:971–978
Fujita T, Yasuda Y, Takakura Y, Hashida M, Sezaki H (1990) 11:149–154
Fung W-P, Przybylski M, Ringsdorf H, Zaharko D (1979) J Natl Cancer Inst 62:1261–1264
Gabizon A, Papahadjopoulos D (1988) Proc Natl Acad Sci USA 85:6949–6953
Gabizon A, Shiota R, Papahadjoupoulos D (1989) J Natl Cancer Inst 81:1484–1488
Gabizon AA (1992) Cancer Res 52:891–896
Galloway JA, Fineberg SE, Fineberg NS, Goldman J (1982) Effect of purity and beef content on complications of insulin therapy. In: Gueriguian JL, Bransone ED, Outschoorn AS (eds) Hormone drugs. Proceedings of the FDA-USP workshop on drug and reference standards for insulins, somatotropins and thyroid-axis hormones, Bethesda, MD. USP Convention, pp 244–253
Garcia-Ortega P (1984) Br Med J 288:1271
Gasparini M, Ripamonti M, Seregni E, Regalia E, Buraggi GL (1988) Int J Cancer [Suppl] 2:81–84
Ghose T, Norvell ST, Guclu A, Bodurtha A, Mac Donald AS (1977) J Natl Cancer Inst 58:845–852
Godal A, Fodstad O, Pihl A (1989) Int J Cancer 32:515–521
Goldstein G (1987) Transplant Proc 19[2 Suppl 1]:1–6
Goldstein G, Fuccello AJ, Norman DJ, Shield CF, Colvin RB, Cosimi AB (1986) Transplantation 42:507–511

Goodenough PW, Jenkins JA (1991) Biochem Soc Trans 19:655–662
Goodman GE, Hellstrom I, Brodzinsky L, Nicaise C, Kulander B, Hummer D (1990) J Clin Oncol 8:1093–1092
Gossain VV, Rovner OR, Mohan K (1985) Ann Allergy 55:116–118
Gould BJ, Borowitz MJ, Groves ES, Carter PW, Anthony D, Weiner LM, Frankel AE (1989) J Natl Cancer Inst 81:775–781
Grabowski GA, Barton NW, Pastores G, DAmbrosia JM, Banerjee TK, McKee MA, Parker C, Schiffmann R, Hill SC, Brady RO (1995) Ann Intern Med 122:33–39
Grabowski GA, Pastores G, Brady RO, Barton NW (1993) Pediatr Res 33 (4 pt 2):819a
Grammar LC, Metzger BE, Patterson Rl (1984) JAMA 251:1459–1460
Grammer L (1986) Clin Rev Allergy 4:189–200
Graybill JR, Cravin PC, Taylor RL, Williams DM, Magee WE (1982) J Infect Dis 145: 748–752
Gregoriadis G (ed) (1984) Liposome technology, vol 1–3. CRC Press, Boca Raton
Gregoriadis G, Neerunjun ED (1975) Biochem Biophys Res Commun 65:537–544
Gribben JG, Devereux S, Thomas NSB, Keim M, Jones HM, Goldstone HA, Lynch DC (1990) Lancet 335:434–437
Guy RH, Hadgraft J (1985) Int J Pharm Int 24:267–274
Habeeb AFSA, Cassidy HG, Singer SF (1958) Biochim Biophys Acta 29:587–593
Haller DG (1988) J Clin Oncol 6:1213–1215
Heath TD (1987) Methods Enzymol 149:111–118
Hellstrom KE, Hellstrom I, Brown JP (1984) Med Oncol Tumor Pharmacother 1:143–147
Higaki JN, Fletterick RJ, Craik CS (1992) Trends Biochem Sci 173:100–104
Hirano T, Ohashi S, Morimoto S, Tsuda K, Kobayashi T, Tsukagoshi S (1986) Makromol Chem 187:2815–2824
Hirayama S, Sato F, Oda T, Maeda H (1986) Jpn J Antibiot 39:815–822
Hirsch R, Chatenoud L, Gress RE, Sachs DH, Bach JF, Bluestone JA (1989) Transplantation 47:853–857
Ho DH, Brown NS, Yen A, Holmes R, Keating M, Abuchowski A, Newman RA, Krakoff IH (1986) Drug Metab Disp 14:349
Ho DH, Wang CY, Lin JR, Brown N, Newman RA, Krakoff IH (1988) Drug Metab Disp 16:27
Holder AT, Aston R, Preece MA, Ivanyi J (1985) J Endocrinol 107:R9–R12
Huang SK, Lee K-D, Hong K, Friend DS, Papahadjopoulos D (1992) Cancer Res 52:5135–5143
Hughes M, Vassilakos A, Andrews DW, Hortelano G, Belmont JW, Chang PL (1994) Human Gene Ther 5:1445–1455
Itri LM, Campion M, Dennin RA, Palleroni AV, Gutterman JU, Groopman JE, Trown PW (1987) Cancer 59:668–674
Jaffers GJ, Fuller TC, Cosimi AB, Russell PS, Winn HJ, Colvin RB (1986) Transplantation 41:572–578
Jain RK (1990) Cancer Metast Rev 9:253–266
Johnsson K, Alleman RK, Widmer H, Benner SA (1993) Nature 365:530–532
Jones B, Holland JF, Glidewell O, Jacquillat C, Weil M, Pochedly C, Sinks L, Chevalier L, Mauier HM, Koch K, Falkson G, Patterson R, Seligman B, Sartorius J, Kura F, Haurani F, Stuart M, Bergert EO, Ruymann F, Sawitsky A, Forman E, Pluess H, Truman J, Hakami N (1979) Med Pediatr Oncol 3:387–400
Jones VE, Leskowitz S (1965) Nature 207:596–597
Juliano RL, Stamp D (1976) Nature 261:235–237
Kaarsholm NC, Norris K, Jorgenson RJ, Mikkelsen J, Ludvigsen S, Olsen OH, Sorenson AR, Havelund S (1993) Biochemistry 32:10773–10778
Kaplan SL, August GP, Blethen SL, Brown DR, Hintz RL, Johansen A, Plotnick LP, Underwood LE, Bell JJ, Blizzard RM, Foley TP, Hopwood NJ, Kirkland RJ, Rosenfeld RG, Van Wyk JJ (1986) Lancet i:697–700
Katre NV (1993) Adv Drug Deliv Rev 10:91–114

Keating MJ, Holmes R, Lerner S, Ho DH (1993) Leuk Lymph 10 [Suppl]:153
Kennel SJ, Falcioni R, Wesley JW (1991) Cancer Res 51:1529–1536
Khazaeli MB (1989) Hybridoma 8:231–239
Khazaeli MB, Conry RM, LoBuglio AF (1994) J Immunother 15:42–52
Khazaeli MB, Saleh MN, Wheeler RH, Huster WJ, Holden H, Carrano R, LoBuglio AF (1988) J Natl Cancer Inst 80:937–942
Kinsky SC, Nicoletti RA (1977) Annu Rev Biochem 46:49–67
Konno T, Maeda H, Iwai K, Maki S, Tashiro S, Uchida M, Miyauchi Y (1984) Cancer 54:2367–2374
Konno T, Maeda H, Iwai K, Tashiro S, Maki S, Morinaga T, Mochinaga M, Hiraoki T, Yoyokama I (1983) Eur J Cancer Clin Oncol 19:1053–1065
Kopecek J (1981) In: Williams DF (ed) Systematic aspects of biocompatability, vol II. CRC Press, Boca Raton, pp 159–180
Kreuter J (1983a) Pharm Acta Helv 58:196–209
Kreuter J (1983b) Pharm Acta Helv 58:217–226
Kreuter J (1983c) Pharm Acta Helv 58:242–250
Kubota K, Yamada T, Ogura A, Ishizak T (1990) J Pharm Sci 79:179–184
Kumar D (1993) Horm Metab Res 25:360–364
Kurokawa T, Iwasa S, Kakinuma A (1989) Biotechnology 7:1163–1176
Kurtz AB, Nabarro JDN (1980) Diabetologia 19:329–334
Kurtzberg J, Asselin B, Poplack D, Grebanier A, Chen R, Franklin A, Scudiery D, Fisherman J (1993) Proc Am Assoc Cancer Res 34:1807a
Kurtzman SH, Russo A, Mitchell JB, de Graff W, Sindelar WF, Brechbiel MW, Ganson OA, Friedman AM, Hines JJ, Gamson J, Atcher RE (1988) J Natl Cancer Inst 80:449–452
Land VJ, Shuster JJ, Pullen J, Harris M, Krance RA, Castleberry R, Adabutu J, Barboas JL, Rosen D (1989) Proc Am Soc Clin Oncol 8:215
Larsen GR, Timony GA, Horgan PG, Barone KM, Henson KS, Angus LB, Stoudemire JB (1991) J Biol Chem 266:8156–8161
Larson SM, Carrasquillo JA, Reynolds JC (1984) Cancer Invest 2:363–381
Lau D, Kuzma G, Wei C-M, Livingston DJ, Huiung N (1987) Biotechnology 5:953–958
Le Doussal JM, Gruaz-Guyon A, Martin M, Gautherot E, Delaage M, Barbet J (1990) Cancer Res 50:3445–3452
Lee VHL (1986) Enzymatic barrier to peptide and protein absorption and use of penetration enhancers to modify absorption. In: Davis SS, Illum L, Tomlinson E (eds) Proceedings of NATO advanced research workshop: advanced drug delivery systems for peptides and proteins. Plenum, Copenhagen, p 87
Lee WY, Sehon AH, Akerblom E (1981) Int Arch Allergy Appl Immunol 64:100–114
Lees A, Finkelman F, Inman JK, Witherspoon K, Johnson P, Kennedy J, Mond JJ (1994) Vaccine 12:1160–1166
Leslie D (1977) Br Med J ii:736–737
Lieberman P, Patterson R, Metz R, Lucena A (1984) JAMA 215:1106–1112
Lindner J, MacNeil LW, Marney S, Conway M, Rivier J, Vale W, Rabin D (1981) J Clin Endocrinol Metab 52:267–270
LoBuglio AF, Wheeler RH, Trang J, Haynes A, Rogers K, Harvey EB, Sun L, Ghrayeb J, Khazaeli MB (1989) Proc Natl Acad Sci USA 86:4220–4224
Lockwood DH, Prout TE (1965) Metabolism 14:530–538
Lopez-Berestein G, Fainstein V, Hopper R, Mehta K, Sullivan MP, Keating M, Rosenblum MG, Mehta R, Luna M, Hersh EM, Reuben J, Juliano RL, Bodey GP (1985) J Infect Dis 151:704–710
Lundin K, Berger L, Blomberg F, Wilton P (1991) Acta Paediatr Scand 372:167–169
Machy P, Lerman L (1987) Liposomes. John Libbey Eurotext, Paris
MacLeod JJR (1978) Bull Hist Med 52:295–312
Maeda H (1991) Adv Drug Deliv Rev 6:181–202

Maeda H, Matsumoto T, Konno T, Iwa K, Uida M (1984) J Protein Chem 3:181–193
Maeda H, Seymour LW, Miyamoto Y (1991) Bioconj Chem 3:351–362
Maeda H, Seymour LW, Miyamoto Y (1992) Bioconj Chem 3:351–362
Martin RJ, Denyer SP, Hadgraft J (1987) Int J Pharm 39:23–32
Martis L (1986) 28th Ann Nat Ind Pharm Res Conf Washington
Mashburn LT, Wriston JC (1964) Arch Biochem Biophys 105:450–452
Mayer AM (1982) Diabetes Forecast 35:44–48
Mayer LD, Bally MB, Cullis PB (1986) Biochim Biophys Acta 857:123–126
McGrath KG, Patterson R (1985) Clin Exp Immunol 63:421–426
McGrath KG, Zeffren B, Alexander J, Kaplan K, Patterson RJ (1985) Allergy Clin Immunol 76:453–457
McPherson JM, Livingston DJ (1989) Pharm Tech 13(9):32–42
Mehta M (ed) (1993) Medical economics. PDR Library on CD-ROM
Melino G, Hobbs JR, Radford M, Cooke KB, Evans AM, Castello MA, Forrest DM (1985) Protids Biol Fluids 32:413–447
Melissen PMB, van Vianen W, Bidjal O, van Marion M, Bakker-Woudenberg IAJM (1993) Biotherapy 6:113–124
Meredith R, Khazaeli MB, Plott G, Liu TP, Russell C, Wheeler R, LoBuglio A (1992) Antibody Immunoconj Radiopharmaceut 5:138
Meyer J, Squier CA, Gerson SJ (eds) (1984) The structure and function of oral mucosa. Pergamon, New York
Mirsky IA, Kawamura K (1966) Endocrinology 78:1115–1119
Miyata K, Nakagawa Y, Nakamura M, Ito T, Sugo K, Fujita T, Tomoda K (1988) Agric Biol Chem 52:1575–1581
Moseley AB, Caskey CT (1993) Adv Drug Del Rev 12:131–142
Mumtaz M, Bachhawat BK (1992) Biochim Biophys Acta 1117:174–178
Muranishi S (1990) Crit Rev Ther Drug Carrier Syst 7:1–33
Murphy TM, Hadgraft J (1990) In: Scott RC, Guy RH, Hadgraft J (eds) Predictors of percutaneous penetration: methods, measurement, modeling. IBC Technical Services, London, p 333
Murray GJ, Howard KD, Richards S, Barton NW, Brady RO (1991) J Immunol Methods 137:113–120
Musch K, Wolf AS, Lauritzen C (1981) Clin Chim Acta 113:95–100
Nagai TJ (1985) Contr Release 2:121–134
Needham D, McIntosh TJ, Lasic DD (1992) Biochim Biophys Acta 1108:40–48
Nesbit N, Chard R, Evans A, Karon M, Hammond GD (1979) Am J Pediatr Hematol Oncol 1:9–13
Neuberger MS, Williams GT, Fox RO (1984) Nature 312:604–608
New RRC, Chance ML, Thomas SC, Peters W (1978) Nature 272:55–56
Nishikawa M, Ohtsubo Y, Ohno J, Fujita T, Koyama Y, Yamashita F, Hashida M, Sezaki H (1992) Int J Pharm 85:75–85
Norman C (1985) Science 228:1176–1177
Oppenheim RC (1981) Int J Pharm 8:217–234
Owens DR (1980) Human insulin: clinical pharmacological studies in normal man. MTP Press, Lancaster
Pastores GM, Sibille AR, Grabowski GA (1993) Blood. 82:408–416
Peracchia C (1978) Nature 271:669–671
Petersen SB, Martel P (1998) Protein engineering – new or improved proteins for mankind. In: Cabral J, Best D, Boross L, Tramper J (eds) Applied biocatalysis. Harwood Academic (in press)
Pikal MJ (1990) Pharm Res 7:118–126
Preece MA (1986) Experience of treatment with pituitary derived human growth hormone with special reference to immunological aspects. In: Milner MDG, Flodh H (eds) Immunological aspects of human growth hormone. Medical Education Services, Oxford, pp 9–16
Reed BR, Chen AB, Tanswell P, Prince WS, Wert RM Jr, Glaesle-Schwartz L, Grossbard EB (1990) Thromb Haemost 64:276–280

Reisner C, Moul DJ, Cudworth AG (1978) Br Med J ii:56
Renold AE, Steinke J, Soeldner JS, Antoniades HN, Smith RE (1966) J Clin Invest 45:702–713
Richards SM, Olson TA, McPherson JM (1993) Blood 82:1402–1409
Ritschel WA, Adolph S, Ritschel GB, Schroeder T (1990) Methods Find Exp Clin Pharmacol 11:281–287
Roberts HW, Cromartie R (1984) Prog Clin Biol Res 150:1–18
Robinson GD (1927) J Pharmacol Exp Ther 32:81–88
Roig de Vargas-Lanares CD (1978) In: Hafez ESE, Evans TN (eds) The human vagina. Elsevier/North Holland Biomedical, Amsterdam, p 193
Rosenblum MG, Cheung L, Murray J (1994) Cancer Bull 46:34–39
Sallan SE, Hitchcock-Bryan S, Gelber R, Cassady JR, Frei E III, Nathan DG (1983) Cancer Res 43:6501–5607
Sandhu JS (1992) Crit Rev Biotech 12:437–462
Sato Y, Beutler E (1993) J Clin Invest 91:1909–1917
Schlichtkrull J, Brange K. Christiansen AH, Hallund O, Heding LG, Jorgensen KH (1974) Horm Metab Res 5[Suppl 1]:134–143
Sehon AH, Lee WY (1981) Int Arch Allergy Appl Immunol 66[Suppl 1]:39–42
Sela M (1969) Science 166:1365–1374
Senior J, Delgado C, Fisher D, Tilcock C, Gregoriadis G (1991) Biochim Biophys Acta 1062:77–82
Seymour LW, Ulbrich K, Strohalm J, Kopecek J, Duncan R (1990) Biochem Pharmacol 39:1125–1131
Smith DH, Byrn RA, Marsters SA, Gregory T, Groopman JE, Capon DJ (1987) Science 238:1704
Smith PL, Wall DA, Gochoco C, Wilson G (1992) Adv Drug Del Rev 8:253–290
Spitler LE, del Rio M, Khentigan A, Wedel NI, Brophy NA, Miller LL, Harkonen WS, Rosendorf LL, Lee HM, Mischak RP, Kawahata RT, Stoudemire JB, Fradken LB, Bautista EE, Scannon PJ (1987) Cancer Res 47:1717–1723
Starling JJ, Maciak RS, Hinson NA, Nichols CL, Briggs SL, Laguzza BC (1989) Cancer Immunol Immunother 28:171–178
Steiner DF (1967) Trans NY Acad Sci Ser II 30:60–68
Steiner DF, Hallund O, Rubenstein A, Cho S, Bayliss C (1968) Diabetes 17:725–736
Storm G, Wilms HP, Crommelin DJA (1991) Biotherapy 3:25–42
Stratford RE, Lee VHL (1985) J Pharm Sci 74:731–734
Straus EK (1961) Proc Soc Exp Biol Med 106:617–621
Takahaski T, Yamaguchi T, Kitamura K, Suzuyama H, Honda M, Yokota T, Kotanagi H, Takahashi M, Hashimoto H (1988) Cancer 61:881–888
Takakura T, Fujita T, Hashida M, Maeda H, Sezaki H (1989b) J Pharm Sci 78:219–222
Takakura T, Kaneko Y, Fujita T, Hashida M, Maeda H, Sezaki H (1989a) J Pharm Sci 78:117–121
Takakura Y, Fujita T, Hashida M, Maeda H, Sezaki H (1989) J Pharm Sci 78:219–222
Takakura Y, Kaneko Y, Fujita T, Hashida M, Maeda H, Sezaki H (1989) J Pharm Sci 78:117–121
Takakura Y, Masuda S, Tokuda H, Nishikawa M, Hashida M (1994) Biochem Pharmacol 47:853–858
Tallal L, Tan C, Oettgen HE, Wollner N, McCarthy M, Helson L, Burchenal J, Karnofsky D, Murphy ML (1970) Cancer 25:306–320
Tepper RI, Mule JJ (1994) Human Gene Ther 5:153–164
Thistlethwaithe JR Jr, Stuart JK, Mayes JY, Gaber AO, Woodle S, Buckingham MR, Stuart FP (1988) Am J Kidney Dis 11:112–119
Thistlewaithe JR Jr, Cosimi AB, Delmonico FL, Rubin RH, Talkoff-Rubin N, Nelson PW, Fang L, Russell PS (1984) Transplantation 38:695–701
Tomlinson E (1988) Anal Proc 25:293–295
Tomlinson E (1989) Drug News Perspect 2:5–14
Torchilin VP, Klibanov AL, Huang L, O'Donnell S, Nossiff ND, Khaw BA (1992) FASEB J 6:2716–2719

Trouet A, Baurain R, Deprez-De Campeneere D, Masquelier M, Pirson M (1982) Targeting of antitumoral and antiprotozolaldrugs by covalent linkage to protein carriers. In: Grogoriadis G, Senior J, Trouet A (eds) Targeting of drugs. Plenum, New York, pp 19–30
Trozak DJ (1985) J Am Acad Derm 13:247–251
Trubetskoy VS, Torchilin VP, Kennel S, Huang L (1992) Biochim Biophys Acta 1131: 311–313
Tuchscherer G, Servis C, Corradin G, Blum U, Rivrier J, Mutter M (1992) Protein Sci 1:1377–1386
Tuft L (1928) Ann J Med Scr 176:707–720
Underwood LE, Voina SJ, Van Wyk JJ (1974) J Clin Endocrinol Metab 38:288–297
Veillard MM, Longer MA, Martens TW, Robinson JR (1987) J Contr Release 6:123–132
Vigeral P, Chkoff N, Chatenoud L, Campos H, Lacombe M, Droz D, Goldstein G, Bach JF, Kreis H (1986) Transplantation 41:730–733
Walles M, Daniels M, Ray KP, Cottingham JD, Aston R (1987) Biochem Biophys Res Commun 149:187–193
Wang BS, Szewczyk E, Shieh HM, Hart IC (1990) J Endocrinol 127:481–485
Weiner LM, O'Dwyer J, Kitson J, Comis RL, Frankel AE, Bauer RJ, Konrad MS, Groves ES (1989) Cancer Res 49:4062–4067
Weissman G, Bloomgarden D, Kaplan R, Cohen C, Hoffstein S, Collins T, Gotlieb A, Nagle D (1975) Proc Natl Acad Sci USA 72:88–92
Werier J, Cheung AH, Matas AJ (1991) Lancet 337:1351
White GC, Taylor RE, Blatt PM (1983) Blood 62:141–145
Widder KJ, Marion PA, Morris RM, Howard FP, Poore GA, Senyei AE (1983) Eur J Cancer Clin Oncol 19:141–147
Widder KJ, Senyei AE, Ranney DF (1979) Magnetically responsive microspheres and other carriers for the biophysical targeting of antitumor agents. In: Garattini S et al (eds) Advances in pharmacology and chemotherapy. Academic, New York, pp 213–217
Widder KJ, Senyei AE, Sears B (1982) J Pharm Sci 71:379–387
Wileman T, Bennett M, Lilleymann J (1983) J Pharm Pharmacol 32:762–765
Wileman TE, Foster RL, Elliott PNC (1986) J Pharm Pharmacol 38:264–271
Wiles PG, Guy R, Watkins SM, Reeves WG (1983) Br Med J 287:531
Williams AC, Barry BW (1992) Crit Rev Ther Drug Carrier Syst 9:305–353
Williams JJ (1922) Metab Res 2:729–751
Wilton, Gunnarson (1988) Acta Paediatr Scand 343:95–101
Windsor E, Cronheim GE (1961) Nature 190:263–264
Winter G, Harris WJ (1993) Trends Pharmacol Sci 14:139–143
Witkowski A, Witkowska HE, Smith S (1994) J Biol Chem 269:379–383
Woodle ES, Thistlewaithe JR, Jolliffe LK, Zivin RA, Collins A, Adain JA, Bodmer M, Athwal D, Alegre ML, Bluestone JA (1992) J Immunol 148:2756–2763
Yang R, Lee C, Hsu H, Shorr RGL (1994) FASEB J 8:A94
Yasuda Y, Fujita T, Takakura Y, Hashida M, Sezaki H (1990) Chem Pharm Bull 38: 2053–2063
Yoyokama M (1992) Crit Rev Ther Drug Carrier Sys 9:213–248
Zuckerman JE, Murray JL, Rosenblum MG (1987) Proc Am Assoc Cancer Res 28: 1522

CHAPTER 5

Targeted Toxin Hybrid Proteins

R.J. KREITMAN and I. PASTAN

A. Introduction

I. Protein Toxins That Inhibit Protein Synthesis

The goal of targeted therapy is to kill cells bearing specific receptors. This is in contradistinction to cytotoxic chemotherapy, which depends on biochemical differences between normal and target cells. Many types of malignant cells, including those resistant to cytotoxic chemotherapy, display unique proteins on the cell surface, making such cells potentially sensitive to targeted therapy. Unlike small chemotherapeutic agents, which enter the cell by passing though the membrane, targeted therapy must enter through the specific receptors or antigens on the cell surface. Because such sites number only thousands per cell, the targeted agent must be extremely potent. Typically, protein toxins are used since they kill cells by catalytic mechanisms. In fact, it has been shown with both bacterial and plant toxins that only one molecule in the cytoplasm of cells is sufficient to kill the cell (CARRASCO et al. 1975; YAMAIZUMI et al. 1978; WILLINGHAM, FITZGERALD and PASTAN unpublished data).

1. Plant Toxins

Protein toxins from eukaryotic organisms, such as plants and fungi, typically inhibit protein synthesis by catalytically inactivating ribosomes. Examples include ricin, abrin, pokeweed antiviral protein, gelonin and saporin (UCKUN 1993; VITETTA et al. 1987). Ricin and abrin are composed of a binding domain and a catalytic domain connected through a disulfide bond. Other plant toxins contain only the catalytic domain.

2. Bacterial Toxins

Bacterial toxins, which include diphtheria toxin (DT) and Pseudomonas exotoxin (PE), are secreted as single chains. They inhibit protein synthesis by catalytically ADP ribosylating elongation factor 2 in the cytoplasm of target cells. Bacterial toxins contain a binding domain, to bind to target cells, and a translocation domain, to gain access to the cytoplasm.

II. Structure and Function of Pseudomonas Exotoxin

1. Definition of Domains and Mechanism of Intoxication

PE (Fig. 1) is composed of 613 amino acids and contains several domains which are structurally and functionally distinct (ALLURED et al. 1986; HWANG et al. 1987). Domain Ia (amino acids 1–252) binds to the $\alpha2$ macroglobulin receptor, which is commonly expressed on the surface of animal cells (KOUNNAS et al. 1992). After internalization into the target cell, domain II (amino acids 253–365) undergoes proteolytic cleavage between Arg279 and Gly280, and the remaining C-terminal portion of domain II functions to translocate the 37-kDa fragment (amino acids 280–613) into the cytosol (OGATA et al. 1990, 1992). The function of domain Ib (amino acids 365–399) is unknown, and much of it (amino acids 365–380) can be removed without loss of activity (SIEGALL et al. 1989a; KREITMAN et al. 1993a). Domain III (amino acids 400–613) contains the ADP ribosylating activity. Amino acids 603–613 are not needed for enzymatic activity (CHAUDHARY et al. 1990b), but the carboxyl terminal amino acids, REDLK, are needed for cytotoxic activity. These can be replaced by the endoplasmic reticulum (ER) retention sequence KDEL (CHAUDHARY et al. 1990b), suggesting that REDLK functions as an ER retention sequence to transport the 37-kDa fragment of PE to the ER, the site from which it can translocate to the cytosol.

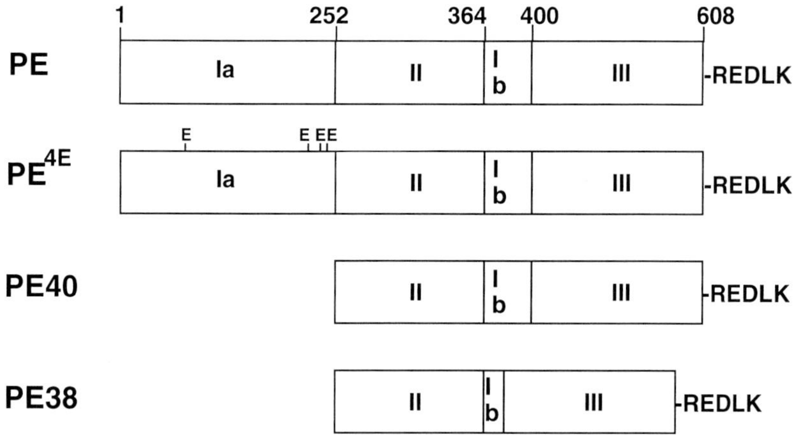

Fig. 1. Schematic diagram of Pseudomonas exotoxin (PE) and truncated forms. Domains 1a, II, and Ib are represented by the indicated *boxes*. Box III represents domain III up until amino acid 608, and amino acids 609–613, the C-terminus of PE, is indicated. In PE^{4E}, the basic amino acids at positions 57, 246, 247 and 249 are replaced with glutamate residues to decrease cell binding. In PE40, domain Ia is missing from PE, and in PE38, domain Ib is shortened by removing amino acids 365–380 of PE

2. Mutants Lacking Cell Binding

To prevent targeting of PE to normal cells, domain Ia is either mutated or deleted, as shown in Fig. 1. PE^{4E} is a mutated form of PE where four basic residues at positions 57, 246, 247 and 249 in domain Ia are converted to glutamate (Chaudhary et al. 1990c). PE40 is the 40-kDa form of PE where domain Ia is deleted. PE40 can be further shortened to PE38 by removing amino acids 365–380, which contain a disulfide bond.

III. Types of Toxins Made with PE

1. Chemical Conjugates vs Recombinant Fusions

PE or PE derivatives were first targeted by chemically coupling the toxin to monoclonal antibodies or growth factors (Pirker et al. 1985; Kondo et al. 1988; FitzGerald et al. 1983). More recently, using recombinant-DNA techniques, ligands have been fused directly to the amino termini of PE derivatives (Fig. 2). The activity of these fusions is made possible by the protease site in PE, which allows an intracellular protease to separate the ligand from the fragment of PE (amino acids 280–613), which translocates to the cytosol (Ogata et al. 1992). An analogous protease site is present for diphtheria toxin (Williams et al. 1990). The absence of such a site in plant toxins like ricin makes effective fusions of ligands to those toxins more difficult to make (Cook et al. 1993).

2. Fusion Toxins Containing Transforming Growth Factor-α

One of the earliest PE-containing fusion toxins was transforming growth factor α (TGFα)-PE40, composed of TGFα fused to PE40 (Fig. 2). This toxin was very active against cells displaying the epidermal growth-factor receptor (EGFR), with half-maximal protein synthesis inhibition (IC_{50}) achieved at concentrations as low as 0.1 ng/ml (Siegall et al. 1989b). Because EGFRs are present on normal liver cells in humans and mice, the single-dose LD_{50} in mice was low (1.5 μg intraperitoneally), but antitumor effects in mice bearing human EGFR-positive tumors were nevertheless observed (Pai et al. 1991; Kreitman et al. 1992b). To improve the binding about 15-fold, TGFα was fused to a derivative of PE40 lacking cysteines (Fig. 2), thus eliminating the Cys265–Cys287 and Cys372–Cys379 disulfide bonds (Edwards et al. 1989). The resultant molecule, TP40, also showed antitumor in animal models (Heimbrook et al. 1990, 1991; Kunwar et al. 1993).

Because of its specific binding to EGFR-bearing hepatocytes, TP40 was brought to clinical trial to be used nonsystemically, as intravesical therapy for bladder carcinoma. Forty-three patients with superficial bladder cancer received between 0.15 mg/week and 9.6 mg/week. No dose-dependent local or systemic toxicity was observed. Of 13 patients with carcinoma in situ, 9 showed histopathologic resolution of disease post-treatment, although responses were

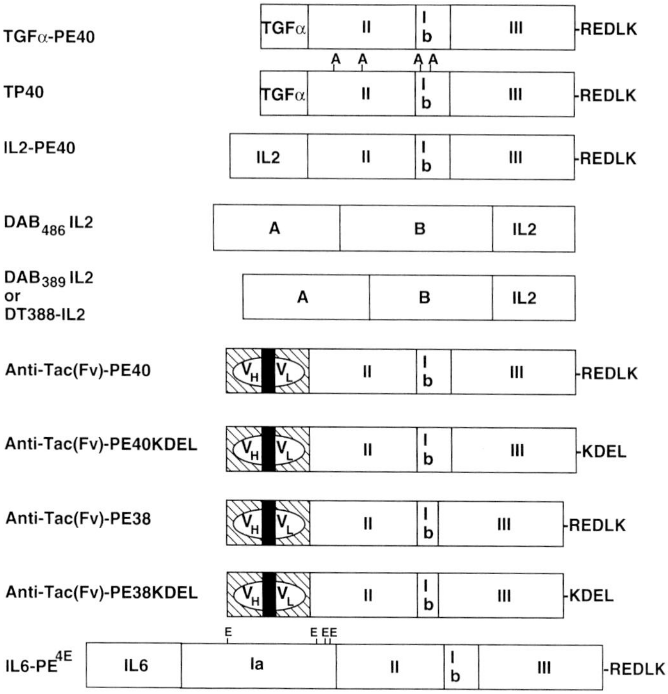

Fig. 2. Schematic diagram of chimeric toxins made with Pseudomonas exotoxin (PE) or diphtheria toxin (DT). Ia, II, Ib and III represent the respective domains of PE, as in Fig. 1. *Boxes A* and *B* represent the A chain and truncated B chain of DT, respectively. $DAB_{486}IL2$ consists of methionine followed by the first 486 amino acids of DT, followed by human interleukin (IL) 2. Both DT388-IL2 and $DAB_{389}IL2$ consist of methionine followed by the first 388 amino acids of DT, followed by human IL2. Between Thr386 of DT and Pro2 of IL2, DT388-IL2 contains the extra amino acids RPHMAD (CHAUDHARY et al. 1991), while $DAB_{389}IL2$ contains HA (WILLIAMS et al. 1990). Also IPEGEA follows amino acid 131 of IL2 in DT388-IL2. In the recombinant immunotoxins, anti-Tac(Fv) is composed of V_H and V_L separated by a flexible linker of sequence $(G_4\,S)_3$

not seen in evaluable patients with more advanced disease. Thus, TP40 was effective against low-volume disease in this compartmental protocol.

3. Fusion Toxins Containing Interleukin 2

To target hematologic tumors or autoimmune disorders, where the target cells display interleukin2 receptors (IL2Rs), IL2 was fused to PE40 (Fig. 2). IL2-PE40 was cytotoxic against IL2R-bearing cell lines from patients with adult T-cell leukemia (ATL) and was active in murine models of autoimmune disease, but was poorly cytotoxic toward human-activated T-cells or fresh ATL cells from patients (LORBERBOUM-GALSKI et al. 1988a, 1989, 1990; KREITMAN et al. 1990). Similarly, the diphtheria toxin-containing fusion protein $DAB_{486}IL2$

(Fig. 2), which was very cytotoxic toward ATL lines, required high concentrations to kill fresh ATL cells (WILLIAMS et al. 1987; KIYOKAWA et al. 1989). A possible reason for this phenomenon is that fresh target cells have lower affinity sites (IL2Rα–IL2Rβ complex, K_d~10 pM) than cell lines (KODAKA et al. 1990; WALDMANN 1991). While fresh target cells still have ample low-affinity sites (IL2Rα alone, K_d~10 nM), the binding to IL2 is markedly reduced. A desirable alternative was to target low-affinity sites with a high-affinity ligand, such as the anti-Tac antibody.

4. Recombinant Immunotoxins

The monoclonal antibody anti-Tac, which binds to IL2Ra with high affinity (K_d = 100 pM), was originally chemically conjugated to PE derivatives, but the cytotoxic effects were similar to that seen with IL2-PE40 (BATRA et al. 1989; KONDO et al. 1988). To make a single-chain fusion toxin with the binding activity of this antibody, anti-Tac was converted to the single-chain (Fv) form and then anti-Tac(Fv) was fused to PE40 (Fig. 2). The Fv form is composed of the variable heavy (V_H) domain fused to the variable light (V_L) domain through a peptide linker, of sequence $(G_4S)_3$. Anti-Tac(Fv)-PE40 was very cytotoxic toward ATL cell lines, with IC_{50}s as low as 0.15 ng/ml (CHAUDHARY et al. 1989). The recombinant immunotoxin was also very cytotoxic toward activated human T-cells and fresh malignant cells from patients with ATL (CHAUDHARY et al. 1989; BATRA et al. 1990; KREITMAN et al. 1990). More recently, other antibodies have been converted to single-chain immunotoxins, including anti-erbB2 (e23 and FRP5), anti-Le^y (B3 and BR96), anti-ovarian cancer (OVB3), anti-transferrin receptor (HB21 and E6), anti-IL2Rβ (Mik-b1) and anti-prostate carcinoma (PR1) monoclonal antibodies (BATRA et al. 1991, 1992; BRINKMANN et al. 1991, 1993; CHAUDHARY et al. 1990a; FRIEDMAN et al. 1993; KREITMAN et al. 1992c; NICHOLLS et al. 1993; WELS et al. 1992).

B. Preclinical Development of Anti-Tac(Fv) Toxins

I. Background

Anti-Tac(Fv)-PE40KDEL, anti-Tac(Fv)-PE38KDEL and anti-Tac(Fv)-PE38 are mutants of anti-Tac(Fv)-PE40, as shown in Fig. 2. Anti-Tac(Fv)-PE40 was converted to anti-Tac(Fv)-PE40KDEL, by replacing the carboxy terminal amino acids REDLK with KDEL. This improved cytotoxic activity, and increased animal toxicity several-fold (SEETHARAM et al. 1991). In addition, anti-Tac(Fv)-PE40 and anti-Tac(Fv)-PE40KDEL were each shortened slightly to anti-Tac(Fv)-PE38 and anti-Tac(Fv)-PE38KDEL, respectively, by removing amino acids 365–380 of PE (SIEGALL et al. 1989a; KREITMAN et al. 1993a). This also removed the Cys372–Cys379 disulfide bond. We have determined that the KDEL mutation increases cytotoxic activity, but that removing amino acids 365–380 of PE does not change cytotoxic activity (KREITMAN et al. 1993a). As

Table 1. Cytotoxicity of anti-IL2R toxins on cell lines. ATAC-4 cells are A431 epidermoid carcinoma cells transfected with the gene encoding human IL2Rα. Cells were incubated with toxins for 18–24h and pulsed with [^{3}H]leucine to determine inhibition of protein synthesis

Cell line (type)	Sites/cell IL2Rα	IC_{50} (mg/ml) IL2Rβ	Anti-Tac(Fv)-PE38KDEL	Anti-Tac(Fv) PE40	DT388-IL2	IL2-PE40
HUT-102 (ATL)	500,000	18,000	0.06	0.2	0.1	3
MT-1 (ATL)	500,000	Absent	0.02	0.2	>100	80
ATAC-4 (epiderm)	200,000	Absent	0.025	0.06	200	23

The IC_{50} is the concentration of toxin necessary for 50% inhibition of protein synthesis. Most of the binding and cytotoxicity data was reported previously (LORBERBOUM-GALSKI et al. 1988b, 1990; CHAUDHARY et al. 1989, 1990a; KREITMAN et al. 1993a, 1994)
ATL, adult T-cell leukemia; IL, interleukin; PE, Pseudomonas exotoxin

shown in Table 1, the cytotoxic activity of anti-Tac(Fv)-PE38KDEL is higher than that of anti-Tac(Fv)-PE40 toward several IL2R-bearing cell lines. It is also notable that anti-Tac(Fv)-PE38KDEL is much more cytotoxic than the IL2 toxins IL2-PE40 and DT388-IL2 toward cell lines with low or undetectable levels of IL2Rβ, which is needed to form the high-affinity IL2R.

II. Efficacy Data on Relevant Human Cells

1. Human Activated T-Lymphocytes

To determine whether anti-Tac(Fv) toxins could target cells that mediate autoimmune diseases, we tested the recombinant immunotoxins against activated human T-lymphocytes. Cells obtained from normal peripheral blood were activated using either phytohemagglutinin (PHA) or with the mixed lymphocyte reaction (MLR). In the latter case, cells from two normal individuals were incubated together until they became IL2R-bearing activated T-lymphoblasts. Anti-Tac(Fv) toxins were very cytotoxic against such cells, with IC_{50}s of as low as 0.025 ng/ml. In contrast, normal resting lymphocytes were resistant, even at 1000 ng/ml (KREITMAN et al. 1993a). The cytotoxic activity of anti-Tac(Fv)-PE38KDEL toward activated human T-cells was >1000-fold higher than that of IL2-PE40 and was also greater than that of either DT388-IL2 or DAB_{389}IL2. Both of the latter fusion proteins contain IL2 following the methionine and the first 388 amino acids of DT (CHAUDHARY et al. 1991; WILLIAMS et al. 1990). While DT388-IL2 and DAB_{389}IL2 differ slightly in junctional amino acids, in our assays, their cytotoxic activity was similar. DAB_{389}IL2 is currently undergoing clinical testing. As shown in Fig. 3, the difference between the cytotoxicity of anti-Tac(Fv)-PE38KDEL and that of DT388-IL2 or DAB_{389}IL2 on the activated human T-cells ranged from several-

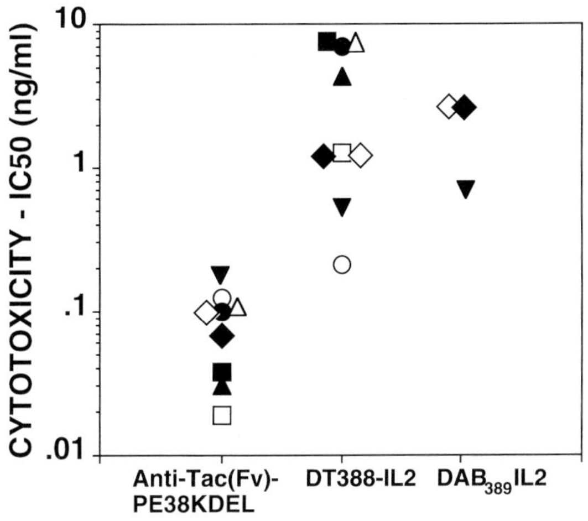

Fig. 3. Comparison of recombinant toxins against activated human T-cells. Human T-cells were activated using either the mixed lymphocyte reaction (●), or phytohemagglutinin (other markers) as described previously (KREITMAN et al. 1993a; PURI et al. 1994). Each type of point marker refers to a separate assay

fold to over 100-fold, depending on different states of T-cell activation (Fig. 3). We have found that towards both MLR and PHA-activated T-cells, the cytotoxic activity of anti-Tac(Fv)-PE40, anti-Tac(Fv)-PE38 and anti-Tac(Fv)-PE40KDEL is similar to that of anti-Tac(Fv)-PE38KDEL (KREITMAN et al. 1993a). We propose two possible reasons for the improved activity of anti-Tac(Fv) toxins over IL2 toxins. The first is that while the ATL cell line HUT-102 contains up to 18,000 high-affinity IL2Rs/cell, activated T-cells contain only 2000–4000 (ROBB et al. 1984). Alternatively, after internalization into a target cell, IL2 may direct the toxin to be destroyed in lysozomes (LOWENTHAL et al. 1986), whereas anti-Tac(Fv) may allow more toxin molecules to reach the ER, where they can access the cytoplasm.

2. Fresh Adult T-Cell Leukemia Cells

In terms of activity against malignant target cells, we did not believe that activity against ATL cell lines would necessarily be predictive of the potential therapeutic utility of anti-Tac(Fv) toxins. One reason for this is that cells from patients may have much lower numbers of IL2Rs than do cell lines. Another reason is that the intracellular transport and processing of the toxin within the target cells in patients might be so slow that the toxin is degraded before reaching the cytoplasm. Thus, we tested the recombinant immunotoxin against fresh cells from patients with ATL and chronic lymphocytic leukemia (CLL).

To study the effect of recombinant immunotoxin against cells from patients with ATL, the Ficoll-purified cells of 40 ATL patients were incubated

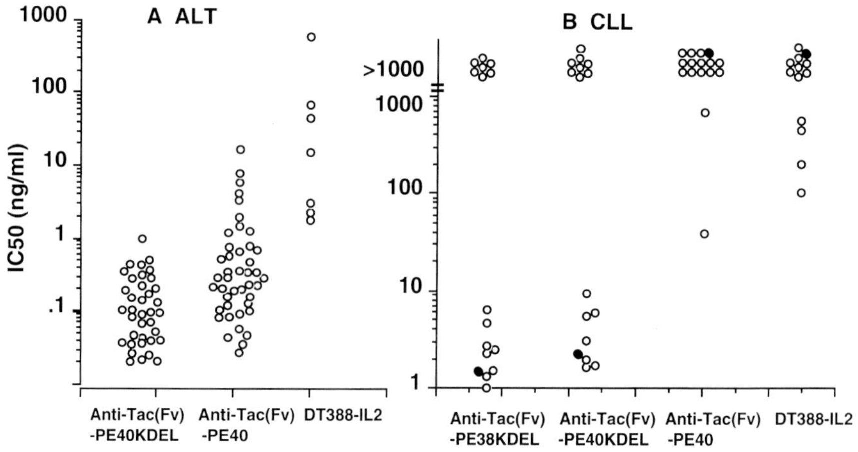

Fig. 4A,B. Cytotoxicity of recombinant toxins against fresh human cells. Fresh adult T-cell leukemia (ATL) (**A**) or chronic lymphocytic leukemia (CLL) (**B**) cells were incubated with the indicated recombinant toxin for 60 h and pulsed with [^{3}H]leucine to determine inhibition of protein synthesis, as described previously (Kreitman et al. 1992a, 1993b; Saito et al. 1994). The *solid circles* indicate a patient with the prolymphocytic variant of CLL

with toxins for 2–3 days and then labeled with [^{3}H]-leucine. All samples showed sensitivity to anti-Tac(Fv)-PE40 as well as anti-Tac(Fv)-PE40KDEL and we have yet to find a resistant sample. The activities of anti-Tac(Fv)-PE38 and anti-Tac(Fv)-PE38KDEL were essentially identical to those of anti-Tac(Fv)-PE40 and anti-Tac(Fv)-PE40KDEL, respectively. (Kreitman et al. 1993a). Figure. 4A shows a scatter plot of the IC_{50}s. The activity of anti-Tac(Fv)-PE40KDEL was usually several-fold to 40-fold higher than that of anti-Tac(Fv)-PE40 (Kreitman et al. 1993b; Saito et al. 1994), but the effect of the KDEL terminus was negligible towards cells from some patients. Also, the soluble IL2R levels present in the serum of ATL patients, who have recently obtained partial response to chemotherapy, did not completely block the cytotoxic activity of the immunotoxin. The response of fresh ATL cells to DT388-IL2 was usually lower (Fig. 4A). This is probably because fresh cells from ATL patients display 1600 to >20,000 IL2Rα sites/cell, but high-affinity sites usually number only a few hundred (Kodaka et al. 1990).

3. Fresh Chronic Lymphocytic Leukemia Cells

CLL cells proved to be much different in immunotoxin sensitivity than ATL cells. Just over half of the cells tested were found to be sensitive to anti-Tac(Fv)-PE38KDEL, with IC_{50}s <10 ng/ml, but the remainder were resistant (Fig. 4B). The resistant cells often displayed well over 1000 IL2Rα sites/cell (Kreitman et al. 1992a). Moreover, without the KDEL carboxyl terminal mutation, anti-Tac(Fv)-PE40 showed poor reactivity with all of the CLL samples (Fig. 4B). This data indicated that CLL cells, perhaps as a result of terminal differentiation, had lost the ability to transport the intracellular toxin containing REDLK, and some samples had even lost the ability to recognize the KDEL-ER localization signal. The poor reactivity of CLL cells with DT388-IL2 was expected, since such cells usually display <100 high-affinity IL2R sites/cell (YAGURA et al. 1990). An important question is whether the less differentiated proliferating cells of patients with CLL, which are the most important therapeutic targets, have not yet lost sensitivity to REDLK or even IL2-containing toxins. If so, patients with CLL, and similar tumors such as lymphomas, would not necessarily require a KDEL-containing toxin to respond.

III. Efficacy Data in an Animal Model of IL2R-Bearing Cancer

1. Production of the Human ATAC-4 Line

We were limited in testing anti-Tac(Fv) toxins in animal models of leukemia or autoimmune disease because anti-Tac does not bind to murine IL2Rα. Therefore, A431 cells, which grow well in animals, were transfected with a gene encoding the human IL2Rα. The resultant ATAC-4 cells expressed 200,000 IL2Rα sites/cell and showed high sensitivity to anti-Tac(Fv) toxins, with IC_{50}s <0.1 ng/ml (KREITMAN et al. 1993a, 1994). ATAC-4 cells were considered an appropriate model for human ATL, despite the fact that the number of IL2Rα sites on ATAC-4 is over 10-fold higher, because ATAC-4 cells are devoid of the IL2Rβ chain that is thought to be important for high rates of internalization (ROBB and GREEN 1987).

2. Toxicity of Anti-Tac(Fv) Toxins in Mice

To determine the maximum tolerated dose (MTD) of anti-Tac(Fv) toxins in mice, they were tested at different doses intravenously (i.v.). When given daily for 3 days (QD × 3), the maximum nonlethal dose of anti-Tac(Fv)-PE38KDEL was 0.1 mg/kg × 3. Twice this amount could be given safely every other day for three doses (QOD × 3). The cause of death in the mice was hepatocellular necrosis, probably due to nonspecific uptake of the toxin in hepatocytes. This was observed in mice receiving the MTD, but not in those receiving half this dose (KREITMAN et al. 1994). It should be added, however, that because of the high variability of transaminases in normal mice and high normal levels in mice, the mice receiving lower doses could have had several-fold elevations of transaminase that were undetected.

3. Pharmacokinetics in Mice

To determine how long cytotoxic levels of anti-Tac(Fv) toxin would remain in the circulation of mice, 1 μg (0.05 mg/kg) of anti-Tac(Fv)-PE38KDEL was administered i.v., and serum levels were determined by testing the cytotoxic

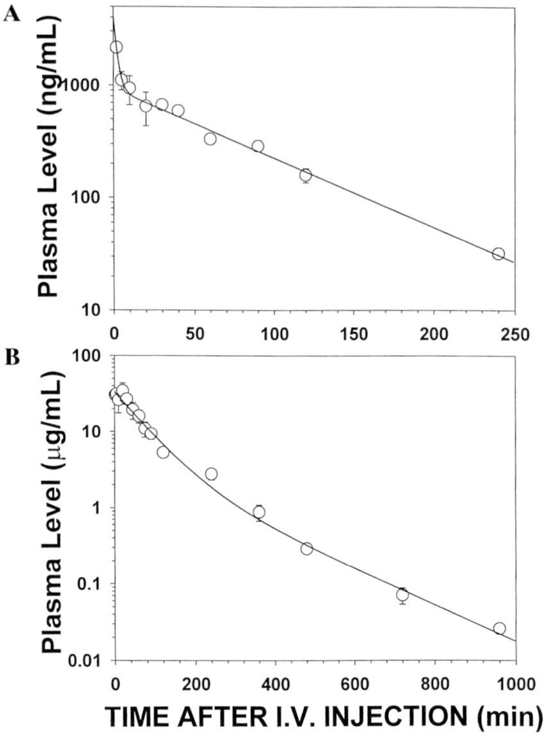

Fig. 5A,B. Pharmacokinetics of recombinant immunotoxins. **A** Anti-Tac(Fv)-PE38KDEL (0.05 mg/kg) was injected i.v. into mice and serum levels determined. Points shown are averages of duplicate mice. **B** Anti-Tac(Fv)-Pseudomonas exotoxin (PE) 38 (0.75 mg/kg) was injected i.v. into Cynomolgus monkeys. Serum levels are the means of triplicate monkeys. Standard deviations are indicated by error bars, when larger than the size of the point markers

activity of the serum. Figure 5A shows that the $T_{1/2}\alpha$ was ~1 min, and the $T_{1/2}\beta$ was 44 min. Thus, while the half life of the 63-kDa anti-Tac(Fv)-PE38KDEL was short, the levels even 4 h after injection were >1000-fold higher than the IC_{50} toward ATAC-4 cells.

4. Antitumor Activity in Tumor-Bearing Mice

To examine the antitumor activity of anti-Tac(Fv) toxins, several different nonlethal doses of anti-Tac(Fv)-PE38KDEL, anti-Tac(Fv)-PE38 or anti-Tac(Fv)-PE40 were administered to nude mice 4 days after injection with ATAC-4 cells, when tumors approximately 75 mm^3 had formed. Figure 6A shows that antitumor activity was dose dependent and was significant, even at 1% of the LD_{50}. The antitumor effect was specific, since treatment with the anti-Tac(Fv)-PE38KDEL^{Asp553} mutant lacking ADP-ribosylation activity, or the Mik-β1(Fv)-PE38KDEL toxin lacking IL2Rα-binding activity showed no

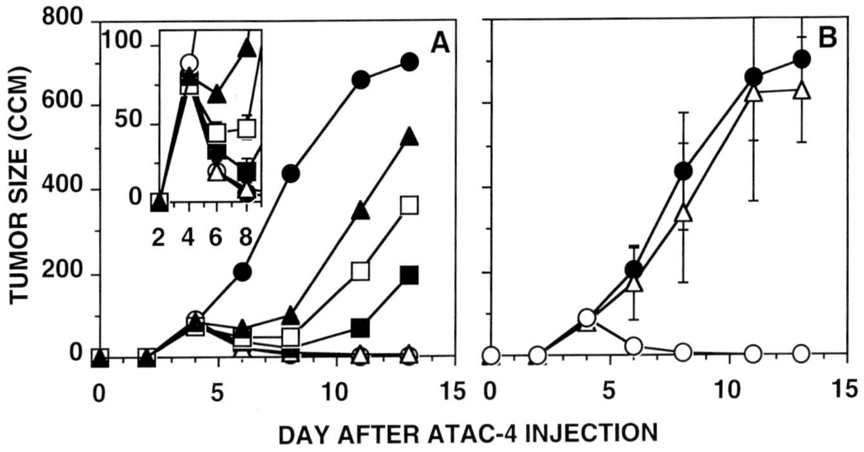

Fig. 6A,B. Antitumor activity of anti-Tac(Fv)-PE38KDEL in mice. Nude mice were injected subcutaneously with ATAC-4 cells (3×10^6), were treated on day 0 and, on day 4, began three daily doses of immunotoxin given i.v. **A** Mice received anti-Tac(Fv)-PE38KDEL at dose levels of 2.5 μg/kg (▲), 5 μg/kg (❐), 10 μg/kg (■), 50 μg/kg (Δ) or 100 μg/kg (○). **B** The antitumor effect of anti-Tac(Fv)-PE38KDEL 100 μg/kg (○) was contrasted with that of Mik-β1(Fv)-PE38KDEL 100 μg/kg (○), which does not bind to interleukin (IL)2Rα. **A, B** The control group was treated with anti-Tac(Fv)-PE38KDEL^{Asp553} 100 μg/kg (●), which has no ADP ribosylation activity due to conversion of Glu^{553} of PE to Asp (Kreitman et al. 1994)

antitumor activity (Fig. 6B) (Kreitman et al. 1994). In repeated experiments, complete remissions were observed in 16 of 19 mice receiving 7% of the LD_{50}. Complete remissions were often durable, lasting over 3 months, until the mice were sacrificed. Alternatively, mice relapsing after high doses were found to be carrying tumors that had lost expression of the IL2Rα-encoding plasmid (Kreitman et al. 1994). This was attributed to the fact that ATAC-4 cells that express IL2Rα could be selected in vitro using neomycin-containing media, but not in vivo.

IV. Primate Testing

The low levels of IL2Rα on resting lymphocytes and high levels of IL2Rα on activated T-cells from non-human primates react with anti-Tac. Therefore, toxicity in primates should be useful in predicting targeted toxicity in humans, and possibly in predicting toxicity due to nonspecific internalization of the toxin in normal cells.

1. Pharmacokinetics in Cynomolgus Monkeys

The average serum levels of anti-Tac(Fv)-PE38 in three Cynomolgus monkeys after injection of 0.75 mg/kg i.v. are shown in Fig. 5B. Prior to injection, the monkeys were screened for the absence of neutralizing antibodies to the toxin, and serum levels were determined by their cytotoxic activity, as in Fig. 5A. The terminal half-life was 114 min., over twice the level found in mice. The difference is probably due to delayed excretion in primates, since the serum levels are too high to be explained by toxin binding to the low levels of cell-associated or soluble-intravascular IL2Rα. A prolonged half-life should enhance antitumor activity, provided it does not increase toxicity and limit the maximum tolerated dose.

2. Toxicity in Cynomolgus Monkeys

The toxicity of anti-Tac(Fv)-PE38 and anti-Tac(Fv)-PE38KDEL was assessed by injecting Cynomolgus monkeys i.v. every other day for 3 days. The every other day administration schedule was chosen for several reasons. First, the nonspecific toxicity of the immunotoxins was evident in mice within 2 days after the dose, so that a patient could be spared the second and third doses if hypertransaminasemia or other laboratory abnormalities appeared by day two after the first injection. Second, a treatment course of at least 1 week could be given before the onset of anti-toxin neutralizing antibodies.

Table 2 shows the immunotoxin levels and transaminase elevations following various doses of anti-Tac(Fv)-PE38 and anti-Tac(Fv)-PE38KDEL in Cynomolgus monkeys. Anti-Tac(Fv)-PE38 resulted in transaminase elevations that were roughly proportional to both the dose and the 10-min post-injection serum level. Interestingly, the 1 mg/kg $\times$ 3 dose was over fivefold higher than the MTD of anti-Tac(Fv)-PE38 in mice when given as three daily doses (KREITMAN et al. 1994). The lowest dose of 0.02 mg/kg i.v. QOD $\times$ 3 is considered the no-effect dose. Data is also shown for monkeys treated with three doses of anti-Tac(Fv)-PE38KDEL, either 0.02 mg/kg or 0.2 mg/kg. Mild but significant transaminase elevations were observed in most of these monkeys, and one of four monkeys treated with the low dose of anti-Tac(Fv)-PE38KDEL developed a grade-III elevation of ALT. The monkeys showed no evidence of renal insufficiency, capillary-leak syndrome or hypersensitivity reactions. Based on this data, we chose anti-Tac(Fv)-PE38 (LMB-2) for a phase-I trial in patients with CD25+ hematologic malignancies, since this immunotoxin consistently appeared relatively nontoxic at low doses. At this time, the maximum tolerated dose of anti-Tac(Fv)-PE38 remains unknown after treatment of 22 patients with doses up to 40 μg/kg i.v. QOD $\times$ 3. The most common toxicities include fever and transaminase elevations not associated with decreases in hepatic function. Several partial responses have been observed, and most of the patients have not developed immunogenicity to the first cycle.

V. Production Issues

A method for the large-scale production of anti-Tac(Fv)-PE38KDEL, which utilizes an IL2Rα-affinity column, was recently reported (SPENCE et al. 1993).

Table 2. Toxicity of anti-Tac(Fv) containing immunotoxins in monkeys

	Dose given i.v. days 1, 3, 5	Pre	Day 1	Day 3	Day 5
Mean serum or plasma levels 10 min post (ng/ml)					
1. Anti-Tac(Fv)-PE38	0.02 mg/kg (n = 4)		390 ± 10	270 ± 70	280 ± 45
2. Anti-Tac(Fv)-PE38	0.3 mg/kg (n = 4)		8000 ± 1000	5900 ± 540	5400 ± 880
3. Anti-Tac(Fv)-PE38	0.75 mg/kg (n = 1)		27	28	23
4. Anti-Tac(Fv)-PE38	1.0 mg/kg (n = 1)		37	34	27
6. Anti-Tac(Fv)-PE38KDEL	0.02 mg/kg (n = 4)		310 ± 10	300 ± 20	290 ± 40
6. Anti-Tac(Fv)-PE38KDEL	0.2 mg/kg (n = 2)		2900 ± 400	2300 ± 200	1200 ± 600
Transaminase grades					
1. Anti-Tac(Fv)-PE38	0.02 mg/kg (n = 4)	0–I	0–I	0–I	0–I
1. Anti-Tac(Fv)-PE38	0.3 mg/kg (n = 1)	0–I	0–II	I–II	I–II
1. Anti-Tac(Fv)-PE38	0.75 mg/kg (n = 1)	0	III	III	III
2. Anti-Tac(Fv)-PE38	1.0 mg/kg (n = 1)	0	II	IV	III
3. Anti-Tac(Fv)-PE38KDEL	0.02 mg/kg (n = 4)	0–I	0–I	I–II	I–III
3. Anti-Tac(Fv)-PE38KDEL	0.2 mg/kg (n = 2)	0–I	0–II	I–II	I–II

Cynomolgus monkeys were treated with the indicated immunotoxin diluted into 1 ml/kg HSA-PSB as a 1-min bolus i.v. infusion on days 1, 3 and 5. Serum levels were determined by the cytotoxic activity of dilutions of monkey serum drawn 10 min after each dose, using a standard curve generated from purified immunotoxin. Blood for chemistry determinations was drawn prior to the dose on days 1, 3 and 5. Grade I, II, III and IV toxicity is defined as either transaminase being <2.5-fold, 2.6- to 5-fold, 5.1- to 20-fold and >20-fold the upper limit of normal, respectively, which for each transaminase is 50 U/l. PE, Pseudomonas exotoxin; i.v., intravenous

We have recently published a more conventional method not utilizing protein-containing columns, which also produces the protein in high yield and with low endotoxin content (Buchner et al. 1992; Kreitman and Pastan 1993). DNA encoding the protein is placed in an expression vector utilizing the T7 promoter (Studier and Moffat 1986). *E. coli* cells containing the plasmid are grown in a fermenter, and induced by the addition of a lactose analog for 90 min. The crude insoluble protein (inclusion bodies) is isolated by repeated washings in the detergent triton-X-100. These washings remove most of the endotoxin from the protein. The crude protein is then denatured and reduced in 6 M guanidine solution containing dithioerythritol and refolded in a redox buffer containing oxidized glutathione. After concentrating and dialyzing the refolded protein, monomeric immunotoxin is isolated by anion exchange and size-exclusion chromatography. The yield of fully active protein is currently 3.5–5% of total recombinant protein, or 170–250 mg from a 10-l fermentation induced at an OD_{650} of 6–8. This method has been found to give very good yields when applied to the production of other proteins, such as DT-containing chimeric toxins, IL4 and toxins containing IL2, IL4 or IL6 (KREITMAN and PASTAN 1993).

C. Preclinical Development of IL6-PE^{4E}

I. Background

IL6Rs have been found on multiple myeloma and hepatocellular carcinoma cells (KAWANO et al. 1988; SIEGALL et al. 1990b). While hepatoma is an important cause of cancer throughout the world, prostate cancer incidence is increasing in the United States, and such cells also express IL6R (SIEGALL et al. 1990c). Moreover, in a recent study, 11 of 13 malignant prostates from patients contained readily detectable levels of IL6R messenger RNA (mRNA) (SIEGSMUND and PASTAN 1994). In 1988, it was shown that IL6R-bearing cells could be targeted using a recombinant fusion of IL6 with PE40 (SIEGALL et al. 1988). IL6-PE^{4E}, a chimeric toxin having nearly tenfold the cytotoxic activity of IL6-PE40, was obtained by fusing IL6 to the amino terminus of PE^{4E} (SIEGALL et al. 1990a). IL6-PE^{4E} was found to be cytotoxic to hepatocellular, epidermoid and prostatic carcinoma, and myeloma cell lines with IC_{50}s as low as 1 ng/ml (SIEGALL et al. 1990a, 1990c). IL6-PE^{4E} was found to have antitumor activity against a human IL6R-bearing hepatocellular carcinoma when administered to nude mice carrying these tumors (SIEGALL et al. 1991). While this agent is being explored for in vivo therapy of IL6R-bearing tumors in humans, a more immediate goal is to use this agent for purging marrow of patients with multiple myeloma prior to autologous bone-marrow transplantation (ABMT).

II. Study of IL 6-PE^{4E} for Ex Vivo Marrow Purging in Multiple Myeloma

1. Rationale

Multiple myeloma causes 10% of the hematologic malignancies and 10,000 deaths in the United States per year (ALEXANIAN and DIMOPOULOS 1994). While conventional chemotherapy is not curative and rarely induces complete responses, ablative chemoradiotherapy, followed by autologous bone-marrow transplantation, induces complete remissions in close to 50% of patients. However, long-term survival is uncommon because of the high frequency of relapse. While it is unclear whether relapse is primarily due to reinfusion of tumor cells or due to failure of the ablative chemotherapy to kill sufficient tumor in vivo, purging bone marrow prior to reinfusion with monoclonal antibodies and complement has recently been attempted (ANDERSON et al. 1993). Because recombinant toxins can kill receptor-bearing cells without requiring complement, IL6-PE^{4E} was developed as a potential purging agent. The normal hematopoietic cells to be treated would include not only bone marrow, but also stem cells isolated from the blood of the patient (FERMAND et al. 1993; JAGANNATH et al. 1992).

The goal in developing IL6-PE^{4E} for this purpose was to determine whether the recombinant toxin was safe ex vivo as well as effective. This required a number of studies, including determining whether IL6-PE^{4E} could kill myeloma cells in bone marrow, and whether IL6-PE^{4E} would spare normal hematopoietic precursors, which are needed for engraftment. Finally, in lieu of extensive non-human primate toxicology studies, it was necessary to show the absence of significant carryover of active toxin into patients during reinfusion.

2. Efficacy Against Fresh Marrow Cells from Myeloma Patients

We did not believe it was relevant to mix myeloma cell lines with normal bone marrow and determine if the myeloma cells were killed in order to show that IL6-PE^{4E} could kill multiple myeloma cells in the bone marrow. The reason for this is that, while myeloma cell lines express 10,000–20,000 IL6Rs/cell, fresh cells from the bone marrow of patients express only about 500 sites/cell (KAWANO et al. 1988). Therefore, we took Ficoll-purified fresh bone-marrow aspirates from 15 patients and tested them for sensitivity to IL6-PE^{4E}. It was found that samples from five patients were very sensitive, with IC_{50}s ranging from <1 ng/ml to 6 ng/ml; three patients showed intermediate sensitivity, with IC_{50}s ranging from 30 ng/ml to 140 ng/ml; and the remaining seven patients showed little or no sensitivity (KREITMAN et al. 1992d). It was found that the marrow samples that were the most active, the ones showing the most [^{3}H]-leucine uptake in the absence of toxin, had the highest sensitivity to IL6-PE^{4E}. Since all samples were contaminated with normal bone marrow, which was found to be resistant to IL6-PE^{4E}, we could not exclude the possibility that in some of the "resistant" samples, the malignant cells were actually killed. However, the signal was too low to detect because of low [^{3}H]-leucine incorporation, even in the absence of toxin.

3. Safety Toward Fresh Normal Marrow Cells

We have tested several bone-marrow specimens from normal patients, who had bone marrow donated for the purpose of allogeneic bone-marrow transplantation. No sensitivity to IL6-PE^{4E} at 1000 ng/ml was detected (KREITMAN et al. 1992d). However, since the stem cells constitute a very small percentage of normal bone marrow, selective toxicity to these cells could not be ruled out by this study.

4. Safety of IL6-PE^{4E} Toward Normal Hematopoietic Progenitors

To determine whether IL6-PE^{4E} would kill normal CFU-GM and BFU-E populations from patient marrow, these cells were isolated from the marrow of four multiple-myeloma patients and one normal patient. After incubating these progenitor cells with 1000 ng/ml IL6-PE^{4E}, no significant change in viability was observed (KREITMAN et al. 1992d).

5. Lack of Prevention of Bone Marrow Engraftment

In vitro assays are limited in their ability to demonstrate the safety of a purging agent, because one cannot, at this time, be sure that the progenitor cells tested are identical to, or representative of, stem cells. Because murine IL6Rs bind human IL6, we decided to expose murine bone marrow to high levels of IL6-PE^{4E}, and determine whether the toxin prevented bone-marrow engraftment in the mice. Accordingly, syngeneic-donor bone marrow was exposed to 300 μg/ml IL6-PE^{4E} under purging conditions, and the marrow was infused into syngeneic lethally irradiated mice. Conditions were chosen so that 100% of lethally irradiated mice would die without transplant, and 50% of the irradiated mice would survive if transplanted with marrow not treated with toxin. No significant difference in survival was observed, whether the marrow was treated or not with IL6-PE^{4E}. This study strongly indicates that ex vivo IL6-PE^{4E} would not prevent marrow engraftment in humans.

6. Carryover of IL6-PE^{4E} In Vivo

The rationale for the use of IL6-PE^{4E} for ex vivo purging of bone marrow, rather than in vivo treatment of multiple myeloma, is to avoid the potential in vivo toxicity of the drug, at least until the potential in vivo toxicity of IL6-PE^{4E} to primates is better understood. However, it is clear that 100% of the drug will not be washed from the bone-marrow cells prior to reinfusion into the patient. While it is not clear that cell-associated IL6-PE^{4E} could dissociate from the cells in vivo and cause toxicity, the possibility nevertheless exists. IL6-PE^{4E} could become cell associated by two mechanisms. The first is by binding to IL6Rs on the malignant (or possibly normal) cells. The second is by non-specifically binding with the cells.

To determine the maximum carryover by specific binding, we would assume that $3\text{–}10^8$ cells/kg of body weight (2.1×10^{10} cells for a 70 kg patient) are exposed to 80 μg/ml IL6-PE^{4E} in a volume of 260 ml. If the cells were 1% myeloma cells and each contained 1000 sites/cell, which is just above the maximum number of 840 sites/cell for fresh myeloma cells reported previously

(Kawano et al. 1988), a maximum of 2.1×10^{11} molecules, 3.5×10^{-13} mol, or 30 ng of IL6-PE^{4E}, would carryover into the patient. This would amount to a maximum of 0.4 ng/kg single dose, which represents less than 1/1,000,000 of the LD_{50} in mice of 450 μg/kg i.v. (Kreitman and Pastan 1993).

The maximum carryover by nonspecific binding must be determined empirically. We needed to determine the amount of IL6-PE^{4E} that might stay bound nonspecifically to bone-marrow stem cells, even after the washing steps, and after reinfusion into the patient disassociate from the cell and represent active IL6-PE^{4E} in vivo. To do this, human bone-marrow or stem cells (6.7×10^7/ml) that were exposed to IL6-PE^{4E} at 80 μg/ml and washed according the typical purging protocol, were added to IL6R-bearing U266 cells. The amount of active toxin could then be determined using a standard cytotoxicity curve with purified IL6-PE^{4E} on U266 myeloma cells. It was found that 6.7×10^4 washed bone-marrow cells contained no detectable cytotoxic activity. Since the IC_{50} of purified IL6-PE^{4E} on U266 cells in the same assay was 0.6 ng/ml, this indicates that a 70-kg patient receiving 2×10^{10} washed autologous cells would receive <18 μg active toxin or <0.26 μg/kg. This would represent <1/1700 of the mouse LD_{50}.

III. Production of IL6-PE^{4E}

We have recently reported a procedure for the large-scale production of IL6-PE^{4E}, which is nearly identical to that described for anti-Tac(Fv)-PE38KDEL (Kreitman and Pastan 1993). Interestingly, the yield of purified monomer was 20% of total recombinant inclusion body protein, which was at least four times the yield of anti-Tac(Fv)-PE38KDEL. Moreover, the expression levels of IL6-PE^{4E} were high, so that 100 mg purified monomer could be obtained from 20 g *E. coli* cell paste, obtained after inducing a 2-l culture at an OD_{650} of 2.

D. Summary

Anti-Tac(Fv)-PE38 and Anti-Tac(Fv)-PE38KDEL are cytotoxic in tissue culture toward ATL cell lines as well as fresh cells from patients with ATL and CLL, and activated human T-lymphocytes. These molecules are well tolerated in non-human primates at levels that induce complete regression of IL2R-bearing carcinoma xenografts in mice. Responses to anti-Tac(Fv)-PE38 have already been seen in a phase-I trial in patients with CD25+ hematologic malignancies, and the trial is yet to be completed.

IL6-PE^{4E} is cytotoxic in tissue culture toward epidermoid, hepatocellular and prostatic carcinoma and multiple myeloma cell lines, and also fresh bone marrow samples from patients with multiple myeloma. It shows no evidence of toxicity toward hematopoietic progenitors, and even at 300 mg/ml does not prevent normal murine bone marrow from engrafting into syngeneic lethally irradiated mice. After washing the cells, no significant active toxin should carry over in vivo after bone marrow reinfusion. A clinical trial is being considered for multiple myeloma patients using myeloablative therapy which incorporates IL6-PE^{4E} as an agent to purge autologous bone marrow or stem cells isolated from blood prior to reinfusion in the patient. Non-human primate toxicology studies are continuing to determine the safety of IL6-PE^{4E} used in vivo to treat multiple myeloma and possibly prostate and hepatocellular carcinoma.

References

Alexanian R, Dimopoulos M (1994) The treatment of multiple myeloma. New Engl J Med 330:484–489

Allured VS, Collier RJ, Carroll SF, McKay DB (1986) Structure of exotoxin A of Pseudomonas aeruginosa at 3.0-Ångstrom resolution. Proc Natl Acad Sci USA 83:1320–1324

Anderson KC, Andersen J, Soiffer R, Freedman AS, Rabinowe SN, Robertson MJ, Spector N, Blake K, Murray C, Freeman A, Coral F, Marcus KC, Mauch P, Nadler LM, Ritz J (1993) Monoclonal antibody-purged bone marrow transplantation therapy for multiple myeloma. Blood 82:2568–2576

Batra JK, FitzGerald DJ, Chaudhary VK, Pastan I (1991) Single-chain immunotoxins directed at the human transferrin receptor containing Pseudomonas exotoxin A or diphtheria toxin: anti-TFR(Fv)-PE40 and DT388-anti-TFR(Fv). Mol Cell Biol 11:2200–2205

Batra JK, FitzGerald D, Gately M, Chaudhary VK, Pastan I (1990) Anti-Tac(Fv)-PE40: a single chain antibody Pseudomonas fusion protein directed at interleukin 2 receptor bearing cells. J Biol Chem 265:15198–15202

Batra JK, Jinno Y, Chaudhary VK, Kondo T, Willingham MC, FitzGerald DJ, Pastan I (1989) Antitumor activity in mice of an immunotoxin made with anti-transferrin receptor and a recombinant form of Pseudomonas exotoxin. Proc Natl Acad Sci USA 86:8545–8549

Batra JK, Kasprzyk PG, Bird RE, Pastan I, King CR (1992) Recombinant anti-erb B2 immunotoxins containing Pseudomonas exotoxin. Proc Natl Acad Sci USA 89:5867–5871

Brinkmann U, Gallo M, Brinkmann E, Kunwar S, Pastan I (1993) A recombinant immunotoxin that is active on prostate cancer cells and that is composed of the Fv region of monoclonal antibody PR1 and a truncated from of Pseudomonas exotoxin. Proc Natl Acad Sci USA 90:547–551

Brinkmann U, Pai LH, FitzGerald DJ, Willingham M, Pastan I (1991) B3(Fv)-PE38KDEL, a single-chain immunotoxin that causes complete regression of a human carcinoma in mice. Proc Natl Acad Sci USA 88:8616–8620

Buchner J, Pastan I, Brinkmann U (1992) A method for increasing the yield of properly folded recombinant fusion proteins: single-chain immunotoxins from renaturation of bacterial inclusion bodies. Anal Biochem 205:263–270

Carrasco L, Fernandez-Puentes C, Vazquez D (1975) Effects of ricin on the ribosomal sites involved in the interaction of the elongation factors. Eur J Biochem 54:499–503

Chaudhary VK, Batra JK, Gallo MG, Willingham MC, FitzGerald DJ, Pastan I (1990a) A rapid method of cloning functional variable-region antibody genes in Escherichia coli as single-chain immunotoxins. Proc Natl Acad Sci USA 87:1066–1070

Chaudhary VK, FitzGerald DJ, Pastan I (1991) A proper amino terminus of diphtheria toxin is important for cytotoxicity. Biochem Biophys Res Commun 180:545–551

Chaudhary VK, Gallo MG, FitzGerald DJ, Pastan I (1990a) A recombinant single-chain immunotoxin composed of anti-Tac variable regions and a truncated diphtheria toxin. Proc Natl Acad Sci USA 87: 9491–9494

Chaudhary VK, Jinno Y, FitzGerald D, Pastan I (1990b) Pseudomonas exotoxin contains a specific sequence at the carboxyl terminus that is required for cytotoxicity. Proc Natl Acad Sci USA 87:308–312

Chaudhary VK, Jinno Y, Gallo MG, FitzGerald D, Pastan I (1990c) Mutagenesis of Pseudomonas exotoxin in identification of sequences responsible for the animal toxicity. J Biol Chem 265:16306–16310

Chaudhary VK, Queen C, Junghans RP, Waldmann TA, FitzGerald DJ, Pastan I (1989) A recombinant immunotoxin consisting of two antibody variable domains fused to Pseudomonas exotoxin. Nature 339:394–397

Cook JP, Savage PM, Lord JM, Roberts LM (1993) Biologically active interleukin 2-ricin A chain fusion proteins may require intracellular proteolytic cleavage to exhibit a cytotoxic effect. Bioconjug Chem 4:440–447

Edwards GM, DeFeo-Jones D, Tai JY, Vuocolo GA, Patrick DR, Heimbrook DC, Oliff A (1989) Epidermal growth factor receptor binding is affected by structural determinants in the toxin domain of transforming growth factor-alpha-Pseudomonas exotoxin fusion proteins. Mol Cell Biol 9:2860–2867

Fermand J-P, Chevret S, Ravaud P, Divine M, Leblond V, Dreyfus F, Mariette X, Brouet J-C (1993) High-dose chemoradiotherapy and autologous blood stem cell transplantation in multiple myeloma: results of a phase II trial involving 63 patients. Blood 82:2005–2009

FitzGerald DJP, Padmanabhan R, Pastan I, Willingham MC (1983) Adenovirus-induced release of epidermal growth factor and Pseudomonas toxin into the cytosol of KB cells during receptor-mediated endocytosis. Cell 32:607–617

Friedman PN, McAndrew SJ, Gawlak SL, Chace D, Trail PA, Brown JP, Siegall CB (1993) BR96 sFv-PE40, a potent single-chain immunotoxin that selectively kills carcinoma cells. Cancer Res 53:334–339

Heimbrook DC, Stirdivant SM, Ahern JD, Balishin NL, Patrick DR, Edwards GM, DeFeo-Jones D, FitzGerald DJ, Pastan I, Oliff A (1990) Transforming growth factor α-Pseudomonas exotoxin fusion protein prolongs survival of nude mice bearing tumor xenografts. Proc Natl Acad Sci USA 87:4697–4701

Heimbrook DC, Stirdivant SM, Ahern JD, Balishin NL, Patrick DR, Edwards GM, Defeo-Jones D, FitzGerald DJ, Pastan I, Oliff A (1991) Biological activity of a transforming growth factor-alpha-Pseudomonas exotoxin fusion protein in vitro and in vivo. J Indust Microbiol 7:203–208

Hwang J, FitzGerald, DJ, Adhya S, Pastan I (1987) Functional domains of Pseudomonas exotoxin identified by deletion analysis of the gene expressed in *E. coli.* Cell 48:129–136

Jagannath S, Vesole DH, Glenn L, Crowley J, Barlogie B (1992) Low-risk intensive therapy for multiple myeloma with combined autologous bone marrow and blood stem cell support. Blood 80:1666–1672

Kawano M, Hirano T, Matsuda T, Taga T, Horii Y, Iwato K, Asaoku H, Tang B, Tanabe O, Tanaka H, Kuramoto A, Kishimoto T (1988) Autocrine generation and requirement of BSF-2/IL-6 for human multiple myelomas. Nature 332:83–85

Kiyokawa T, Shirono K, Hattori T, Nishimura H, Yamaguchi K, Nichols JC, Strom TB, Murphy JR, Takatsuki K (1989) Cytotoxicity of interleukin 2-toxin toward lymphocytes from patients with adult T-cell leukemia. Cancer Res 49:4042–4046

Kodaka T, Uchiyama T, Ishikawa T, Kamio M, Onishi R, Itho K, Hori T, Uchino H, Tsudo M, Araki K (1990) Interleukin-2 receptor β-chain (p70–75) expressed on leukemic cells from adult T cell leukemia patients. Jpn J Cancer Res 81:902–908

Kondo T, FitzGerald D, Chaudhary VK, Adhya S, Pastan I (1988) Activity of immunotoxins constructed with modified Pseudomonas exotoxin A lacking the cell recognition domain. J Biol Chem 263:9470–9475

Kounnas MZ, Morris RE, Thompson MR, FitzGerald DJ, Strickland DK, Saelinger CB (1992) The α_2-macroglobulin receptor/low density lipoprotein receptor-related protein binds and internalizes Pseudomonas exotoxin A. J Biol Chem 267:12420–12423

Kreitman RJ, Pastan I (1993) Purification and characterization of IL6-PE^{4E}, a recombinant fusion of interleukin 6 with Pseudomonas exotoxin. Bioconjug Chem 7:581–585

Kreitman RJ, Chaudhary VK, Waldmann T, Willingham MC, FitzGerald DJ, Pastan I (1990) The recombinant immunotoxin anti-Tac(Fv)-Pseudomonas exotoxin 40 is cytotoxic toward peripheral blood malignant cells from patients with adult T-cell leukemia. Proc Natl Acad Sci USA 87:8291–8295

Kreitman RJ, Chaudhary VK, Kozak RW, FitzGerald DJP, Waldmann TA, Pastan I (1992a) Recombinant toxins containing the variable domains of the anti-Tac monoclonal antibody to the interleukin-2 receptor kill malignant cells from patients with chronic lymphocytic leukemia. Blood 80:2344–2352

Kreitman RJ, Chaudhary VK, Siegall CB, FitzGerald DJ, Pastan I (1992b) Rational design of a chimeric toxin: an intramolecular location for the insertion of transforming growth factor α within Pseudomonas exotoxin as a targeting ligand. Bioconjug Chem. 3:58–62

Kreitman RJ, Schneider WP, Queen C, Tsudo M, FitzGerald DJP, Waldmann TA, Pastan I (1992c) Mik-β1(Fv)-PE40, a recombinant immunotoxin cytotoxic toward cells bearing the β-chain of the IL-2 receptor. J Immunol 149:2810–2815

Kreitman RJ, Siegall CB, FitzGerald DJP, Epstein J, Barlogie B, Pastan I (1992d) Interleukin-6 fused to a mutant form of Pseudomonas exotoxin kills malignant cells from patients with multiple myeloma. Blood 79:1775–1780

Kreitman RJ, Batra JK, Seetharam S, Chaudhary VK, FitzGerald DJ, Pastan I (1993a) Single-chain immunotoxin fusion between anti-Tac and Pseudomonas exotoxin: relative importance of the two toxin disulfide bonds. Bioconjug Chem 4:112–120

Kreitman RJ, Chaudhary VK, Waldmann TA, Hanchard B, Cranston B, FitzGerald DJP, Pastan I (1993b) Cytotoxic activities of recombinant immunotoxins composed of Pseudomonas toxin or diphtheria toxin toward lymphocytes from patients with adult T-cell leukemia. Leukemia 7:553–562

Kreitman RJ, Bailon P, Chaudhary VK, FitzGerald DJP, Pastan I (1994) Recombinant immunotoxins containing anti-Tac(Fv) and derivatives of Pseudomonas exotoxin produce complete regression in mice of an interleukin-2 receptor-expressing human carcinoma. Blood 83:426–434

Kunwar S, Pai LH, Pastan I (1993) Cytotoxicity and antitumor effects of growth factor-toxin fusion proteins on human glioblastoma multiforme cells. J Neurosurg. 79: 569–576

Lorberboum-Galski H, Barrett LV, Kirkman RL, Ogata M, Willingham MC, FitzGerald DJ, Pastan I (1989) Cardiac allograft survival in mice treated with IL-2-PE40. Proc Natl Acad Sci USA 86:1008–1012

Lorberboum-Galski H, FitzGerald D, Chaudhary V, Adhay S, Pastan I (1988a) Cytotoxic activity of an interleukin 2-Pseudomonas exotoxin chimeric protein produced in Escherichia coli. Proc Natl Acad Sci USA 85:1922–1926

Lorberboum-Galski H, Kozak RW, Waldmann TA, Bailon P, FitzGerald DJP, Pastan I (1988b) Interleukin 2 (IL2) PE40 is cytotoxic to cells displaying either the p55 or p70 subunit of the IL2 receptor. J Biol Chem 263:18650–18656

Lorberboum-Galski H, Garsia RJ, Gately M, Brown PS, Clark RE, Waldmann TA, Chaudhary VK, FitzGerald DJP, Pastan I (1990) IL2-PE66^{4Glu}, a new chimeric protein cytotoxic to human-activated T lymphocytes. J Biol Chem 265:16311–16317

Lowenthal JW, MacDonald HR, Iacopetta BJ (1986) Intracellular pathway of interleukin 2 following receptor-mediated endocytosis. Eur J Immunol 16:1461–1463

Nichols PJ, Johnson VG, Andrew SM, Hoogenboom HR, Raus JCM, Youle RJ (1993) Characterization of single-chain antibody (sFv)-toxin fusion proteins produced in vitro in rabbit reticulocyte lysate. J Biol Chem 268:5302–5308

Ogata M, Chaudhary VK, Pastan I, FitzGerald DJ (1990) Processing of Pseudomonas exotoxin by a cellular protease results in the generation of a 37,000-Da toxin fragment that is translocated to the cytosol. J Biol Chem 265:20678–20685

Ogata M, Fryling CM, Pastan I, FitzGerald DJ (1992) Cell-mediated cleavage of Pseudomonas exotoxin between Arg^{279} and Gly^{280} generates the enzymatically active fragment which translocates to the cytosol. J Biol Chem 267:25396–25401

Pirker R, FitzGerald DJP, Hamilton TC, Ozols RF, Laird W, Frankel AE, Willingham MC, Pastan I (1985) Characterization of immunotoxins active against ovarian cancer cell lines. J Clin Invest 76:1261–1267

Puri RK, Mehrotra PT, Leland P, Kreitman RJ, Siegel JP, Pastan I (1994) A chimeric protein comprised of IL-4 and Pseudomonas exotoxin (IL4-PE4 E) is cytotoxic for activated human lymphocytes. J Immunol (in press)

Robb RJ, Greene WC (1987) Internalization of interleukin 2 is mediated by the β chain of the high-affinity interleukin 2 receptor. J Exp Med 165:1201–1206

Robb RJ, Greene WC, Rusk CM (1984) Low and high affinity cellular receptors for interleukin 2. J Exp Med 160:1126–1146

Saito T, Kreitman RJ, Hanada S, Makino T, Utsunomiya A, Sumizawa T, Arima T, Chang CN, Hudson D, Pastan I, Akiyama S (1994) Cytotoxicity of recombinant Fab and Fv immunotoxins on adult T-cell leukemia lymph node and blood cells in the presence of soluble interleukin-2 receptor. Cancer Res 54:1059–1064

Seetharam S, Chaudhary VK, FitzGerald D, Pastan I (1991) Increased cytotoxic activity of Pseudomonas exotoxin and two chimeric toxins ending in KDEL. J Biol Chem 266:17376–17381

Siegall CB, Chardhary, VK, FitzGerald DJ, Pastan I (1988) Cytotoxic activity of an interleukin 6-Pseudomonas exotoxin fusion protein on human myeloma cells. Proc Natl Acad Sci USA 85:9738–9742

Siegall CB, Chaudhary VK, FitzGerald DJ, Pastan I (1989a) Functional analysis of domains II, Ib, and III of Pseudomonas exotoxin. J Biol Chem 264:14256–14261

Siegall CB, Xu Y-H, Chaudhary VK, Adhya S, FitzGerald D, Pastan I (1989b) Cytotoxic activities of a fusion protein comprised of TGFα and Pseudomonas exotoxin. FASEB J 3:2647–2652

Siegall CB, FitzGerald DJ, Pastan I (1990a) Cytotoxicity of IL6-PE40 and derivatives on tumor cells expressing a range of interleukin 6 receptor levels. J Biol Chem 265:16318–16323

Siegall CB, Nordan RP, FitzGerald DJ, Pastan I (1990b) Cell-specific toxicity of a chimeric protein composed of interleukin-6 and Pseudomonas exotoxin (IL6-PE40) on tumor cells. Mol Cell Biol 10:2443–2447

Siegall CB, Schwab G, Nordan RP, FitzGerald DJ, Pastan I (1990c) Expression of the interleukin 6 receptor and interleukin 6 in prostate carcinoma cells. Cancer Res 50:7786–7788

Siegall CB, Kreitman RJ, FitzGerald DJ, Pastan I (1991) Antitumor effects of interleukin 6-Pseudomonas exotoxin chimeric molecules against the human hepatocellular carcinoma, PLC/PRF/5 in mice. Cancer Res 51:2831–2836

Siegsmund MJ, Pastan I (1994) Interleukin 6 receptor mRNA in prostate carcinomas and benign prostate hyperplasia. J Urol (in press)

Studier FW, Moffatt BA (1986) Use of bacteriophage T7 RNA polymerase to direct selective high-level expression of cloned genes. J Mol Biol 189:113–130

Uckun FM (1993) Annotation: immunotoxins for the treatment of leukaemia. Br J Haematol 85:435–438

Vitetta ES, Fulton RJ, May RD, Till M, Uhr JW (1987) Redesigning nature's poisons to create anti-tumor reagents. Science 238:1098–1104

Waldmann TA (1991) The interleukin-2 receptor. J Biol Chem 266:2681–2684
Wels W, Harwerth I-M, Hueller M, Groner B, Hynes NE (1992) Selective inhibition of tumor cell growth by a recombinant single-chain antibody-toxin specific for the erbB-2 receptor. Cancer Res 52:6310–6317
Williams DP, Parker K, Bacha P, Bishai W, Borowski M, Genbauffe F, Strom TB, Murphy JR (1987) Diphtheria toxin receptor binding domain substitution with interleukin-2: genetic construction and properties of a diphtheria toxin-related interleukin-2 fusion protein. Protein Eng 1:493–498
Williams DP, Wen Z, Watson RS, Boyd, J, Strom TB, Murphy JR (1990) Cellular procesing of the interleukin-2 fusion toxin DAB_{486}-IL-2 and efficient delivery of diphtheria fragment A to the cytosol of target cells requires Arg^{194}. J Biol Chem 265:20673–20677
Yagura H, Tamaki T, Furitsu T, Tomiyama Y, Nishiura T, Tominaga N, Katagiri S, Yonezawa T, Tarui S (1990) Demonstration of high-affinity interleukin-2 receptors on B-chronic lymphocytic leukemia cells: functional and structural characterization. Blut 60:181–186
Yamaizumi M, Mekada E, Uchida T, Okada Y (1978) One molecule of diphtheria toxin fragment A introduced into a cell can kill the cell. Cell 15:245–250

CHAPTER 6

SB 209763: A Humanized Monoclonal Antibody for the Prophylaxis and Treatment of Respiratory Syncytial Virus Infection

T.G. PORTER, S.G. GRIEGO, T.K. HART, D.E. EVERITT, and S.B. DILLON

A. Introduction

Respiratory syncytial virus (RSV) is a major and highly contagious respiratory pathogen that most people are exposed to within the first year or two of life. It is the dominant cause of severe lower respiratory-tract disease (bronchiolitis and pneumonia) in infants and young children, leading to approximately 100,000 hospitalizations each year in the United States alone (HEILMAN 1990). There are two major antigenically distinct subgroups of RSV, A and B, which can circulate simultaneously and are responsible for predictable winter epidemics each year. In adults and older children, symptoms resemble the common cold and are normally mild and confined to the upper respiratory tract. However, in infants and young children, RSV infection often progresses to the lower respiratory tract, where it can lead to serious respiratory complications requiring hospitalization and, in rare cases, may even result in death. Infants with underlying cardiac or pulmonary disease, and some premature/low-birth-weight infants are at considerably higher risk of more severe disease and increased mortality from RSV infection, and are candidates for prophylactic intervention during the winter months. Although not well studied, immunosuppressed adult populations (the elderly, organ-transplant recipients) are also prone to life-threatening infections.

Significant safety and efficacy hurdles continue to face the development of an effective RSV vaccine (HALL 1994) and the only treatment for RSV that is available, ribavirin (Virazole, ICN), has not been widely used by the medical community. A clear unmet medical need for effective RSV intervention exists. Currently, passive antibody administration represents the most promising approach for the treatment and prevention of RSV disease and is well supported by epidemiological, animal-model and clinical studies (HEMMING and PRINCE 1992; GROOTHUIS 1994). Recently, Respigam (RSVIG), a pooled human serum immunoglobulin (HSIg) preparation containing high levels of neutralizing antibodies to RSV, became the first product approved by the United States Food and Drug Administration (FDA) for the prevention of RSV disease in children under 24 months of age with bronchopulmonary dysplasia (BPD) or a history of prematurity (≤35 weeks gestation). In a pivotal phase-III clinical study, administration of high doses of Respigam (0.75 g/kg), by intravenous (IV) infusion at monthly intervals throughout the RSV season, resulted in significantly fewer RSV lower respiratory-tract infections and a lower incidence of hospitalization due to RSV infection. The approval of Respigam provides strong conceptual support for second-generation monoclonal antibody (mAb) approaches. In addition to retaining the long circulating half-life and immunocompatability of RSVIG, anti-RSV mAbs, such as SB 209763, promise clinical advantages over HSIg preparations, including reduced safety concerns surrounding blood-derived products and, with much lower doses expected, easier methods of administration.

B. Early Challenges in the Development of SB 209763

I. Selection of Target Antigen

In the discovery phase of any antiviral mAb program, the selection of the appropriate viral antigen is critical and, ideally, will be guided by clinical epidemiological and/or animal-model data correlating protection or a favorable clinical outcome to antibody levels against specific viral antigens. Owing to their ready accessibility, viral surface proteins are common, although not exclusive, antibody targets.

During natural infection by RSV, the induction of neutralizing antibodies appears to be limited to the two major surface glycoproteins, G and F, which are responsible for virus attachment and membrane fusion, respectively. The fusion (F) protein is essential for viral entry, mediating membrane fusion between the virus and the host cell. By inducing membrane fusion between infected cells with adjacent uninfected cells to form multinucleated giant cells (syncytia), the F protein is also implicated in later stages of infection, promoting cell-to-cell spread of the virus. The F protein is highly conserved between strains of both A and B subtypes of RSV (COLLINS 1994) and is the most important viral antigen for inducing cross-protective immunity against both subtypes of RSV. The G glycoprotein, which is responsible for attachment of RSV to the host cell, is much less conserved, and major antigenic differences between the A and B subtypes are linked to this protein. Animal studies with neutralizing mAbs to F and G confirmed that passive administration in advance of challenge can completely prevent infection (TAYLOR et al. 1984; WALSH et al. 1984). However, mAbs directed to the G protein would not necessarily be expected to prevent cell-to-cell spread of the virus through syncytia formation; thus, antibodies to the F protein may be more relevant for clearing established infection. Recent studies show that mAbs directed to the F protein are clearly more effective than mAbs to the G protein, when administered therapeutically after initiation of infection (TAYLOR 1993). These key factors, high conservation across both subtypes of the virus and its integral role in both viral entry and cell to cell spread, have made the RSV F protein a preferred target antigen for current anti-RSV mAbs in development.

II. Molecular Engineering of SB 209763

The clinical utility of murine monoclonal antibodies has been severely restricted by their short circulating half-lives (2–3 days or less) and inherent immunogenicity in humans. Over the last decade, a key goal in the molecular engineering of murine mAbs has been to overcome their immunogenicity and to extend their plasma half-life so that sustained therapeutic levels of antibody can be achieved. mAb "humanization" technologies have been introduced to minimize the murine component of the antibody (Adair 1992; Mark and Padlan 1994). A common strategy involves the direct transplantation of the complementarity determining regions (CDRs) responsible for antigen binding from a parent murine antibody into recipient human heavy-chain and light-chain variable-region (VH) frameworks. SB 209763, which is in development by SmithKline Beecham, under a license agreement with Scotgen Biopharmaceuticals Ltd., was derived from a well-characterized murine anti-RSV mAb using CDR grafting technology (see Tempest et al. 1991 where SB 209763 is referred to as HuRSV19HFNSNK). The parent murine mAb selected for humanization, mAb 19, recognized a defined epitope on the RSV F protein, had potent neutralizing and fusion-inhibition activity in vitro, recognized a panel of RSV subgroup A and B isolates, and both prevented and cleared RSV infection in mice (Tempest et al. 1991; Taylor et al. 1991). As reported for other mAbs, direct transplant of the mAb 19 heavy- and light-chain CDRs into human VH frameworks resulted in a humanized mAb, RSHZ00, with very poor binding affinity. A comparison of the murine and human heavy-chain VH sequences revealed a difference in the framework region immediately adjacent to CDR3, in which an arginine, normally present at position 94 in the majority of mouse and human VHs, was absent in mAb 19. This arginine is normally able to form a salt bridge with aspartic acid at position 101, contributing to the conformation of CDR3, and it was speculated that a salt bridge had been imposed in RSHZ00, thus altering the conformation of CDR3 and affecting its interaction with antigen. Accordingly, a modified mAb, SB 209763 (RSHZ19), was constructed with amino acids 91–94 (PheCysAsnSer) from the mAb 19 being used to replace the human VH framework amino acids (Fig. 1). These changes resulted in restoration of antigen-binding ability and the antiviral properties of the parent murine monoclonal antibody.

During construction of the humanized mAb, an IgG_1 isotype was selected for several reasons: (1) it has a long circulating half-life (approximately 3 weeks in humans), making it suitable for long-term prophylaxis and/or chronic therapy, (2) it retains effector functions that can aid in clearing an infected cell, and (3) importantly, there were established large-scale production and purification procedures for IgGs. Other antibody isotypes, such as IgA and IgM, have much shorter half-lives (5–6 days), are potentially more difficult to produce at large-scale and may not access sites of infection as effectively when administered systemically.

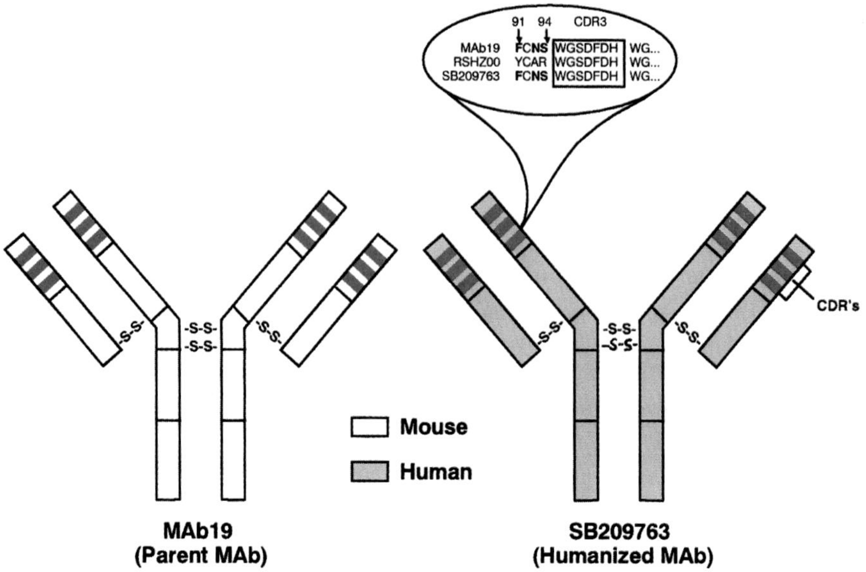

Fig. 1. SB 209763: a humanized monoclonal antibody (mAb) specific for respiratory syncytial virus (RSV). The hyper-variable complementarity determining regions (CDRs) of mAb19, a well-characterized murine mAb specific for the RSV-fusion protein, were grafted into a human immunoglobulin G1 (IgG1) kappa mAb framework. The resulting reshaped antibody, RSHZ00, showed weak affinity for RSV; however, an additional alteration to the framework region immediately flanking CDR3 of the heavy chain (inset) restored binding affinity and antiviral activity (Tempest et al. 1991)

III. Production

As for many protein agents, cost of goods and scale of production were important commercial factors to consider early in the development of SB 209763. A manufacturing process would need to produce multi-kilogram amounts of mAb and be commercially cost effective. Since secreted glycoproteins require correct postranslational processing, mammalian cell lines remain the expression system of choice for manufacturing mAbs, and commercial production is routinely carried out in large-scale bioreactors. Associated costs of mammalian cell culture can often be high, placing emphasis on the generation of a stable, high-expressing manufacturing cell line together with an efficient downstream purification process that would ensure safety of the product through removal and/or inactivation of extraneous contaminants. At the outset of our studies on SB 209763, there was little clinical information, especially in non-immunocompromized patient populations, on the antigenicity, pharmacokinetic and biological properties of human or humanized mAbs produced by recombinant DNA technology. Most human mAbs had been produced by adaptations of traditional hybridoma technology, involving immortalization of human B cells with heteromyeloma fusion partners. Our choice of an amplifiable Chinese Hamster Ovary (CHO) cell-expression system for production of recombinant SB 209763 was based on favorable past experiences demonstrating that CHO cells are capable of high-level heterologous protein expression and are well suited to large-scale cell culture (TRILL et al. 1995). An expression plasmid containing the SB 209763 coding sequences was co-transfected with a plasmid containing dihydrofolate reductase (dhfr) into a dhfr-deficient CHO cell line. Following selection and co-amplification in media containing increasing concentrations of methotrexate, a stable, high-level-production cell line expressing SB 209763 at levels >20 pg/cell/day was generated that was amenable to large-scale culture in stirred-tank bioreactors. An efficient purification process was also established that removed contaminants derived from the host cell, cell-culture medium and raw materials used in the process, and yielded a clinical good manufacturing practices (GMP) grade SB 209763 (SHADLE et al. 1995). The principles of fermentation and purification of recombinant DNA (rDNA)-derived mAbs, such as SB 209763, will be discussed further in Chap. 2 by LUBINIECKI and coworkers.

IV. Primary Structure Analysis

All antibodies contain at least one N-linked carbohydrate in their constant regions that can be important in the maintenance of antibody–effector functions (TAO and MORRISON 1989). As a humanized antibody expressed in an unnatural cell line, it was important to evaluate any changes in the primary structure of SB 209763 resulting from differences in post-translational processing of the mAb. Accordingly, the primary amino acid and carbohydrate structure of SB 209763 expressed in CHO cells was extensively characterized using an integrated approach that combined mass spectrometry with conventional analytical methods (ROBERTS et al. 1995). The study verified more than 99% of the light- and heavy-chain amino acid sequences and confirmed the N and C termini of both chains. The mAb was found to be exclusively glycosylated at the Asn296 site of the CH2 domain and there was no evidence for nonglycosylated forms of SB 209763 or for *O*-linked carbohydrate. The carbohydrate structure of CHO-expressed SB 209763 consisted predominantly of biantennary, core fucosylated carbohydrates lacking sialic acid. This was similar but less complex than that observed in a pool of carbohydrate structures reported for IgGs purified from human serum (TANDAI et al. 1991), which contained both terminal sialic acid and bisecting N-acetyl glucosamine (GlcNAc).

C. Preclinical Evaluation Prior to Testing in Humans

Preclinical pharmacology and safety studies are routinely designed to evaluate and estimate potential adverse events associated with drug administration and to identify a safe starting dose and dose-escalation regimen for early clinical trials (FDA 1994). As a protein agent intended for both acute therapy and longer term prophylaxis of a respiratory virus, particular issues facing SB 209763 included: (1) appropriate efficacy and potency testing in vitro and in vivo, (2) antigenic variation and viral resistance, (3) immunogenicity after single and repeat administration, (4) demonstration of appropriate pharmacokinetics and biodistribution, and (5) a well-characterized preclinical safety profile.

I. Fusion Inhibition: An In Vitro Correlate of Protection

Several antibody-mediated effector mechanisms can contribute to viral clearance in vivo (DIMMOCK 1993). As part of their preclinical characterization, antiviral antibodies are typically evaluated for: (1) in vitro neutralizing activity against prototype and wild-type serologically diverse virus isolates, (2) complement fixation, (3) antibody-dependent cell-mediated cytotoxicity (ADCC), (4) affinity for target antigen, (5) epitope specificity, and (6) activity in animal models, where available. Owing to species differences in recognition, the role of antibody Fc effector functions in vivo is often hard to elucidate. However, the demonstrated ability of Fab and Fab_2 antibody fragments lacking Fc to neutralize virus in vitro and in animal models of RSV infection (PRINCE et al. 1990; CROWE et al. 1994) suggests that effector functions, such as complement fixation and ADCC, may not be essential for the activity of mAbs, such as SB 209763, although their contribution to clearance of RSV in human infection is unknown.

Standard viral neutralization assays, such as plaque reduction, are commonly used for screening and quantifying the antiviral activities of candidate antibodies in vitro. In this format, where immune serum or mAbs are incubated together with virus in advance (1 h) of exposure to permissive cells, the assay typically measures antibodies that block cell attachment and/or penetration of cell-free virus. In cases such as RSV, where infection can be spread through contact of infected with uninfected cells, standard neutralization assays may fail to identify antibodies that interfere with such later stages of viral infection and, consequently, may not always be the best correlate of protection in vivo. Studies conducted on large panels of mAbs directed to different epitopes of the RS virus F protein have shown that only a subset of neutralizing mAbs also interfere with virus/cell or cell/cell membrane fusion (BEELER and COELINGH 1989). The epitopes recognized by the fusion-inhibiting mAbs tended to be more conserved in wild-type strains, suggesting a requirement of these regions for viral infectivity. The significance of this finding was further defined by TAYLOR et al. (1992; ARBIZA et al. 1992) in studies that profiled in vitro versus in vivo biological properties of a panel of murine- and bovine-neutralizing and fusion-inhibiting mAbs. The results of these studies demonstrated that neither IgG isotype, ability to fix complement or neutralizing titer correlated with the ability of the mAbs to passively protect mice from RSV challenge. Only those mAbs that inhibited the formation of multinucleated giant cells (syncytia) in RSV-infected cell cultures correlated directly with high-level protection in vivo. The ability of SB 209763 to inhibit RSV-induced giant-cell formation in Vero cell monolayers is illustrated in Fig. 2. Whereas numerous large multinucleated cells are evident in RSV-infected cultures

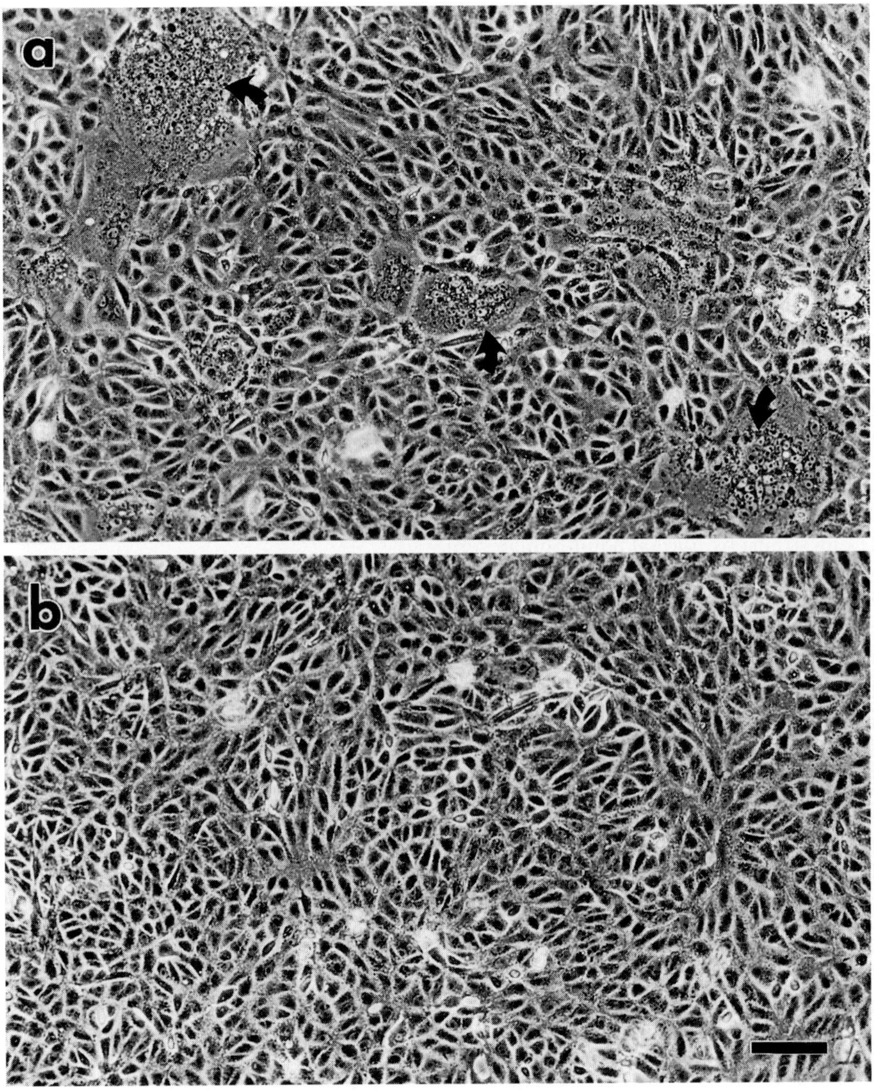

Fig. 2a,b. Inhibition of giant cell formation by SB 209763. Phase-contrast light micrographs of VERO cultures infected with the long strain of RSV. Cultures were incubated with virus for 4 h and then washed. Media with or without SB 209763 was added and the cultures allowed to incubate for 3–4 days. **a** Control well containing numerous multi-nucleated syncytia (*arrows*). **b** Well that contained 10 μg/ml SB 209763 in the incubation medium. SB 209763 completely blocked the formation of syncytia. *Bar* = 100 μm grown in the absence of mAb, the number of these giant cells is drastically reduced when the RSV-infected cells are grown in the presence of SB 209763.

For preclinical characterization of SB 209763 and other anti-RSV mAbs, we used a RSV fusion-inhibition (FI) assay that differs from the traditional neutralization assay by allowing cells to be infected by RSV prior to exposure with the mAb. mAbs are not added to cell cultures until 4–18 h post-infection and remain in culture for the duration of the assay. RSV-infected cells are then detected utilizing anti-F protein antiserum together with a horseradish peroxidase (HRP)-labeled secondary antibody. Color development was shown to be directly proportional to spread of infection and equivalent to directly scoring cell monolayers for syncytia formation by microscopy. An alternative "microneutralization" assay combines features of the plaque-reduction assay (antibody is premixed with virus) and the FI assay (antibody remains for the duration of the assay), thereby measuring all forms of virus neutralization.

As shown in Table 1, SB 209763 showed high potency and equivalent activity in both the FI and neutralization assays (EC50 of 1.3 ± 0.5 μg/ml and 0.9 ± 0.5 μg/ml, respectively). Interestingly, unlike SB 209763, the ratios of neutralization to FI titers in a group of normal adult subjects ranged from 10 to more than 50 (Table 1), indicating that, in all cases, fusion-inhibiting anti-

Table 1. Fusion inhibition and neutralization of type A respiratory syncytial virus (Long strain) by normal adult human sera

Donor	Fusion-inhibition titer (dilution $\times 10^{-1}$)	Neutralization titer (dilution $\times 10^{-1}$)	NT/FI ratio
TT	99	1250	12.6
FD	87	1219	14.0
DH	77	2041	26.5
CW	115	5943	51.7
BL	130	2404	18.5
MC	94	2015	21.4
GM	18	765	42.5
BS	41	1490	36.3
GS	11	109	9.9
PK	545	6409	11.8
CDP	106	880	8.3
JS	276	2647	9.6
DL	10	85	8.5
SB 209763	1.3 ± 0.5 μg/ml	0.9 ± 0.5 μg/ml	0.7
Gammar S19610	12.5–25 μg/ml	65–130 μg/ml	5

Sera was obtained from healthy adults (male and female). Five-fold dilutions of serum were added to VERO cells infected with 96 $TCID_{50}$ RS Long virus according to either the neutralization or fusion inhibition format described in the text. Titers represent the reciprocal dilution, or concentration of reference standard, which caused a 50% reduction in enzyme-linked immunosorbent assay (ELISA) signal based on regression analysis

FI, fusion inhibition titer; NT, neutralization titer

bodies represented a much lower fraction of the overall neutralization activity in immune serum. The ratio of neutralizing to FI titers in a commercially available HSIg preparation, Gammar™, was also approximately fivefold higher, potentially accounting for the much higher neutralization titers (>1:250) in circulation required to provide protection in animal models of RSV, compared with mAbs, such as SB 209763, where titers of less than 1:50 appear to be protective (WYDE et al. 1995). If FI is truly a correlate for protection, the relatively low FI titers observed may contribute to the frequent reinfection of adults by RSV throughout life.

II. Antigenic Variation

Antigenic variation and the production of viral escape mutants is a potential issue facing all anti-viral agents, including mAbs. It was, therefore, very important to investigate the conservation of the SB 209763 epitope across serologically diverse strains of RSV. Furthermore, as significant changes in the antigenic properties of culture-adapted prototypic laboratory virus strains were possible, it was essential to include fresh clinical isolates that had undergone minimal adaptation in vitro. By immunofluorescence assay, SB 209763 recognized 56 clinical virus isolates from nasopharyngeal aspirates of infected children collected between 1984 and 1995 from diverse geographical regions in both the northern and southern hemispheres. In addition, the mAb also demonstrated potent neutralization activity (EC50's 0.4–3 μg/ml) against a subgroup of these isolates representing both A and B subgroups of RSV (WYDE et al. 1995). The data indicate that the epitope region of SB 209763 is well conserved; however, the generation of escape mutant viruses in response to treatment with SB 209763 is still of theoretical concern and will be addressed by active monitoring of viral isolates obtained during clinical trials.

III. Animal Models of Respiratory Syncytial Virus Infection

Several issues surround the choice of animal models for testing anti-viral agents, including antibodies, for RSV infection. Following the failure of a formalin-inactivated RSV vaccine and subsequent enhancement of disease upon natural RSV infection in vaccinated children (KIM et al. 1969), the potential for disease exacerbation by any RSV immunotherapy needs to be evaluated. Many laboratory animal species can be infected experimentally with human RSV; however, clinical signs of disease, if any, are usually limited to the upper respiratory tract, and pulmonary pathological effects are minimal. The chimpanzee and calf represent disease models, as they develop clinical signs of RSV illness, such as profuse rhinorrhea, cough and sneezing. However, the cost and limited availability of seronegative chimpanzees (an endangered species) and the large size of calves have precluded extensive use of these models. Owl monkeys, which exhibit only mild rhinorrhea, have also been used for the evaluation of vaccines and therapeutic agents (Hemming et al. 1985).

The most extensively used and well-characterized small-animal models of human RSV infection are cotton rats (*Sigmodon hispidus*) or Balb/c mice. Human RSV replicates for 7–9 days in the upper and lower respiratory tract of these animals (10^4–10^5 PFU/g tissue), causes mild to moderate pulmonary histopathologic changes in the absence of overt upper or lower respiratory-tract disease, and induces neutralizing antibody titers, which correlate with resistance to reinfection. Protective serum antibody titers (1:200–1:400), achieved by convalescent serum or administration of HSIg, which are associated with protection from lung infection in cotton rats, roughly correlate to levels of maternally derived antibodies in young infants with relative resistance to serious RSV infection, and to titers associated with efficacy in RSVIG treated patients.

SB 209763 was effective in reducing lung virus titers in both Balb/c mouse and cotton rat models of RSV infection. Complete clearance of virus was observed in mice treated with 5 mg/kg SB 209763 (Tempest et al. 1991), and SB 209763 completely protected cotton rats from lower respiratory-tract infection at 10 mg/kg (Wyde et al. 1995). In the latter study, SB 209763 was shown to be greater than 100-fold more potent than Gammar™. In the same study, histopathology in cotton rats was proportional to RSV titers in the lung and inversely proportional to SB 209763 dose, providing no evidence for disease exacerbation.

IV. Safety and Pharmacokinetics

Preclinical safety and pharmacokinetic studies were designed to support the IV and intramuscular (IM) administration of SB 209763. The selection of animal species for toxicity testing of monoclonal antibodies is generally guided by the results of a human tissue-binding study, in which the cross-reactivity of SB 209763 with normal human tissues is examined. By immunohistochemical analysis, SB 209763 did not bind to human tissues; hence, we chose to conduct our toxicology assessment of SB 209763 in non-human primates because of the homology of primate immunoglobulins (Lewis et al. 1993) and the likelihood of reduced antigenicity of a humanized antibody in monkeys. Interestingly, examples of antiviral mAbs that cross-react with human tissue, potentially as a result of viral mimicry of host proteins, have been reported (Fujinami et al. 1983) that would necessitate the need for screening of animal tissues from a variety of hosts for a similar epitope-expression pattern. By conducting studies in monkeys, we were able to assess the long-term safety and repeat-dose regimens of SB 209763 that would have been precluded in lower species, where a high incidence of immunogenicity following administration of a human or humanized mAb was observed.

SB 209763 was well tolerated by the monkeys in all acute and chronic studies. No cardiovascular effects, adverse clinical effects or clinical pathology were recorded over a 3-month observation period following single or repeat doses of SB 209763, administered IV or IM, to cynomolgus macaques at doses up to 200 mg/kg, a 20-fold multiple of the clinical dose (Davis et al. 1995). Furthermore, in an age group more representative of the primary target population (infants), SB 209763 was equally well tolerated when administered as five daily doses of 10 mg/kg and 40 mg/kg to infant baboons. SB 209763 was antigenic in the macaques; however, the incidence of responses was low (2/14). The initial antibody responses were primarily directed toward the human framework of the humanized mAb and not the murine-derived CDRs. Antibody responses were seen following a single IV or IM dose and repeated administration did not increase the incidence of responses(Davis et al.1995).

Results of pharmacokinetic studies of SB 209763 in non-human primates were equally encouraging (Davis et al. 1995) and supported both IV and IM administration of the humanized mAb. SB 209763 exhibited essentially identical pharmacokinetics in adult macaques and infant baboons, demonstrating a typical biphasic decline in plasma concentration with a long terminal half-life of 21–24 days, which is consistent with the expected half-life for an endogenous IgG_1 (around 23 days) in humans. Absorption of SB 209763 after IM administration was rapid and complete (bioavailability >82%). Owing to the potential for pre-exiting antibodies to RSV in human plasma that could compete for binding to F protein, a standard antigen-based enzyme-linked immonosorbent assay (ELISA) was unsuitable for the detection of circulating levels of SB 209763 in clinical samples. To avoid this problem, an ELISA was developed and evaluated during preclinical studies (Davis et al. 1995) that utilized two bovine anti-idiotypic mAbs, B11 and B12, raised to the parent murine mAb (mAb 19), which recognized only the CDR domains of SB 209763. The format of the anti-idiotypic sandwhich ELISA is shown schematically in Fig. 3. Following capture of SB 209763 from a plasma sample by one

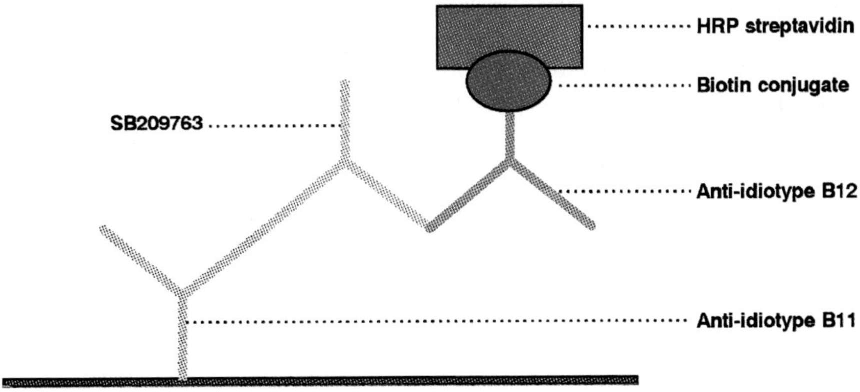

Fig. 3. Double anti-idiotype "sandwich" enzyme-linked immunosorbent assay (ELISA) used to measure SB 209763 levels in monkey and human plasma

anti-idiotypic mAb, B11, bound to a microtiter plate, the complex was probed with biotinylated B12, then detected with HRP-conjugated streptavidin. The anti-idiotype sandwich ELISA provided a specific and highly sensitive assay for both preclinical and clinical studies with a lower level of detection, around 1 ng/ml, in undiluted human plasma.

D. Challenges for the Early Clinical Development of SB 209763

The early clinical development of SB 209763 presented the dual challenge of introducing a novel protein into human use and designing an appropriate sequence of studies to show efficacy in a frail pediatric population. The target indications required that pivotal studies of treatment efficacy be conducted in hospitalized infants infected with RSV. The studies of prophylaxis needed to enroll infants at high risk of serious RSV disease, who had prematurity, underlying congenital heart disease or BPD. Among the issues we needed to address included: (1) the study population for first administration of SB 209763, with accompanying safety considerations, (2) the usefulness of pharmacodynamic markers of pharmacologic effect, (3) the choice of formulations for IV or IM administration, (4) the extent of concern and surveillance for an immune response to SB 209763, (5) the transition into the appropriate pediatric populations, and (6) the choice of the dose range.

Clinical development of SB 209763 proceeded after preclinical development criteria had been satisfied; the completed toxicology studies described above confirmed that: (1) there was no significant toxicity associated with either IV or IM administration to primates, (2) the pharmacokinetics in primates were as expected for IgG, and (3) there was no binding to human tissue in vitro. The clinical plan called for parallel development of the IV formulation for treatment of acute RSV infection and the IM formulation for prophylaxis of infants and young children at risk of serious infection. Each indication was to begin with a single dose, dose-rising safety and pharmacokinetic study in healthy volunteers, followed by a phase-IIa safety study in the target pediatric populations. Following safety and pharmacokinetic checkpoints, pivotal trials could then proceed.

I. Selection of the Initial Study Population and Safety Considerations

It was necessary to move as efficiently as possible into large-scale pivotal studies in the appropriate pediatric populations, but only after adequate safety had been demonstrated. Early meetings with pediatric investigators and members of pediatric Institutional Review Boards (IRBs) made it clear that some evidence of safety in adults was necessary prior to subjecting infants to studies of a novel investigational recombinant biologic product. A traditional approach would have been to begin with adults or older adolescents who had an indication for the therapy, and then to study successively younger cohorts of subjects. However, an appropriate older population at reasonable risk of serious RSV disease that would be suitable for a first clinical study could not be identified.

Healthy volunteers offered many advantages for a first study, including the ability to give informed consent, a competent immune system so that any potential antigenicity of the product could be monitored, and efficiency of recruitment and follow-up. However, at that time, we were aware of only one mAb, a human mAb to cytomegalovirus (Azuma et al. 1991) that had been tested in early studies in human volunteers. Draft guidelines (FDA 1994) for the development of mAbs published by the FDA Center for Biologics Evaluation and Research (CBER) suggested that "... healthy volunteers are generally not appropriate." for clinical studies, and this was related to the "... potential immunogenicity of most mAbs studied to date (guidelines issued in 1994) and the usual paucity of detailed toxicity data in healthy animals that possess the target antigen or in an animal disease model ...". It was also suggested that a research subject might not be able to receive mAb therapy in the future if anti-mAb antibodies developed. Other toxicity-related concerns included the consequences of targeting a native receptor, such as antibodies aimed at lymphocytes that may cause a cytokine-release syndrome. Additionally, the expected long half-life of a mAb would preclude the rapid withdrawal of exposure if adverse effects occurred.

Nevertheless, the guidelines noted situations in which healthy volunteers might be used in phase I, including circumstances when the risks of studying a new agent in the index population are too high, or when the index population is too ill to give interpretable safety data. After reviewing relevant literature and toxicology results, we reasoned that SB 209763 should not pose significantly more risk to a healthy volunteer than pooled native immune globulins. SB 209763 would be given in a relatively small volume, would lack the potential for passing blood-borne infectious agents and would present a very low risk for an adverse immune response. Murine mAbs clearly elicit an immune response (Sequeira and Cutler 1992), but adverse clinical consequences are rarely seen when murine mAbs are administered, even in the presence of known high human anti-mouse antibody (HAMA) concentrations (Henry et al. 1992; Russell et al. 1992). Furthermore, SB 209763 did not target any native tissue receptors and was not expected to form antigen–antibody complexes in healthy, uninfected individuals. Since the beginning of our clinical program, several further examples have been published of healthy volunteers being used for early studies of mAbs or mAb fragments (Stephens 1994). Concerns about immunogenicity expressed in the CBER draft guidelines were based primarily on early experience with murine or chimeric antibodies that had been evaluated prior to 1994.

A further concern about the use of healthy volunteers for the initial clinical study was that the pharmacokinetics in adults might be misleading and would not correspond to those in a pediatric group, which would include low-weight premature infants and RSV-infected infants that may have circulating antigen. However, literature describing the kinetics of IV immune globulin (IVIG) identified age- and weight-related kinetic differences only for infants weighing less than about 1 kg. We, thus, felt reassured that an early study in adults could give relevant information.

II. Pharmacodynamic Markers to Establish Pharmacologic Effect

As described, RSV infection in mouse and cotton rat models of disease served as reasonable models to measure virus clearance, neutralization and FI titers, although no available model exists that closely mimics human disease. The preclinical studies showed that dose correlated closely with plasma levels of SB 209763, FI titers and clearance of RSV from lung tissue. In humans, the ultimate endpoints for the prophylaxis indication would be prevention of serious infections requiring hospitalization, and for treatment, reduction in measures of length and severity of infection in infants hospitalized with severe infection. As no accurate or acceptable method for quantifying lower respiratory-tract virus during clinical trials is available, FI titers appeared to be the most useful potential marker of pharmacologic effect that might ultimately predict clinical efficacy.

III. Formulation Considerations for Clinical Studies

Studies of RSVIG (Groothuis et al. 1993) have been hampered by the large volume (up to 700 ml per dose) required by this preparation. A key advantage of mAb products would be a significant reduction in dose (both concentration and volume) requirements. An initial formulation of 10 mg/ml was considered adequate for the therapy trials of IV administration of SB 209763. At this concentration, the highest dose (10 mg/kg) could easily be delivered IV to a 10 kg infant in a volume of only 10 ml. It was recognized that an IM dose form would be more acceptable for prophylactic outpatient administration of SB 209763, although the volume would need to be minimized. Formulation development led to a high-concentration formulation of SB 209763 that could be administered as IM injections of less than 1 ml, a volume expected to be well tolerated, even in infants.

IV. Surveillance for Anti-SB 209763 Antibodies

As noted previously, antigenicity is a major concern for new proteins given to humans. An immune response could lead to rapid clearance of the therapeutic antibody. Adverse clinical effects could include serum sickness, glomerular pathology from antigen–antibody complexes or anaphylaxis. Antibodies to

SB 209763 were analyzed using an ELISA method (Davis et al. 1995) for 8–10 weeks after the final dose of the humanized mAb.

V. Transition to the Target Pediatric Population and Choice of Dose

The phase-I checkpoints for both the treatment and prophylaxis indications required that minimal toxicity and no clinically relevant antigenicity be identified. In addition, the terminal half-life of SB 209763 needed to be long enough so that a single dose would be adequate for therapy and that a dosing interval of 1 month or greater would be sufficient for prophylaxis. The initial dose, 0.025 mg/kg given IV, was chosen to be about 1/1000 of the dose that was clearly nontoxic in primates. Preclinical pharmacology studies, discussed previously, guided the selection of the maximum dose, 10 mg/kg. For prophylaxis, pharmacokinetic modeling based on the results of the adult volunteer studies predicted that concentrations that cleared virus in animal species could be maintained with this dose for at least 8 weeks following a single dose.

A further consideration for clinical development was that the RSV season spans only about half of the year and is juxtaposed across hemispheres. Thus, an efficient program was devised to conduct sequential studies in both the northern and the southern hemispheres. As the initial study of healthy adult volunteers was concluding in the United States, preliminary results allowed the start of a phase-IIa study in Australia, early in the RSV season of the southern hemisphere.

VI. Results of Early Clinical Studies

In the first clinical study, SB 209763 or placebo was administered IV to 26 healthy adult men. Single ascending doses ranging from 0.025 mg/kg to 10.0 mg/kg ($n = 16$) or placebo (saline; $n = 10$) were given over 30 min (Everitt et al. 1995; 1996). Subjects had serial blood samples taken over 10 weeks of follow-up for the determination of pharmacokinetics, antibodies to SB 209763 and FI titers. SB 209763 was safe and well tolerated, and no vital sign or laboratory abnormalities were considered related to study drug. Consistent with the long half-life of human IgG1 in humans, SB 209763 had a low plasma clearance and a long elimination half-life of approximately 23 days (Fig. 4). The pharmacokinetics of SB 209763 appeared to be dose proportional over the 400-fold dose range of 0.025–10 mg/kg. Low intersubject variability in total plasma clearance, steady-state volume of distribution and elimination half-life was observed. There was no evidence of an antibody response. Pre-existing anti-RSV antibodies made it difficult to quantify changes in FI titers in this study. Nevertheless, modest increases in RS virus FI activity of approximately threefold were observed, even in the presence of pre-existing anti-RS antibody, when 10 mg/kg SB 209763 was administered (4/4 subjects) and when 5 mg/kg SB 209763 was administered (2/4 subjects).

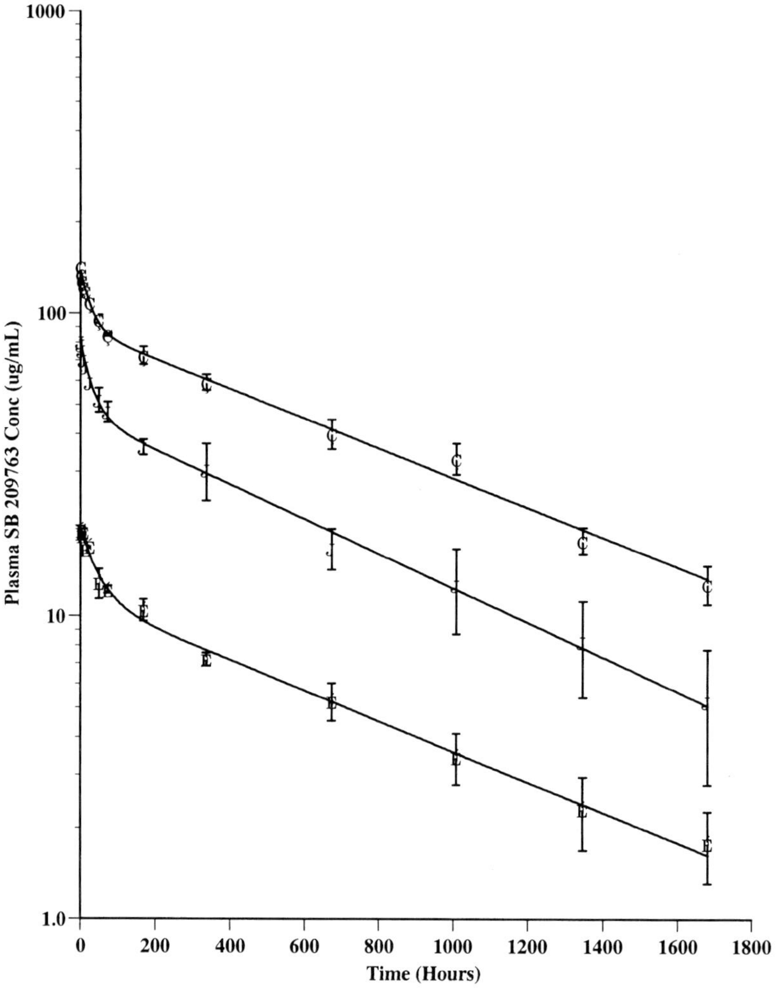

Fig. 4. Plasma SB 209763 concentration versus time profile. Mean (standard deviation) plasma SB 209763 concentration versus time profiles following a single intravenous infusion to healthy adult male volunteers ($n = 4$/group). E = 1.25 mg/kg, J = 5.0 mg/kg, C = 10.0 mg/kg

This study was followed by a single IV-dose study in previously healthy or stable infants aged 1 month to less than 36 months, who were hospitalized with RSV infection (HOGG 1996). The design and dosing scheme were the same as that described for the initial study of healthy adult volunteers. There were 26 patients with an average age of 9 months (range 1–28 months) that completed the study. SB 209763 was well tolerated and no antibodies developed to the humanized mAb. The mean terminal half-life of SB 209763 was independent of dose and averaged 28.4 days (range 17.6–35.3 days), similar to the half-life noted above in healthy adults. Pharmacokinetic area under the curve (AUC) values varied minimally (<two-fold) within each dose group and were approximately dose proportional.

In the second clinical study in healthy volunteers, 26 adult men received either a single dose or repeated IM doses of placebo or SB 209763 (Thompson 1996). Single doses of placebo or 0.021 mg/kg, 0.25 mg/kg or 1.25 mg/kg of SB 209763 were received by six, two, two, and four subjects, respectively. Two doses of placebo, 2.5 mg/kg or 5.0 mg/kg of SB 209763 were administered 8 weeks apart to each of four subjects, who were followed for 8 weeks after each dose. The top dose was limited by the 5-ml volume required in each of two injection sites at this dose. SB 209763 was well tolerated with no significant injection site irritation. No evidence of an antibody response was observed following a single or repeat dose of SB 209763. Maximum plasma concentration (C_{max}) and AUC increased in an approximately dose-proportional manner. The apparent terminal elimination half-life averaged 34 days (range 18–48 days) and appeared to be dose independent.

E. Conclusion

In summary, IV and IM doses of SB 209763 were well tolerated in healthy volunteers, infants and young children. In addition, no clinically relevant antigenicity was identified in phase-I and phase-IIa studies. Doses up to 10 mg/kg can be administered to young infants in a formulation and volume that caused no local reactivity following IM administration. The pharmacokinetics of SB 209763 in humans is similar to that of native IgG. The initial use of healthy volunteers proved to be an efficient way to gain early data on safety. The pharmacokinetics of SB 209763 were remarkably similar in healthy adults, premature infants and infants acutely infected with RSV. This approach allowed RSV-infected infants to be studied with the assurance that SB 209763 should be safe, not strongly antigenic, and with a sufficient terminal half-life that should allow clinical efficacy. Pivotal trials of ultimate clinical efficacy have proceeded following the early preclinical and clinical program that has been described.

References

Adair JR (1992) Engineering antibodies for therapy. Immunological Reviews, No. 130:5–39

Arbiza J, Taylor G, Lopez JA, Furze J, Wyld S, Whyte P, Stott EJ, Wertz G, Sullender W, Trudel M, et al. (1992) Characterization of two antigenic sites recognized by neutralizing monoclonal antibodies directed against the fusion glycoprotein of human respiratory syncytial virus. J Gen Virol 73:2225–2234

Azuma J, Kurimoto T, Tsuji S, et al. (1991) Phase I study on human monoclonal antibody against cytomegalovirus: pharmacokinetics and immunogenicity. J Immunotherapy 10:278–285

Beeler JA, Coelingh K (1989) Neutralization epitopes of the F glycoprotein of respiratory syncytial virus: effect of mutation upon fusion function. J Virol 63:2941–2950

Collins PL (1994) The molecular biology of respiratory syncytial virus (RSV) of the genus pneumovirus, The Paramyxoviruses, Kingsbury DW (ed), Plenum Press, New York, chapter 4, 103–162

Crowe JE, Murphy BR, Chanock RM, Williamson RA, Barbas CF, Burton DR (1994) Recombinant human respiratory syncytial virus (RSV) monoclonal antibody Fab is effective therapeutically when introduced directly into the lungs of RSV-infected mice. Proc Natl Acad Sci 91:1386–90

Davis CB, Hepburn TW, Urbanski JJ, Kwok DC, Hart TK, Herzyk DJ, Demuth SG, Leland M, Rhodes GR (1995) Pre-clinical pharmacokinetic evaluation of the respiratory syncytial virus-specific virus-specific reshaped human monoclonal antibody SB 209763. Drug Metabolism and Disposition 23:1028–1036

Dimmock NJ (1993) Neutralization of animal viruses. Curr Top Microbiol Immunol 183:1–149

Everitt DE, Davis CB, Thompson K, Boike SC, DiCicco R, Ilson B, Demuth SG, Herzyk DJ, Jorkasky DK (1996) The pharmacokinetics, antigenicity and fusion-Inhibition activity of SB 209763, a humanized monoclonal antibody to respiratory syncytial virus, in healthy volunteers. J Infect Dis 174:463–469

Everitt D, Thompson K, DiCicco R, Davis C, Demuth S, Herzyk DJ, Ilson B, Jorkasky D (1995) Safety, pharmacokinetics, antigenicity & fusion inhibition activity of SB 209763, a reshaped human monoclonal antibody to respiratory syncytial virus. Pediatric Research 37:173A

FDA (1994) Draft points to consider in the manufacture and testing of monoclonal antibody products for human use. Center for Bioligics Evaluation and Research, FDA.

Fujinami RS, Oldstone MBA, Wroblewska Z, Frankel ME, Koprowski H (1983). PNAS 80:2346–2350

Groothuis JR, Simoes EAF, Levin MJ, Hall CB, Long CE, Rodriguez WJ, Arrobio J, Meissner HC, Fulton DR, Welliver RC, Tristram DA, Siber GR, Prince GA, Van Raden M, Hemming VG (1993) Prophylactic administration of respiratory syncytial virus immune globulin to high-risk infants and young children, The New England Journal of Medicine 329:1524

Groothuis JR (1994) The role of RSV neutralizing antibodies in the treatment and prevention or respiratory syncytial virus infection in high risk children. Antiviral Research 23:1–10

Hall CB (1994) Prospects for a Respiratory Syncytial Virus Vaccine. Science 265:1393–1394

Heilman CA (1990) From the National Institute of Allergy and Infectious Diseases and the World Health Organization. Respiratory syncytial and parainfluenza viruses. J Infect Dis 161:402–406

Hogg G, Rodruiguez W, Groothuis, Mills J, Thompson K, Nichols A, Everitt D, Miller A, Herzyk D, Jorkaskyk D (1996) Pharmacokinetics & Safety of RSHZ19, a Humanized Monoclonal Antibody to Respiratory Syncytial Virus, in RSV Infected Infants. Presented at the 36th Interscience Conference of Antimicrobial Agents and Chemotherapy, New Orleans, LA, 1996

Hemming VG, Prince GA, Horswood RL, London WT, Murphy BR, Walsh EE, Fischer GW, Weisman LE, Baron PA, Chanock RM (1985) Studies of passive immunotherapy for infections of respiratory syncytial virus in the respiratory tract of a primate model. J Infect Dis 152:1083–1087

Hemming VG, Prince GA (1992) Respiratory Syncytial Virus: Babies and Antibodies. Infectious Agents and Disease 1:24–32

Henry R, Begent J, Pedley RB (1992) Monoclonal antibody administration: current clinical pharmacokinetic status and future trends. Clin Pharmacokinet 23:85–89

Kim HW, Canchola JG, Brandt CD, Pyles G, Chanock RM, Jensen K, Parrot RH (1969) Respiratory syncytial virus disease in infants despite prior administration of antigenic inactivated vaccine. Am. J. Epidemiol. 89:422–434

Lewis AP, Barber KA, Cooper HJ, Sims MJ, Worden J, Crowe JS (1993) Cloning and sequence analysis of kappa and gamma cynomolgus monkey immunoglobulin cDNAs. Dev Comp Immunol 17:549–560

Mark GE, Padlan EA (1994) Humanization of Monoclonal Antibodies. Handb Exp Pharm 113:105–134

Prince GA, Hemming VG, Horswood RL, Baron PA, Murphy BR, Chanock RM (1990) Mechanism of antibody-mediated viral clearance in immunotherapy of respiratory syncytial virus infection of cotton rats. J Virol 64:3091–3092

Roberts GD, Johnson WP, Burman S, Anumula K, Carr SA (1995) An integrated strategy for structural characterization of the protein and carbohydrate components of monoclonal antibodies: application to anti-respiratory syncytial virus Mab. Analytical Chemistry 67:3613–3625

Russell SJ, Llewelyn MB, Hawkins RE (1992) Principles of antibody therapy. BMJ 305:1424–1429

Sequeira LA, Cutler RE (1992) Muromonab CD3 (Orthoclone OKT3): part 2, adverse effects, clinical trials, and therapeutic use. Dial Transplant June, 353–368

Shadle PJ, Erickson JC Scott RG, Smith TM (1995) Antibody Purification. US Patent No. 5,429,746.

Stephens S, Emtage S, Vetterlein, et al. (1994) Clinical experience with CDP571, an engineered anti-TNFα antibody. J Cellular Biochem, Supp D, Supp D:216

Tandai M, Endo T, Sasaki S, Masuho Y, Kochibe N, Kobata A. (1991) Structural study of the sugar moieties of monoclonal antibodies secreted by human-mouse hybridoma. Arch. Biochem. Biophys. 291:339–348

Tao M-H, Morrison SL (1989) Studies of aglycosylated mouse-human IgG. Role of carbohydrate in the structure and effector functions mediated by the human IgG constant region. J Immunol 143:2595–2601

Taylor G, Stott EJ, Hughes M, Collins AP (1984) Respiratory syncytial virus infection in mice, Infection and Immunity 43:649

Taylor G, Furze J, Tempest PR, Bremner P, Carr FJ, Harris WJ (1991) Humanised monoclonal antibody to respiratory syncytial virus. Lancet 337:1411–1412

Taylor G, Stott EJ, Furze J, Ford J, Sopp P (1992) Protective epitopes on the fusion protein of respiratory syncytial virus recognized by murine and bovine monoclonal antibodies. J Gen Virol 73:22–27

Tempest PR, Brenner P, Lambert M, Taylor G, Furze JM, Car FJ, Harris WJ (1991) Reshaping a human monoclonal antibody to inhibit human respiratory syncytial virus infection in vivo. Biotechnology 9:266–271

Thompson KA, Everitt DE, Chapelsky M, Miller AK, Herzyk DJ, Jorkasky D (1996) Safety, pharmacokinetics and antigenicity of single and repeat IM injections of a reshaped human monoclonal antibody (SB 209763) in healthy volunteers. Clin Pharmacol Ther 59:144

Trill JJ, Shatzman AR, Ganguly S (1995) Production of monoclonal antibodies in COS and CHO cells. Curr. Opin. Biotechnol. 6:553–560

Walsh E, Schlesinger JJ, Bradriss MW (1984) Protection from respiratory syncytial virus infection in cotton rats by passive transfer of monoclonal antibodies, Infection and Immunity 43:756.

Wyde PR, Moore DK, Hepburn T, Silverman CL, Porter TG, Gross M, Taylor G, Demuth SG, Dillon SB (1995) Evaluation of the protective efficacy of reshaped human monoclonal antibody RSHZ19 against respiratory syncytial virus in cotton rats. Pediatric Research 38:543–550

CHAPTER 7

Preclinical Development of Antisense Therapeutics

A.A. LEVIN, S.P. HENRY, C.F. BENNETT, D.L. COLE, G.E. HARDEE, and G.S. SRIVATSA

A. Introduction

Phosphorothioate oligodeoxynucleotides appear to be an important new class of human therapeutic agents and are the first compounds shown to have the properties required of antisense drugs. They have well-demonstrated antisense modes of drug action, are stable in vitro and have very acceptable half-lives in vivo. When administered by injection, phosphorothioate oligodeoxynucleotides distribute rapidly to tissues and show excellent pharmacokinetics. To date, they have exhibited mild and manageable toxicities at projected and demonstrated therapeutic doses. Phosphorothioate oligodeoxynucleotides are readily taken up by cells in vivo without need for added uptake enhancers. At lengths of 20–21 nucleotides or less, they exhibit messenger RNA (mRNA) affinities sufficient to inhibit gene expression at doses which, in turn, can provide good therapeutic indices and acceptable cost to the patient.

B. Pharmacology of Antisense Oligodeoxynucleotides

I. Molecular Pharmacology of Antisense Oligodeoxynucleotides

Antisense drugs are nucleic acid oligomer analogs that bind target mRNA receptors through Watson–Crick base pairing. In our experience, screening an adequate number of candidate oligomers for antisense activity is extremely important to the design of these drugs (BENNETT et al. 1995; CHIANG et al. 1991; DEAN and MCKAY 1994). Superficially, it is a simple matter to design an antisense molecule to target mRNA at a specific base sequence. In practice, however, mRNAs are highly structured molecules. Secondary and tertiary mRNA structure and mRNA-bound proteins can all limit drug access to a target base sequence. Thus, screening multiple oligodeoxynucleotide sequences aids identification of accessible sites on a given target mRNA.

In this manner, virtually thousands of unique oligodeoxynucleotides could be made and tested against one target mRNA. As an alternative to that resource-intensive exercise, the initial activity screen is often limited to 15–30 unique oligodeoxynucleotide sequences. The length of screened oligodeoxynucleotides depends, in part, on their chemistry and base composition. For first-generation phosphorothioate oligodeoxynucleotides, 17-mers to 21-mers are sufficiently long to hybridize target mRNA under physiological conditions. If the target sequence is particularly rich in adenosine or thymidine residues, longer oligomers may be needed. Modification of native DNA or RNA structure can afford increased affinity and permit use of shorter oligomers.

At this time, it is not possible to predict which regions of an mRNA will be most accessible to antisense oligodeoxynucleotides. For this reason, oligodeoxynucleotides should be designed to hybridize to various regions of the target mRNA, including the 5′-untranslated region, the coding region and the 3′-untranslated region. Commercial computer programs can aid mRNA target selection. Because predicted secondary structure for large RNAs may not be accurate, however, the programs' main use is to identify regions that might form poorly accessible hairpins or other self-hybridized structures.

In addition to antisense drug action, oligodeoxynucleotides, like all other drugs, may produce other effects when added to cells (STEIN and CHENG 1993; STEIN and KRIEG 1994; STEIN 1995; CROOKE and BENNETT 1996). Several examples of non-antisense effects of oligodeoxynucleotides have recently been described, including inhibition of cell proliferation, attachment to plastic substrates, inhibition of phospholipase A_2 and inhibition of human immunodeficiency virus (HIV) binding to CD4 (CROOKE and BENNETT 1996; WATSON et al. 1992; STEIN et al. 1991; BENNETT et al. 1994). Therefore, it is necessary that direct inhibition of the targeted protein be demonstrated before claiming an antisense mechanism for an oligodeoxynucleotide drug. It is also important to include control oligodeoxynucleotides in the assay to reveal any non-sequence-specific oligodeoxynucleotide effects. There is no single best control oligodeoxynucleotide for this use, as each oligodeoxynucleotide sequence may take a unique three-dimensional shape and can, thus, potentially have unique non-antisense pharmacological effects. Ideally, several control oligodeoxynucleotides are used, including some with various numbers of base-pairing mismatches with the target mRNA sequence. Finally, to demonstrate specificity of an oligodeoxynucleotide drug for inhibiting the expression of its targeted protein, effects on expression of protein coded by closely related genes should be examined. As an example of this latter approach, and of the specificity that has been achieved with first-generation phosphorothioate antisense oligodeoxynucleotides, complete selectivity for inhibiting expression of either the single-base mutated form or the wild-type form of the ha-*ras* gene product has been recently demonstrated (MONIA et al. 1992).

Using proper controls, it has been possible to consistently demonstrate that oligodeoxynucleotides inhibit targeted gene expression by an antisense mechanism of action, and it has been shown that resulting pharmacological effects can be attributed to reduction in targeted protein concentration. It has been further demonstrated that phosphorothioate oligodeoxynucleotides targeted to human intercellular adhesion molecule 1 (ICAM-1) mRNA may inhibit protein expression by multiple antisense mechanisms of action, depending in part on the region of mRNA targeted; oligodeoxynucleotides that bind the 3′-untranslated region of ICAM-1 mRNA achieved reduction in mRNA concentration by an RNase H-dependent mechanism, whereas oligodeoxynucleotides targeting the 5′ untranslated region inhibited ICAM-1 expression by an RNase-independent mechanism (CHIANG et al. 1991). This work was later extended to two additional cell adhesion molecules, vascular cell adhesion molecule 1 (VCAM-1) and E-selectin (BENNETT et al. 1995). Similar observations were made for oligodeoxynucleotides targeting protein kinase C-α (PKCα) expression (DEAN et al. 1994). Taken together, these studies demonstrate that there are several molecular level antisense mechanisms by which mRNA-bound oligodeoxynucleotides may reduce gene expression.

A classic drug-design goal is inhibition of protein function via drug binding to the target protein itself or, perhaps, to a protein involved in biosynthesis of the target protein. In contrast, the goal of antisense drug design is inhibition of protein production, through drug binding to mRNA. Targeting protein production instead of protein function affords a distinct advantage to antisense drugs over conventional drugs; antisense drugs can be designed to inhibit a single member of a closely related family of isoenzymes. In fact, antisense oligodeoxynucleotides can be designed to selectively discriminate among spliced mRNA variants of the same protein. Thus, antisense drugs can be used to explore the molecular pharmacology of specifically reduced protein isotype levels, a therapeutic goal virtually inaccessible through classical drug design.

The PKCs, for example, are a family of serine–threonine protein kinases, which play important roles in transmitting extracellular signals to the cell nucleus. Currently, there are 12 known, genetically unique, isoenzymes of PKC. An oligodeoxynucleotide, ISIS 3521, which selectively inhibited PKCα isotype expression in human-lung carcinoma cells without affecting expression of other PKC isoenzymes, has been identified (DEAN et al. 1994). Cells treated with ISIS 3521 for prolonged periods exhibited reduced ICAM-1 expression in response to phorbol-ester treatment, but not tumor necrosis factor treatment. This demonstrated that at least two independent mechanisms exist for regulating ICAM-1 expression, one dependent on PKCα, the other not requiring PKCα. Similar application of antisense oligodeoxynucleotides will be very useful in determining roles of other isoenzymes in normal and abnormal or disease physiological processes.

Oligodeoxynucleotides targeting various viral gene products have demonstrated reductions in viral replication (AGRAWAL 1992; AGRAWAL et al. 1988b; AZAD et al. 1993, 1995; LISZIEWICZ et al. 1993). In the case of viral assays, however, it has been difficult to demonstrate that oligodeoxynucleotides inhibit replication solely by an antisense mechanism of action. ISIS 2922 is a

21-mer phosphorothioate oligodeoxynucleotide, targeting the immediate early gene product of human cytomegalovirus (CMV) (AZAD et al. 1993). This oligodeoxynucleotide was selected from a large number of candidate oligodeoxynucleotides based on its potency in inhibiting replication of CMV. Characterization of molecular mechanisms of action suggests that ISIS 2922 is capable of inhibiting immediate early gene expression through an antisense mechanism of action, when the gene products are expressed exogenously in cells (ANDERSON et al. 1996). In in vitro viral replication assays, the oligodeoxynucleotide inhibits replication of CMV in a manner not entirely consistent with an antisense mechanism of action (ANDERSON et al. 1996; AZAD et al. 1993). Thus, this antiviral oligodeoxynucleotide is probably working by a dual mechanism of action, in which the antisense mechanism plays a role. Similar findings have been reported for an antiviral oligodeoxynucleotide targeted to HIV.

II. In Vivo Pharmacology of Antisense Oligodeoxynucleotides

An early task facing antisense oligodeoxynucleotide researchers was the demonstration that oligodeoxynucleotides would inhibit expression of targeted gene products when administered in vivo. Direct intracerebral injections of antisense oligodeoxynucleotides were among the earliest successful in vivo applications, demonstrating inhibition of gene expression in the brain (CHIASSON et al. 1992; CROOKE and BENNETT 1996; HOOPER et al. 1994; WEISS et al. 1993; ZHOU et al. 1994). A number of investigators subsequently demonstrated reduction in target protein expression, following direct intracerebral injection and characterized the resulting pharmacological effects.

DEAN and MCKAY were the first to demonstrate that systemic administration of an antisense phosphorothioate oligodeoxynucleotide targeted to a normal host gene resulted in reduced targeted gene expression in normal tissue, thus, demonstrating that phosphorothioate oligodeoxynucleotides are capable of reaching the desired sites of action when administered systemically, and do so without formulation enhancements (DEAN and MCKAY 1994). In this case, they used an oligodeoxynucleotide targeting murine PKCα to selectively reduce expression of PKCα in the liver (DEAN and MCKAY 1994).

An antisense oligodeoxynucleotide targeting PKCα mRNA, ISIS 3521, has been used to treat nude mice bearing human-tumor xenografts. Marked reduction in tumor growth has been observed in the model for glioblastoma (YAZAKI et al. 1996). Animals tolerated treatment with the human PKCα antisense oligodeoxynucleotide. ISIS 3521 is specific for human PKCα and, thus, does not affect PKCα expression in murine cells or tissues; therefore, mice were treated with murine-specific antisense PKCα oligodeoxynucleotide in parallel studies. These studies demonstrated that prolonged treatment with a murine PKCα phosphorothioate oligodeoxynucleotide, ISIS 4189, did not produce adverse effects, which could be correlated with the resulting reduction in PKCα levels, despite direct demonstration of reduction of PKCα in murine tissue (DEAN and MCKAY 1994). These results suggest that in normal tissues, PKCα is not critical for cell survival or tissue function, possibly due to compensatory mechanisms provided by other PKC isoenzymes. In contrast, for some transformed cells, PKCα appears to have a critical role in maintaining cell proliferation, possibly due to increased demands placed on the cell by the transformation process. Systematically administered first-generation phosphorothioate oligodeoxynucleotides can produce pharmacological effects in a matter consistent with an antisense mechanism of action and demonstrate the value of selective isotypic pharmacology.

In addition to cancer and viral diseases, antisense oligodeoxynucleotides have also shown therapeutic potential for the treatment of inflammatory diseases. An antisense oligodeoxynucleotide targeting murine ICAM-1 significantly prolonged survival of a heterotopic heart allograft when administered by continuous intravenous (i.v.) infusion for either 7 days or 14 days (STEPKOWSKI et al. 1994). Histological examination of heart allografts from animals treated with the antisense oligodeoxynucleotide revealed marked reduction in leukocyte infiltration and tissue damage in animals treated with the antisense oligodeoxynucleotides. Two control oligonucletides failed to prolong survival of the allograft, demonstrating that the effect of the ICAM-1 antisense oligodeoxynucleotide was sequence specific. The anti-ICAM-1 oligodeoxynucleotide exhibited synergistic effects on allograft survival when combined with other immunosuppresive therapies, and induced tolerance when combined with monoclonal antibodies to LFA-1. These studies have been extended to pancreatic islet grafts with similar effects (KATZ et al. 1995).

More recently, reductions in ICAM-1 expression in lung tissue following endotoxin challenge were demonstrated in mice treated with a murine-specific oligodeoxynucleotide, directly demonstrating that the ICAM-1 antisense oligodeoxynucleotide reduces expression in vivo (KUMASAKA et al. 1996). The ICAM-1 antisense oligodeoxynucleotide reduced neutrophil trafficking into the airways of mice treated with endotoxin, to an extent similar to monoclonal antibodies to ICAM-1, demonstrating that similar pharmacological effects may be obtained by two different antisense methods for interfering with ICAM-1 function (KUMASAKA et al. 1996). The murine ICAM-1 antisense oligodeoxynucleotide showed activity in murine models of colitis (BENNETT et al. 1997), demonstrating a range of anti-inflammatory activity.

These studies strongly suggest that first-generation phosphorothioate oligodeoxynucleotides can inhibit expression of the targeted gene product, both in vitro and in vivo. Thus, oligodeoxynucleotides are capable of distributing to target tissues, and are taken up by cells in the tissues in a manner such that they have access to targeted mRNA. In addition, antisense inhibition of targeted gene expression can produce profound pharmacological activity, which may be exploited therapeutically. As discussed below, there are some limitations to first-generation phosphorothioate oligodeoxynucleotides, many of which can be alleviated either through formulation or novel medicinal-chemical modifications.

C. Pharmacokinetics and Toxicity of Oligodeoxynucleotide Therapeutics

The first generation of antisense therapeutic agents are nearly all phosphorothioate oligodeoxynucleotides. A number of different sequences have been examined in preclinical studies and the pattern of plasma and tissue distribution, as well as the toxic effects, have been remarkably similar for all, suggesting that the pharmacokinetic profiles and toxicologic effects are sequence independent (AGRAWAL et al. 1988b; COSSUM et al. 1993; SANDS et al. 1994; SRINIVASAN and IVERSEN 1995). It has become clear, as our experience with this class of compounds increases, that this high degree of predictability, from sequence to sequence, is conducive to drug development, because the existing information for the class is relevant to each new oligodeoxynucleotide. Thus, a significant body of knowledge on oligodeoxynucleotide pharmacokinetics and toxicity is developing.

The toxicity and the kinetics of the disposition of oligodeoxynucleotides may be influenced by oligodeoxynucleotide–protein interactions, making it possible that there could be specific sequence motifs that might alter the toxicity of some oligodeoxynucleotides, resulting in differences in potency or toxic effects. However, the experience to date suggests that only quantitative differences exist and, qualitatively, the toxicities are similar (MCINTYRE et al. 1993). In addition, it is also possible that exaggerated pharmacologic activity associated with inhibition of expression of specific proteins could be associated with sequence-specific toxicities. Thus, while it is possible that there could be subtleties in some aspects of the kinetics and toxicity that are sequence dependent, it is clear that most of the toxicities and the pharmacokinetic profiles (observed to date) are independent of sequence.

Existing data on the preclinical pharmacokinetics and toxicity, as well as data on the pharmacokinetics from early clinical trials, support the continued use of these compounds in clinical trials and their eventual use as marketed therapeutic agents.

I. Pharmacokinetics and Metabolism

The toxicity and pharmacokinetics of phosphorothioate oligodeoxynucleotides have been characterized in mice, rats and monkeys exposed for as long as 26 weeks. Because the oral bioavailability of the present generation of oligodeoxynucleotide therapeutic agents is limited, existing data on the pharmacokinetics is limited to parenteral routes of exposure, including i.v., intradermal, subcutaneous, intraperitoneal (i.p.) and intravitreal injections (Cossum et al. 1993, 1994; Agrawal et al. 1991, 1995a; Iversen 1991; Leeds et al. 1996b; Sands et al. 1994; Srinivasan and Iversen 1995). There is significant absorption from these types of injections as well as marked systemic exposure. The only possible exception may be intravitreal injections, which limit systemic exposure and maintain local concentrations in the eye for extended periods of time (Leeds et al. 1996a).

Systemically administered phosphorothioate oligodeoxynucleotides circulate in blood and are widely distributed to tissues. The kinetics of phosphorothioate oligodeoxynucleotides in plasma are multiphasic. Initial clearance rates are in the order of 1 h or less, with much longer terminal elimination phases ($t_{1/2}$ $\geq$40 h). Much of the initial rate of clearance from plasma can be explained on the basis of the rapid uptake by tissues. Typically, the half-life of this early phase is in the order of 10–30 min in mice and rats, and 30–60 min in primates and humans (Cossum et al. 1993; Agrawal et al. 1989; Sands et al. 1994; Glover et al. 1997). The distribution kinetics and the half-lives appear to be sequence independent, but we have observed in rats, monkeys and humans that plasma kinetics are dose dependent. With increasing doses of oligodeoxynucleotide administered by i.v. injections, the plasma concentrations increase in a greater-than-dose linear manner. In phase-I clinical trials, plasma area under curves (AUCs) for ISIS 2302, a 20-mer phosphorothioate, increased sixfold over a fourfold range of doses (Glover et al. 1997). In some species, half-lives are seen to increase with dose, but in man, there is a reduction in the volume of distribution with dose. In normal volunteers given various doses of ISIS 2302 by 2-h i.v. infusion, as dose was increased from 0.5 mg/kg to 2.0 mg/kg, there was a change in volume of distribution (steady state) from 155.4 ml/kg to 97.5 ml/kg. These nonlinearities result in AUC values that increase from approximately 250 μg/min/ml to 1560 μg/min/ml as dose was increased from 0.5 mg/kg to 2.0 mg/kg administered as a 2-h infusion (Glover et al. 1997). These data, and data in animals, suggest that there is some saturable component to the plasma disposition and we propose that tissue distribution may be one of the factors that changes with dose.

After systemic exposure, radiolabel derived from phosphorothioate oligodeoxynucleotides can be found in most organs and tissues. The major sites of distribution are the liver, kidney, spleen and bone marrow. The brain and the testes, because of their blood barriers, are sites of little or no accumulation (Cossum et al. 1993, 1994; Agrawal et al. 1991). However, direct intrathecal administration of oligodeoxynucleotides results in significant distribution to the brain (Szklarczyk and Kaczmarek 1995; Whitesell et al. 1993).

After i.v. or intradermal injections, the kidney is the site of the highest concentration of oligodeoxynucleotide, but the liver has significant concentrations and, because of its greater mass, is the major organ for oligodeoxynucleotide deposition. The spleen, lymph nodes, bone and bone marrow are also sites of deposition of phosphorothioate oligodeoxynucleotides, but the concentrations are well below those in liver and kidney. Phosphorothioate oligodeoxynucleotides can be detected in most other tissues as well, but the concentrations are lower. Similar organ-distribution patterns have been observed with a number of different phosphorothioate-oligodeoxynucleotide sequences of varying lengths, suggesting that the distribution is independent of sequence (COSSUM et al. 1993, 1994; AGRAWAL et al. 1991; SANDS et al. 1994).

As doses are increased, there is a change in the pattern of distribution of oligodeoxynucleotide to various organs, with a slightly smaller percentage of the total dose accumulating in the liver and kidney, and a greater percentage accumulating in other organs, such as the lymph nodes and spleen (GEARY et al. 1997). This pattern of changes is consistent with a saturable component in the distribution to the liver and possibly kidney. Because this change in distribution occurs at high doses, or perhaps after long-term administration, the saturable processes may be important for the interpretation of high-dose toxicology study data and chronically administered clinical doses. For instance, when there is repeated low-dose exposure over long periods, organ concentrations may rise to some saturating concentrations. If this is the case, then repeated administration could produce shifts in the steady-state plasma profile and organ distribution, a phenomenon that will be examined in upcoming preclinical studies. Because organ distribution is thought to be an important factor in the plasma kinetics, the saturable nature of the organ distribution is thought to play a role in the nonlinear plasma kinetics (GEARY et al. 1997).

What is the fate of oligodeoxynucleotides in tissues? Techniques such as fluorescent labeling, immunohistochemistry and autoradiography have been employed to localize oligodeoxynucleotides in tissues (RAPPAPORT et al. 1995; PLENAT et al. 1994; SANDS et al. 1994). In these studies, the specific cell types that take up oligodeoxynucleotides after in vivo administration have been characterized. In the livers of mice treated with phosphorothioate oligodeoxynucleotide, Kupffer cells are enlarged and contain basophilic granules, which are thought to contain oligodeoxynucleotide (SARMIENTO et al. 1994). The most remarkable accumulation of oligodeoxynucleotides is within cells of the proximal tubules in the kidney. In proximal tubular cells, it is possible, with routine histology, to visualize granular inclusions that are thought to contain oligodeoxynucleotide (SARMIENTO et al. 1994). With autoradiography, it is possible to identify the granules containing oligodeoxynucleotide-derived material and it is possible to differentiate cortex from medulla because of the significantly higher concentrations in the cortex (proximal tubules) (RAPPAPORT et al. 1995; SANDS et al. 1994). Micropuncture studies have demonstrated that the proximal tubule accumulates phosphorothioate oligodeoxynucleotides as a result of tubular reabsorption (RAPPAPORT et al. 1995; OBERBAUER et al. 1995). The transport mechanisms into cells and the sites of oligodeoxynucleotide binding are under investigation, but may be the result of binding to the scavenger receptor (SAWAI et al. 1995; WU-PONG et al. 1994). The critical information obtained from these studies is the clear evidence of uptake into cells in animals treated with phosphorothioate oligodeoxynucleotides (SANDS et al. 1994; PLENAT et al. 1994; OBERBAUER et al. 1995). At this time, it is not clear whether cellular accumulation of phosphorothioate oligodeoxynucleotide into cytoplasmic granules within tissue results in delivery to a functional compartment for producing antisense effects, or whether it is a nonfunctional cellular uptake mechanism as demonstrated for cultured cells (BENNETT et al. 1992; NESTLE et al. 1994). Studies are currently being conducted to identify functional cellular-uptake pathways in tissues.

A strategy for determining the factors that influence the disposition of oligodeoxynucleotide is to make chemical modifications that alter various physical–chemical properties of oligodeoxynucleotide, such as protein binding or lipophilicity, and studying the disposition characteristics of the modified compounds. For example, by modifying the protein-binding characteristics of a compound, it is possible to examine the role of protein binding in renal disposition. Altering the protein-binding characteristics of oligodeoxynucleotides by chemical modifications could have significant impact on the distribution of oligodeoxynucleotide to the kidney in the following way; reducing the plasma-protein binding would increase the amount of free oligodeoxynucleotide circulating and might increase the amount of oligodeoxynucleotide filtered by the kidney and then reabsorbed in the proximal tubules. In fact, 2′-propoxy modifications to a phosphodiester oligodeoxynucleotide reduced binding to albumin and increased accumulation in the kidney, compared with an unmodified phosphorothioate oligodeoxynucleotide with a similar sequence (CROOKE et al. 1996). Although part of this renal accumulation could have been the result of increased metabolic stability of the modified oligodeoxynucleotide, reduced protein binding cannot be excluded as a factor controlling this change in disposition. It is also possible to add specific substituents to an oligodeoxynucleotide to take advantage of cellular transport mechanisms, not only to enhance cellular uptake, but to direct the distribution to specific tissue. When the lipophilicity of a phosphorothioate oligodeoxynucleotide was markedly increased by the inclusion of a cholesterol modification at the 5′ end, there was a dramatic increase in the accumulation in the liver. It should also be noted that the alteration and disposition could be related to binding to receptors in other cholesterol-binding proteins. Thus, by modifying the physical–chemical properties of the oligodeoxynucleotides, it is possible to obtain an understanding of the factors that control disposition and organ distribution for oligodeoxynucleotides. Clearly, protein binding and lipophilicity are key factors in these processes, just as they are for small molecules.

Once distributed from plasma, phosphorothioate oligodeoxynucleotides are slowly cleared from tissues (COSSUM et al. 1993, 1994; PLENAT et al. 1994; SANDS et al. 1994; SAIJO et al. 1994; AGRAWAL et al. 1991, 1995b). Using ^{14}C-labeled phosphorothioate oligodeoxynucleotides, where the label was on the C-2 position of thymidine, the terminal elimination half-lives from liver, renal cortex, medulla and bone marrow were 62 h, 78 h, 116 h and 156 h, respectively (COSSUM et al. 1993). With half-lives greater than 48 h, every-other-day administration will result in almost constant or increasing tissue levels with multiple-dose regimens. The slow clearance from tissues is the result of the relatively slow rate of metabolism and clearance (through nuclease-mediated processes) of these compounds in tissues. With half-lives greater than 48 h, repetitive dosing on a daily basis is not necessary to maintain tissue concentrations of phosphorothioate oligodeoxynucleotides, and pharmacologically active concentrations might be maintained with less frequent administration.

While the initial disappearance from plasma is associated with distribution to the tissues, the terminal-elimination phase may be associated with the prolonged metabolic clearance of oligodeoxynucleotide by metabolism in plasma, tissues and excretion. The terminal-elimination phase of the plasma distribution has been characterized in studies employing radiolabeled oligodeoxynucleotides. A 20-mer phosphorothioate oligodeoxynucleotide labeled at the C-2 position of thymidine with ^{14}C had a terminal-elimination half-life of 51 h (COSSUM et al. 1993). This long terminal-elimination half-life has also been observed in studies employing ^{35}S-radiolabeled phosphorothioate (AGRAWAL et al. 1991).

Depending on the nature of the radiolabel, the end products of metabolism are eliminated in the expired air (^{14}C labeled) or in the urine(^{35}S-labeled oligodeoxynucleotide). An oligodeoxynucleotide labeled at the C-2 position of thymidine was primarily metabolized to $^{14}CO_2$. Approximately 50% of a 3.7-mg/kg dose was eliminated in the expired air (COSSUM et al. 1993, 1994). In contrast, following i.v. or i.p. injection of 30 mg/kg of a 20-mer phosphorothioate thiated with ^{35}S, approximately 30% of the radiolabel was eliminated in the urine. The apparent differences can be explained based on site of radio labeling. The ^{14}C-2 of thymidine was metabolized to $^{14}CO_2$, while the ^{35}S-radiolabeled phosphorothioate was eliminated in urine, presumably as sulfur-containing metabolites or chain-shortened metabolites. A small, but significant fraction of the administered dose was recovered from urine as intact oligodeoxynucleotide in mice treated with a 29-mer oligodeoxynucleotide ^{35}S-GEM 91, but significant amounts of compound were identified as metabolites of the parent compound (AGRAWAL et al. 1991, 1995b). Independent of the site of radiolabeling, all studies to date conclude that the principle pathway for the metabolism and, therefore, clearance of oligodeoxynucleotides is through nuclease-mediated shortening from the 3′-or 5′-termini.

The plasma and tissue pharmacokinetics of phosphorothioate oligodeoxynucleotides differ from those of naturally occurring phosphodiester oligodeoxynucleotides. One reason for the difference in pharmacokinetics is the increased resistance of the phosphorothioate linkage to nuclease activity, compared with the phosphodiester bond. This resistance is demonstrated in the longer plasma half-lives of intact phosphorothioate oligodeoxynucleotides compared with phosphodiester oligodeoxynucleotides (SANDS et al. 1994). For example, polyacrylamide-gel electrophoretic (PAGE) analysis of the plasma of monkeys dosed with 30 mg/kg of a phosphodiester oligodeoxynucleotide by i.v. injection, demonstrated that within 5 min, more than approximately 50% of parent 25-mer was degraded and within 15 min, little, if any, intact oligodeoxynucleotide could be detected. Thus, the half-life for the degradation of phosphodiester oligodeoxynucleotide is in the order of 5 min. In contrast, administration of phosphorothioate oligodeoxynucleotides to monkeys at a dose of 1 mg/kg or 5 mg/kg had an initial plasma half-life of 0.6–1 h (AGRAWAL et al. 1995b). A series of studies comparing the disposition of phosphorothioate or phosphodiester oligodeoxynucleotides internally labeled on cytosines with a ^{3}H-methyl further illustrate the remarkable differences in metabolic pattern between these two oligodeoxynucleotide backbones. When the livers, spleen and kidneys of mice treated with phosphodiester oligodeoxynucleotides were analyzed by paired ion chromatography, little or no parent compound was observed after 1 min. The only radiolabel extracted from these tissues coeluted with ^{3}H-5-methyl-2′-deoxydeoxymethyl cytosine or tritiated thymidine, suggesting a rapid and complete metabolic degradation of the administered oligodeoxynucleotide. Phosphorothioate oligodeoxynucleotide-treated mice had significant tritium activity associated with a high-performance liquid chromatography (HPLC) peak that coeluted with parent compound, and no radiolabel was associated with ^{3}H-5-methyl-2′-deoxydeoxymethyl cytosine or tritiated thymidine, suggesting that full degradation to the monomeric forms was retarded by the presence of the phosphorothioate backbone. In fact, after 24 h, approximately 20% of the radiolabel in these tissues remained in a form that coeluted with parent compound (SANDS et al. 1994).

Recent advances in analytical methods are allowing us to characterize both the plasma kinetics and the tissue disposition of phosphorothioate oligodeoxynucleotides without radiolabeling. Using capillary gel electrophoresis (CGE), it is now possible to quantitate chain-shortened metabolites of phosphorothioate oligodeoxynucleotides down through at least 8-mers and possibly shorter. These methods have been applied to both preclinical (GEARY et al. 1997) and clinical studies (LEEDS et al. 1996b; GLOVER et al. 1996 submitted). Using CGE to analyze chloroform–phenol extracts of plasma, we can confirm the clearance of parent drug from the plasma with half-lives on the order of 60 min in non-human primates and in man, and it is also possible to determine the clearance rates of the metabolites. The data from these studies suggest that both parent oligodeoxynucleotide and metabolites are cleared from plasma with similar half-lives (GLOVER et al. 1997).

The technique of CGE separates oligodeoxynucleotides on the basis of molecular weights in a manner analogous to size-exclusion chromatography. With this technique, it is possible to demonstrate there is a pattern of metabolites that is consistent with excision of single nucleotides. Typical electropherograms taken from tissues or plasma a few minutes after injection demonstrate the presence of parent oligodeoxynucleotide as a predominant oligodeoxynucleotide species. Peaks that comigrate with parent minus one nucleotide and parent minus two nucleotides (*n*–1, *n*–2, *n*–3, etc.) decrease in concentration with decreased size. At early times, the pattern of metabolites is such that the parent and the long-chain metabolites represent a greater portion of the total oligodeoxynucleotide pool and the smaller metabolites comprise a lesser fraction. In tissue, this pattern shifts to shorter metabolites over the course of several days, suggesting that there is a progressive shortening of the oligodeoxynucleotide, one base at a time, by exonucleases. Because both CGE and the more traditional gel-electrophoresis techniques separate oligodeoxynucleotide metabolites solely on the basis of molecular size (mass) and charge, these techniques provide no information on the site or direction of the exonuclease activity. By modifying the oligodeoxynucleotide, it is possible to determine the site of degradation. The ends of the oligodeoxynucleotide can be protected from nuclease degradation by conjugating various groups on the 2′ position of the sugar(s) of the terminal nucleotides. When organic groups are conjugated to the 3′ end or the 5′ end there is a reduction in the rate of metabolic degradation. However, the stability of phosphorothioate oligodeoxynucleotides is more greatly enhanced by protecting the 3′ end rather than the 5′-end, suggesting that the chain shortening of phosphorothioate oligodeoxynucleotides occurs primarily from the 3′ end. Thus, it appears that the successive removal of bases from the 3′ end of the oligodeoxynucleotide is the major pathway for metabolic degradation (TEMSAMANI et al. 1993). This same pattern of metabolites has been reported for all species from mouse to man, suggesting that the metabolic pathways are similar. The metabolism, kinetics of distribution and the plasma half-lives are scalable from species to species, allowing for good interspecies extrapolations. This interspecies predictability facilitates the development of these compounds as therapeutic agents.

II. Toxicity of Phosphorothioate Oligodeoxynucleotides

The systemic toxicities of phosphorothioate oligodeoxynucleotides have been studied extensively, in both rodents and monkeys, for a number of different sequences (MCINTYRE et al. 1993; SARMIENTO et al. 1994; SRINIVASAN and IVERSEN 1995; CORNISH et al. 1993; GALBRAITH et al. 1994). Like the pharmacokinetic properties discussed above, the toxicity of phosphorothioate oligodeoxynucleotides appears to be more closely related to the class of the molecule, rather than the specific sequence. Thus, it is possible to generalize about the toxicologic profiles of phosphorothioate oligodeoxynucleotides. There are, however, data suggesting that there are some sequence-related differences in the potency of toxicities, but qualitatively, the toxicities of all phosphorothioate oligodeoxynucleotides are similar (McIntyre et al. 1993). Although there is the suggestion of common toxicities for phosphorothioate oligodeoxynucleotides in general, there are notable differences observed in the toxicity profiles between rodents and monkeys, as will be described below.

Administration of phosphorothioate oligodeoxynucleotides to mice and rats produces a series of histologic changes that have been characterized as a general immune stimulation (Branda et al. 1993). This immunostimulation is characterized by splenomegaly, lymphoid hyperplasia and diffuse histocytosis in multiple organs. Marked splenomegaly was observed in animals treated with phosphorothioate oligodeoxynucleotides at doses of 50 mg/kg and higher with spleen weights increasing up to 2.5-fold over controls (Sarmiento et al. 1994). We have observed splenomegaly at doses as low as 20 mg/kg when administered daily for 14 days (unpublished data). The splenomegaly was accompanied by B-cell hyperplasia, histiocytosis in the connective-tissue capsule and marked extramedullary hematopoiesis. Hematology indicated that there were changes in circulating cells associated with the cellular changes in the spleen with increased total leukocytes, especially neutrophil and monocyte counts. Consistent with the increase in the number of monocytes, there were numerous other tissues and organs from mice treated with 24-mer phosphorothioate oligodeoxynucleotides that showed evidence for histiocytic infiltration, including liver, kidney, heart, thymus, pancreas and periadrenal tissues (Sarmiento et al. 1994). Similar changes have been observed with other phosphorothioate-oligodeoxynucleotide sequences at comparable doses (Henry et al. 1997a,b).

A notable difference in potency of immunostimulation was observed in studies in which mice that were dosed with a 24-mer phosphorothioate oligodeoxynucleotide designed to hybridize to *rel a*, the p65 subunit of NF-κB (GAG GGG AAA CAG ATCGTC CAT GGT), or the sense construct to that sequence (ACC ATG GACGAT CTG TTT CCC CTC). Mice injected intraperitoneally daily for 3 days with 25 mg/kg of the sense sequence had increases in spleen weights of greater than twofold over control values. In contrast, the mice injected with the antisense sequence under identical conditions had a 1.3-fold increase in spleen weights. Although potency differed, histologic examination of the spleens demonstrated that the effects in the spleens were qualitatively similar. An explanation for the differences in potencies for immune stimulation between these two compounds is currently unavailable. Possibilities range from an antisense effect, which specifically blocks the immune stimulation by inhibiting the NF-κB signaling pathway, to differences in tertiary structure or sequence motifs between the oligodeoxynucleotides, which influence immune stimulation. Differences in the ability of oligodeoxynucleotides to stimulate B-cell proliferation may be modulated by certain sequence motifs (BRANDA et al. 1993; YAMAMOTO et al. 1994; KRIEG et al. 1995). Although some of these motifs have been described in the literature, for example palindromic sequences and CpG motifs (KRIEG et al. 1995; YAMAMOTO et al. 1994), it is apparent from studies in our laboratories with oligodeoxynucleotides that do not contain CG motifs, that the stimulation of B-cell proliferation is probably dependent on a complex set of rules that control the interactions of oligodeoxynucleotide with lymphocytes and not a single motif (unpublished observations). High-affinity receptors have been proposed that might control these interactions (BENNETT et al. 1985). Obviously, the identification and structural mapping of high-affinity binding sites on the B lymphocytes will aid in the understanding of this interaction.

The occurrence of immune stimulation in rodents treated with phosphorothioate oligodeoxynucleotides might be related to the polyanionic nature of the compounds. Polyanions, including bacterial DNA, are known to stimulate B-cell proliferation (DIAMANTSTEIN et al. 1971; DIAMANTSTEIN and BLISTEIN-WILLINGER 1978; TALMADGE et al. 1985). The polyanionic phosphorothioate oligodeoxynucleotides appear to be behaving in the same way. The immunostimulation in rodents by polyanions and nucleic acids may, in part, be associated with the polyanion-stimulated B-cell hyperplasia and release of cytokines and interferons (KLINMAN et al. 1996; YAMAMOTO et al. 1994).

Studies in cynomolgus monkeys with the phosphorothioate oligodeoxynucleotide ISIS 2302 (GCC CAA GCT GGC ATC CGT CA), at doses up to 50 mg/kg every other day for 28 days, demonstrated that there was little or no evidence for immune stimulation in the monkey. However, the same sequence in mice produced generalized immunostimulation at doses of 20 mg/kg and above (HENRY et al. 1997a). We have examined the pharmacokinetics of this compound and other phosphorothioate oligodeoxynucleotides in rodents and primates, and cannot identify any differences in plasma kinetics or tissue distribution that would explain the species differences in sensitivity to immunostimulation (HENRY et al. 1997b). Therefore, we conclude that rodents are markedly more sensitive to phosphorothioate oligodeoxynucleotide-induced immunostimulation than primates, and we suggest that the immunostimulatory effects may not be a prominent adverse effect in humans. In clinical trials with the ISIS 2302, there have been no indications of immunostimulation at doses of 2 mg/kg infused over 2 h every other day for four doses (GLOVER et al. 1997).

Other toxicologic effects were also noted in rodents treated with phosphorothioate oligodeoxynucleotides. Treatment of rodents with repeated doses of 100 mg/kg or higher of the antisense to *rel a* (see above for sequence) induced significant histopathologic changes in the livers and kidneys. Doses of 100 mg/kg and 150 mg/kg administered three times a week for 2 weeks produced renal proximal tubular degeneration (SARMIENTO et al. 1994). The extent of tubular degeneration in this study was not sufficient to produce changes in blood urea nitrogen or creatine; however, longer duration of treatment or higher doses have induced both histologic and serum-chemistry changes. Note that the doses required to produce such toxicities far exceed doses intended for use in clinical studies that are in the range of 1–2 mg/kg.

In the livers from mice treated with 100–150 mg/kg of a phosphorothioate oligodeoxynucleotide, there was multifocal hepatocellular degeneration, or necrosis, accompanied by hypertrophy of Kupffer cells. This hepatic change was accompanied by serum-chemistry changes consistent with some hepatic dysfunction (SARMIENTO et al. 1994). Tissue-distribution studies have shown that liver and kidney are sites of deposition of phosphorothioate oligodeoxynucleotide and both Kupffer cells, and proximal tubular cells contained basophilic granules (SARMIENTO et al. 1994). These basophilic granules are thought to be composed of oligodeoxynucleotides or material derived from oligodeoxynucleotide, as demonstrated by autoradiographic studies as discussed above (SANDS et al. 1994; PLENAT et al. 1994; SAWAI et al. 1995; YAMAMOTO et al. 1994; OBERBAUER et al. 1995). Changes in the bone marrow were also observed after 2 weeks of treatment with 100–150 mg/kg of the *rel a* antisense. There was a reduction in the number of megakaryocytes accompanied by approximately 50% reduction in circulating platelets (SARMIENTO et al. 1994).

Treatment of primates with phosphorothioate oligodeoxynucleotides produces a pattern of toxicities quite distinct from those described in rodents. In primates, the dose-limiting toxicities reported in the literature are an inhibition of the clotting cascade and the activation of the complement cascade (HENRY et al. 1994; CORNISH et al. 1993; GALBRAITH et al. 1994). Both of these toxicities are thought be related to the polyanionic nature of the molecules and the binding of these compounds to protein factors in plasma. Phosphorothioate oligodeoxynucleotides inhibit the clotting cascade, as indicated by a concentration-dependent prolongation of APTT. The prolongation of APTT is linearly related to plasma concentration, is readily reversed and directly parallels the plasma concentrations. The inhibition of the clotting cascade can be reproduced in vitro and phosphorothioate oligodeoxynucleotides have been reported to produce this inhibition, independent of the nucleotide sequence (HENRY, unpublished observations). At a dose of 2 mg/kg, infused over 2 h, there is an approximate 50% increase in APTT in normal human volunteers (GLOVER et al. 1997). The mechanism of this inhibition, like that of the immunostimulation in rodents, may be related to the polyanionic nature of the compounds. It is well known that the polyanions are inhibitors of clotting, and phosphorothioate oligodeoxynucleotides may be acting through similar mechanisms. If phosphorothioate oligodeoxynucleotides inhibit the clotting cascade as a result of their polyanionic properties, then binding and inhibition of thrombin would be a likely mechanism of action, similar to heparin's mode of action (ROSENBERGER 1989). In fact, there is a novel oligodeoxynucleotide-based anticoagulant in clinical trials that exploits this property (BOCK et al.

1992). The small degree of inhibition of the coagulation cascade and the transient nature of the effect suggests that this effect is not clinically significant (GLOVER et al. 1997). Clinical trials with as many as 11 different oligodeoxynucleotides, using doses up to 2 mg/kg continuously, with no-dose limiting effects on coagulation, have been reported to date.

The binding of phosphorothioate oligodeoxynucleotides to another class of plasma proteins may be responsible for the activation of the complement cascade. Rapid infusions of phosphorothioate oligodeoxynucleotides cause the activation of the complement cascade and the release of complement split products C3a and C5a with physiologic consequences (CORNISH et al. 1993; GALBRAITH et al. 1994; HENRY et al. 1996). The result of this activation and the release of anaphylatoxins in some monkeys treated with high doses over short infusion times results in marked hematologic effects and marked hemodynamic changes. Hematologic changes are characterized by transient reductions in neutrophils, presumably due to margination, followed by neutrophilia as neutrophils are recruited from immature populations by the release of chemotactic factors. Over the same time course in some monkeys, the activation of the complement cascade produces marked reductions in heart rate, blood pressure and, subsequently, cardiac output. This cardiovascular collapse can be lethal in some animals (CORNISH et al. 1993; GALBRAITH et al. 1994). We have established clear relationships between plasma concentrations of oligodeoxynucleotide and the activation of the complement cascade, such that there are well-defined thresholds of oligodeoxynucleotide concentrations that must be exceeded prior to the activation of the cascade. Complement activation in monkeys is common to all phosphorothioate oligodeoxynucleotides studied to date, and similar threshold concentrations have been identified with a series of phosphorothioate oligodeoxynucleotides (HENRY et al. 1997a). The levels at which we observe complement activation by the alternative pathway are approximately four- to sevenfold greater than plasma levels observed in normal volunteers treated with a 2-h infusion of 2 mg/kg of ISIS 2302 (GLOVER et al. 1997).

The toxicities of phosphorothioate oligodeoxynucleotides are being characterized in a number of laboratories. Although there is further characterization to be done, it is apparent that there are well-defined dose–response relationships for all the toxicities. The pattern of toxicity differs little from sequence to sequence, with some slight differences in potency with some sequences (MCINTYRE et al. 1993). An acceptable safety margin exists between toxicities and targeted clinical doses (Figs. 1 and 2). This consistency in toxicity profiles, combined with the predictability in the pharmacokinetics from animals to man, provides a framework within which we can design safe clinical trials and therapeutic dose regimens. Further work is necessary to understand the mechanisms of the toxicologic effects of phosphorothioate oligodeoxynucleotides and it will be important for workers in the field to begin to understand the relationships between tissue concentrations of oligodeoxynucleotides and toxicity. An understanding of the relationships between organ concentrations and toxicity will allow clinicians to design safe and effective dose regimens for chronic administration of this class of therapeutic agents.

D. Chemistry, Manufacture and Control of Phosphorothioate Oligodeoxynucleotide Drugs

Currently, phosphorothioate-oligodeoxynucleotide drugs are synthesized for clinical use by sequential coupling of activated 3′-phosphoramidite nucleoside monomers to a starting nucleoside that is covalently bound to

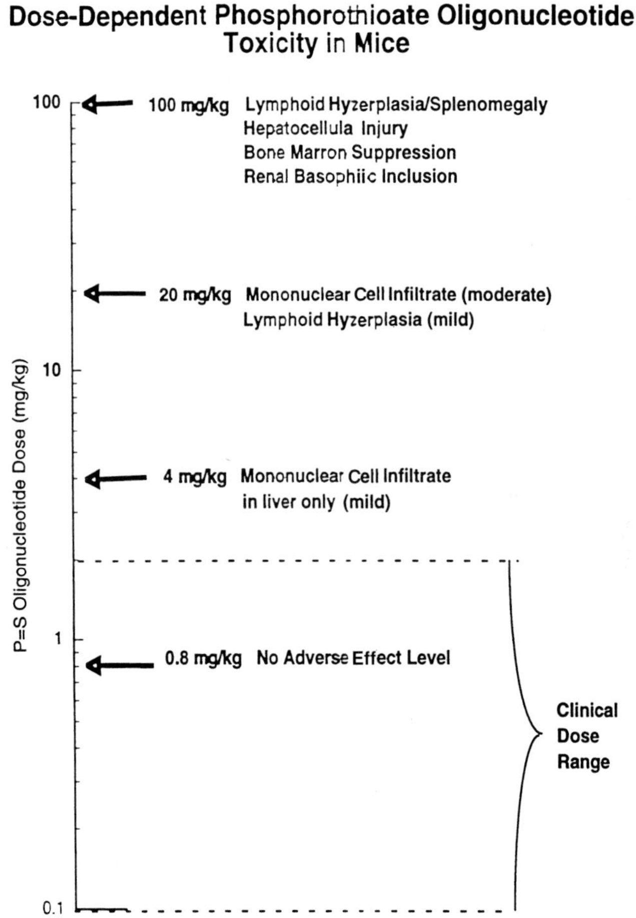

Fig. 1. Dose-response relationship in mice following repeated intravenous injection of a phosphorothicate oligonucleotide every other day for 28 days. Similar dose reponse is observed for other phosphorothioate oligonucleotides although the exact results depend on the dose regimen and route of administration

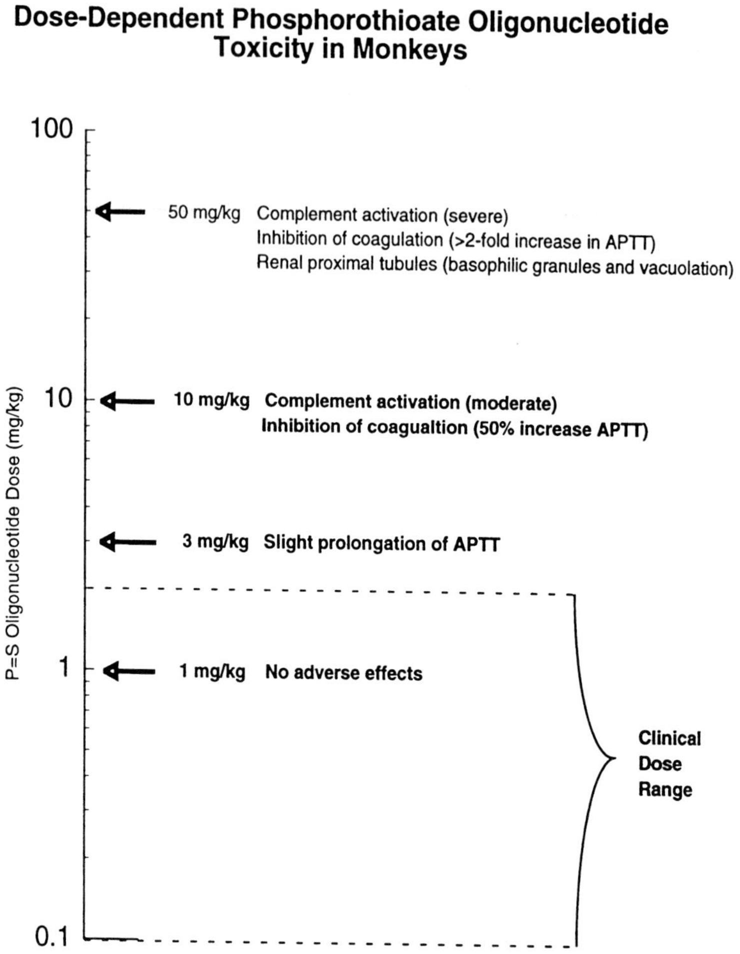

Fig. 2. Dose-response relationship in monkeys following repeated intravenous injection of phosphorothioate oligonucleotide every other day for 28 days. Similar dose response is observed for other phosphorothioate oligonucleotides although the exact results depend on the dose regimen

solid support. These unique trivalent phosphorous monomers are highly reactive and, thus, preferable to less reactive pentavalent phosphorous monomers. Successful sequential synthesis of an oligomeric molecule of defined sequence demands that all coupling steps proceed in very high yield, near unity. To bring antisense phosphorothioate oligodeoxynucleotides into human clinical use, it has been necessary to develop chemistry that allows monomers to couple with stepwise yields of 98.5% and higher. At these extremely high stepwise yields, the manufacture of phosphorothioate DNA oligomers by sequential solid phase-supported synthesis becomes practical and effective (RAVIKUMAR et al. 1995). In this section, we highlight our experience in optimizing the synthesis, purification and quality control of phosphorothioate oligodeoxynucleotides. We also discuss the biological and regulatory implications of the presence of related oligonucleotides that is unique to the development of synthetic phosphorothioate oligodeoxynucleotide therapeutics.

I. Synthesis of Phosphorothioate Oligodeoxynucleotides

1. Chemistry of Elongation

Strategies for sequential organic synthesis of any oligomeric molecule are dictated by synthon-coupling efficiency. In the case of phosphorothioate oligomers, the synthesis starting material is a 5′-protected nucleoside covalently bound to solid support via its 3′-hydroxyl group. Each synthesis elongation cycle for the addition of one monomer to the growing chain begins with removal of a 5′-OH protecting group. Commonly, the 5′-OH protecting group is a 4, 4′-dimethoxytrityl (DMT) group. The DMT group is removed by mild acid treatment, liberating the DMT cation as the counter ion of the acid anion. A dry solvent wash is used to remove DMT salts and excess acid, with efficiency of DMT removal at each step being key to overall synthesis yield. Small amounts of DMT cation may be converted to DMT alcohol by reaction with trace water present in the reaction mixture. The free 5′-hydroxyl groups are next allowed to react with activated 3′-phosphoramidite nucleoside monomer. Following this rapid reaction, unreacted monomer is washed from the support with dry solvent.

A number of unproductive events may occur during the course of an elongation cycle. First, some chains bearing free 5′ hydroxyls may fail to elongate. Chains that have not elongated are subjected to a 5′-hydroxyl "capping" reaction. Capping is typically done by acetylation with acetic anhydride with the aid of an activating nitrogenous base. If the capping reaction proceeds in quantitative yield and the resulting ester is stable throughout the full synthesis, these capped elongation failures (sometimes referred to as "deletion sequences", "shortmers" or "deletionmers") will appear as impurities with lengths of $n-1$, $n-2$, $n-3$, etc., in the crude final product (n-mer); their length depends on how early a failure to elongate occurred in a given shortmer. In practice, reverse-phase HPLC purification of crude trityl-on product limits deletion-sequence impurities in final product to $n-1$ and $n-2$. This class of impurities is sometimes referred to as "terminal deletion sequences". The $n-1$ and $n-2$ terminal deletion sequences have a slightly lower affinity for target mRNA than the full-length sequence and would be expected to display pharmacological activity marginally lower than the full-length drug. The presence of low levels of these terminal deletion sequences is, therefore, relatively inconsequential from the standpoint of antisense biological activity and safety.

It is possible that some chains that fail to elongate, as for steric reasons, will also avoid capping. Some of these chains may then resume reacting with incoming phosphoramidite monomer in subsequent elongation cycles. These, too, will exhibit lengths of n–1, n–2, etc., as final-product impurities, but their nucleoside ("base") deletions will be internal to the chain, as opposed to being at the 5′ end. These related compounds are sometimes referred to as "internal deletion sequences". In addition, some chains may not remain detritylated and, thus, will fail to elongate. The trityl cation is stable in dry solution, as evidenced by its persistent bright-orange color in the reaction mixture. These reactive cations may absorb to the anionic solid support and/or its organic bonded phase of growing oligomer chains. After the wash step is completed, the absorbed DMT cations may desorb during or prior to addition of the next monomer solution. Free DMT cations can then rapidly retritylate 5'-hydroxyl groups and, in the absence of acid, the equilibrium will lie far to the side of the DMT-ether. The retritylated oligomers will then fail to elongate. During subsequent elongation cycles, these chains will have a high probability of being successfully deprotected and elongated. This mechanism, thus, constitutes an additional source of "internal deletion sequences".

From a biological-activity standpoint, these internal deletion sequences, which are of different base sequence than the intended oligoncleotide sequence, would not be expected to bind well to target mRNA and would, therefore, be inactive. They may, however, bind nonspecifically to plasma proteins, resulting in a safety profile similar to that of the oligodeoxynucleotide drug.

In the event that there is a mRNA target complementary to an internal deletion-sequence impurity, hybridization leading to an unintended biological response is possible. However, at the low level of internal deletion sequences present in the purified product, significant pharmacologic effect from antisense action is unlikely. Therefore, the final product may be rationally controlled by measuring full-length content, usually by PAGE or CGE, and setting a control specification for total deletion sequences.

2. Chemistry of Sulfurization

In the synthesis of phosphorothioate DNA oligomers, the sulfur atom at each phosphate linkage is introduced during the elongation cycle. After the coupling reaction, the intermediate linkage is in (trivalent) phosphite triester form. The phosphite is then oxidatively sulfurized to the (pentavalent) phosphorothioate. Yield of this sulfurization reaction is high, in the order of 99.7–99.8%. The remaining 0.2–0.3% comprises phosphites that have reacted with trace water or, in some cases, sulfurization-reaction byproducts, to form phosphodiester linkages identical to those in natural DNA.

For a 20-mer containing 19 internucleotide linkages, the total incidence of full-length "monophosphodiester" oligomers is about 19 times 0.2–0.3%, or about 4–6% of full-length chains. A much smaller fraction óf oligomers contain more than one phosphodiester linkage (higher order partial phosphodiesters). Monophosphodiester oligomers are stable to attack by endonucleases, as endonucleolytic cleavage does not proceed at a significant rates unless a phosphorothioate oligomer contains at least three adjacent phosphodiester linkages (personal communication). If a phosphodiester linkage is at the 3′- or 5′ terminus of an oligomer, one nucleotide may be cleaved from this minor (perhaps 0.5%) impurity, but will produce a stable phosphorothioate oligomer with nearly the same affinity for mRNA target as the parent drug.

The stability of lower order (mono-, di-, tri-) partial phosphodiester sequences is essentially equal to that of the parent drug, and their affinity for the mRNA target is slightly enhanced relative to the parent drug. The partial phosphodiester sequences are, therefore, not impurities at all in the classical sense, but are actually related compounds with similar activity and safety profile as the fully thioated drug. Nonetheless, phosphorous (^{31}P) NMR spectroscopy or weak anion-exchange HPLC is used to determine and control the levels of these related compounds in phosphorothioate drug substances.

3. *O,O*-Linked Phosphorothioate DNA Diastereoisomerism

O,O-linked, phosphorothioate diester internucleotide linkages are chiral; thus, a 20-mer phosphorothioate DNA oligomer could, in principle, contain 524,288 diastereomers. Clearly, it is beyond the selectivity and peak capacities of any existing or anticipated separation technologies to speciate this number of diastereomers in a synthetic bulk drug substance.

It is likewise difficult to conceive of separation methods for isolation of a single pure diastereomer from a reaction mixture containing 524,288 virtually identical compounds. Stereocontrolled synthesis of phosphorothioate DNA oligodeoxynucleotides has been attempted with the highest enantiomeric excesses yet reported (Suska et al. 1993). The net stereoselectivity achieved is low enough, however, that the best short oligomers are only about 60% stereochemically pure. A 20-mer made via this stereocontrolled chemistry would, thus, contain all the 524,288 diastereomers contained in material made by non-stereocontrolled chemistry. In the attempt to make a stereo-pure drug, qualitatively, the same mixture is produced as if no stereo control had been attempted and, at the same time, one cannot know or determine the diastereomeric composition of the product.

Since it is not presently possible to synthesize *O,O*-linked phosphorothioate oligodeoxynucleotides with 100% enantiomeric excess at each linkage, the key issue is whether or not manufacture of the drug substance is under process control and reproducibly delivers drug substance with essentially the same diastereomeric composition, to provide assurance that drug studied in animal toxicity studies has the same diastereomeric composition as drug used in human clinical trials and later manufactured for market. It has been demonstrated that solid-phase synthesis of phosphorothioate oligodeoxynucleotides by phosphoramidite coupling is under inherent process control, such that the enantiomer ratio at each linkage, irrespective of flanking base identity, sequence, and chain position, does not significantly vary from 1:1 (Wyrzykiewicz and Cole 1995). The products of such syntheses, thus, have essentially the same diastereomeric composition, lot to lot. In an incompletely stereocontrolled synthesis, by contrast, individual diastereomers might fall to low levels in a given lot, but cannot be detected using current analytical technology. If a given diastereomer were present at 100-fold higher, but still low level in a later lot, this large percentage change would go undetected, i.e., it is possible to have large variations in the diastereomeric ratios of the final drug substance that are undetected and uncontrolled. From a regulatory and patient-safety point of view, it is vastly preferable to have a reproducible and controllable non-stereoselective process, rather than an inherently uncontrollable stereoselective process.

It is also important to note that it is not presently possible to know which of a half million diastereomeric phosphorothioate 20-mers of identical sequence would have (marginally) the highest affinity for target mRNA and, presumably, the greatest drug activity. It is not possible even to reliably predict whether the all-Sp or the all-Rp compound will be most active, for example, having made and tested such compounds against mixed R/S compounds. Without the ability to predict which single diastereomer might be (marginally) the best antisense drug, there has been no reason to accept the lack of process control inherent in producing a diastereomerically enriched, but not diastereomerically pure product. Finally, the cost of stereocontrolled synthetic oligomers would be high because of the lower process yields involved. For all these technical, safety-related and regulatory reasons, non-stereoselective chemistry is utilized for the manufacture of phosphorothioate oligodeoxynucleotide antisense drugs.

II. Purification of Phosphorothioate Oligodeoxynucleotides

In current solid-phase supported phosphorothioate DNA manufacture, reverse-phase preparative HPLC plays a key role; 5′-DMT-protected full-length oligomer and internal deletionmers are readily separated from less lipophilic capped or free hydroxy-terminal deletionmers, up to and including *n*–1 length terminal deletionmers, which arise from failure to couple in the final elongation step. Reverse-phase HPLC is also highly effective for removing 5′-DMT internal deletionmers of length *n*–2 and shorter from full-length DMT-on product. Preparative reverse-phase HPLC purification typically takes a crude product of about 73% full-length assay to 90% full-length assay in a single step. The method does not, however, achieve complete selectivity for full-length product and *n*–1 internal trityl-on deletionmers. This chromatographic selectivity problem has not been completely solved by either reverse-phase HPLC of 5′-protected intermediates or by anion-exchange HPLC of a fully deprotected product. From a manufacturing standpoint, the key issue is analytical control. Using length-selective analytical methodologies, such as PAGE and CGE, as in-process control checks, it has been possible to ensure that the impurity profiles of manufactured bulk drug-substance lots are constant lot-to-lot in terms of total deletionmer content.

III. Quality Control of Phosphorothioate Oligodeoxynucleotides

Oligodeoxynucleotides intended for use in clinical studies require thorough characterization of their overall quality (Kambhampati et al. 1993). Identity of the oligodeoxynucleotide drug substance is commonly established by a combination of techniques including mass spectrometry for confirmation of molecular weight, ^{31}P NMR spectroscopy for confirmation of backbone phosphorothioate structure, and elemental analysis. The presence of a characteristic ultraviolet (UV) maximum in the vicinity of 260 nm may be used to confirm the presence of a DNA strand. CGE migration time and/or anion-exchange chromatography retention time, when recorded against a reference standard (with a previously established sequence authenticity), may serve as additional confirmation of identity. Other indirect approaches for establishing sequence authenticity include base composition analysis and DNA duplex melting temperature.

Unequivocal identity of the oligodeoxynucleotide drug may only be established by sequencing the sample. Due to their inherent stability to nucleases, the conventional enzyme-based procedures for direct digestion of oligodeoxynucleotides to the nucleobases are not practical for phosphorothioate oligodeoxynucleotides. Chemical conversion of the phosphorothioate to the corresponding phosphodiester prior to enzymatic degradation has been successfully applied to determine both the base composition and sequence (Schuette et al. 1994; Wyrzykiewicz and Cole 1994). More recently, this approach was utilized in establishing the sequence of phosphorothioate oligodeoxynucleotides through application of matrix-assisted laser desorption ionization time of flight (MALDI-TOF) mass spectrometry (Schuette et al. 1995).

The purity of the phosphorothioate drug substance is commonly established by a combination of impurity profile (area–%) and assay (% by weight). The primary objectives of the impurity profile analysis are to separate compounds of closely related structure, as well as any adventitious impurities from the analyte of interest, and to have adequate sensitivity to accurately quantitate these structurally related impurities at low levels. CGE, by virtue of its length-based mechanism of separation, resolves the failure sequences in synthetic oligodeoxynucleotides, but is insensitive to changes in the phosphate backbone, resulting in comigration of the full-length partial (or complete) phosphodiester compounds with the fully thioated sequence. Anion-exchange HPLC has been shown to be selective for differences in the phosphate backbone and has proven to be a useful complementary technique for establishing a comprehensive impurity profile of phosphorothioate oligodeoxynucleotides (COHEN et al. 1993; BERGOT and EGAN 1992; AGRAWAL et al. 1990;WARREN and VELLA 1993; SRIVATSA et al. 1997). The %-w/w assay is performed either by HPLC or electrophoresis against a reference standard of pre-established purity (SRIVATSA et al. 1994; CARLSON et al. 1994). This serves as a final confirmation of mass balance for the drug substance and can demonstrate that the drug substance is free of significant levels of adventitious impurities. Finally, each drug-substance sample is tested by a variety of analytical techniques to assure that it is free of other potential impurities, such as residual organic volatile compounds and heavy metals.

Prior to use in preclinical or clinical applications, it is necessary to assure the microbiological quality of a drug substance. Currently, the limulus amebocyte lysate (LAL) test is used to detect the presence of endotoxins in pharmaceutical preparations. The test is based on an endotoxin-induced clotting of the LAL. Phosphorothioate oligodeoxynucleotides inhibit gelling of the LAL, rendering the detection of endotoxin present in the sample difficult. Dilutions necessary to overcome this inhibition effectively raise the endotoxin limit of detection and reduce accuracy of the endotoxin assay. At the present time, the detection limits are sufficient to support 3 mg/kg i.v. administration of phosphorothioate oligodeoxynucleotides over a 2-h infusion period. Attempts to overcome this inhibition problem include precipitation of the phosphorothioate oligodeoxynucleotide from the sample by utilization of divalent cations prior to LAL analysis. This approach has been shown to lower the endotoxin limit of detection by approximately 1000-fold by significantly reducing the inhibition of the gel formation by the oligodeoxynucleotide matrix (FELDMAN et al. 1995).

In summary, utilization of the most efficient chemistries for synthesis and purification, coupled with state-of-the-art quality-control programs, help assure the high quality of drug substance used in establishing the safety and efficacy of antisense oligodeoxynucleotides.

E. Formulation and Drug Delivery of Oligodeoxynucleotides

I. Physical–Chemical Properties

Due to the presence of a mixture of diastereisomers, phosphorothioate oligodeoxynucleotides are amorphous solids possessing the expected physical properties of hygroscopicity, low-bulk density, electrostatic charge pick up and poorly defined melting point prior to decomposition. Their good chemical stability allows storage in the form of a lyophilized powder, spray-dried powder or a concentrated, sterile solution; more than 3 years of storage is possible at refrigerated temperatures.

Due to their polyanionic nature, phosphorothioate oligodeoxynucleotides are infinitely soluble in neutral and basic conditions. Drug-product concentrations are limited (in select applications) only by an increase in solution viscosity. The apparent solubility is also influenced by the counterion composition, ionic strength and pH. Phosphorothioate oligodeoxynucleotides have an apparent pKa in the vicinity of 2 and will come out of solution in acidic environments, i.e., the stomach. This precipitation is readily reversible with increasing pH or by acid-mediated hydrolysis (unpublished results).

Instability of phosphorothioates has been primarily attributed two degradation mechanisms: oxidation and acid-catalyzed hydrolysis. Oxidation of the P=S bond in the backbone has been observed at elevated temperatures and under intense UV light, leading to partial phosphodiesters (still pharmacologically active) and are readily monitored by anion-exchange HPLC. Under acidic conditions, hydrolysis reactions followed by chain-shortening depurination reactions have been documented by length-sensitive electrophoretic techniques (unpublished results).

Phosphorothioate formulations in simple buffer solutions show remarkable chemical stability which allows for room-temperature stable products with a shelf life of at least 2 years. While the predominant mechanism of degradation is oxidation, it is difficult to detect a statistically significant impact upon the low rate of oxidation by the addition of antioxidants or headspace oxygen control.

II. Formulation

In animal studies, there is overwhelming evidence of biological activity of oligodeoxynucleotides upon i.v. injections of simple solutions. Thus, it has been relatively straightforward to formulate the first-generation drug products in support of early clinical trials. Simple, buffered solutions have been successfully used in clinical studies by i.v., intradermal, subcutaneous and intravitreal injections. These formulations have, thus far, been non-preserved and are amenable to terminal sterilization. Early data are supportive of oral, transdermal and pulmonary routes of delivery using these simple solutions (Fuhrman et al. 1995).

As antisense technology is broadly applicable to any clinical indication involving downregulation of protein expression, it has the potential to access clinical indications that require chronic, systemic dosing. In addition to the obvious demands of high bioavailability, low toxicity etc., these chronic, systemic applications place high demands on pharmaceutical elegance and ease of use. It has been recently reported that tritium-labeled ISIS-3082, a novel 20-mer phoshorothioate that targets murine ICAM-1 mRNA, can be incorporated into a variety of lipid and aqueous formulations. All formulations are capable of moving approximately 6% of the total reactivity to the receptor compartment of a Franz diffusion cell across hairless mouse skin, when driven from an occluded donor compartment. However, cardiolipin phosphatidylcholine cholesterol liposomes increased the dermal retention of radioactivity by 6.5-fold. Subsequently, CGE analysis showed the radioactivity to be correlated with intact, i.e., full-length, oligodeoxynucelotide (OCHELTREE et al. 1996).

As polyanionic species inherently possess a high degree of water solubility, the phosphorothioate oligodeoxynucleotides present obvious opportunities for traditional sustained-release strategies. Complexation and/or non-covalent interactions with other ionic or charge-inducable species have been reported in the literature, including divalent counterions, polycations, polysaccarides, proteins, polyamines and dendrimers. In general, these interactions produce complexes with reduced solubilities, which may be manipulated with respect to dosage-form parameters predicted by solubility (equilibria) and dissolution (kinetic) theories. Therefore, these formulations generally must account for the injection volume, dose, vehicle-buffering effect, counter-ion excess and effective surface area for dissolution. The usefulness of other strategies that rely upon masking the depot from the biological environment (microparticulate systems, implants) will be determined as effective doses are defined in humans.

III. Drug Delivery: Targeting, Uptake and Release

As pointed out earlier in this review, targeting among cell populations is often unnecessary since specificity is inherent. Other considerations, such as dose conservation, non-specific effects and toxicity (previously reviewed) may provide some justification for targeting of phosphorothioate oligonucelotides. Interesting progress has been reported regarding the passive targeting achieved using liposome-encapsulated therapeutics. Accumulation at sites of infection, inflammation and tumor growth have been attributed to increased circulation times of these materials and the leaky vasculatures associated with these processes (LASIC and NEEDHAM 1995; BOMAN et al. 1995). One caution regarding these observations is worth noting. Since the mononuclear phagocyte system (MPS) is largely responsible for clearing these materials from circulation, misleading data regarding circulation time may be obtained in species with less-evolved systems, i.e., rodents.

Until recently, there existed the academic debate of whether oligonucleotides would reach their sites of action in intact cells. This question has stimulated a great deal of research interest in the areas of targeting, cellular uptake and release. It is now an accepted technique to use cationic lipids for cell transfections in in vitro screening assays (FELGNER et al. 1995). Use of these systems in vivo have been limited to date by instability and toxicity problems. However, a system with increased in vitro stability in the presence of serum has been reported (LEWIS et al. 1996). Similarly, the use of the cationic polymer, polyethylenimine, for increasing cellular transfection in culture and in newborn mice has been recently reported (BOUSSIF et al. 1995).

Looking past the question of uptake, a novel approach to releasing endosomal contents into the cytoplasm after uptake has been recently reported (LEE et al. 1996). A 58-kDa protein isolated from Listeria monocytogenes was incorporated into pH-sensitive liposomes. The intracellular fate of this system was monitored with a pH-sensitive fluorescent dye. It could be determined that as soon as the endosome began to acidify, the liposome/endosome contents were released into the cytosol. As with the other delivery systems mentioned above, the eventual usefulness (incremental improvements) of a particular approach will be determined in the near future as we further define the mechanisms and governing restrictions for the inter- and intracellular trafficking of oligonucleotides.

F. Summary

In this chapter, we have summarized the fundamental pharmacology, pharmacokinetics, toxicology, manufacture, quality control and formulation issues that apply to the development of phosphorothioate oligodeoxynucleotides as therapeutic agents. Our experience with this first generation of antisense therapeutics demonstrates that these compounds have marked pharmacologic activities, an attractive pharmacokinetic profile and clearly defined dose-related toxicities that provide ample safety margins for the intended therapeutic dose ranges. In addition, many of the early doubts regarding the ability to manufacture and assay oligodeoxynucleotides have been successfully addressed. With the economies of scale currently achievable, the manufacturing costs of these first-generation oligodeoxynucleotides have been dramatically lowered, making the dream of practical oligodeoxynucleotide therapeutics more of a reality. Based on our preclinical and clinical experience to date, we can confidently rephrase the commonly asked question from "Does antisense technology work?" to "How successfully can antisense drugs be marketed?"

References

Agrawal S (1992) Antisense oligonucleotides as antiviral agents. Trends Biotechnol: 152–158

Agrawal S, Goodchild J, Civeira MP, Thornton AH, Sarin PS, Zamecnik PC (1988b) Oligodeoxynucleoside phosphoramidates and phosphorothioates as inhibitors of human immunodeficiency virus. Proc Natl Acad Sci USA 85:7079–7083

Agrawal S, Ikeuchi T, Sun D, Sarin PS, Konopka A, Maizel J, Zamecnik PC (1989) Inhibition of human immunodeficiency virus in early infected and chronically infected cells by antisense oligodeoxynucleotides and their phosphorothioate analogues. Proc Natl Acad Sci USA 86:7790–7794

Agrawal S, Tang JT, Brown DM (1990) Analytical study of phosphorothioate analogues of oligodeoxynucleotides using high-performance liquid chromotography. J Chromatogr:396–399

Agrawal S, Temsamani J, Galbraith W, Tang J (1995b) Pharmacoketics of antisense oligonucleotides. Clin Pharmacokinet 28:7–16

Agrawal S, Temsamani J, Tang JY (1991) Pharmacokinetics, biodistribution, and stability of oligodeoxynucleotide phosphorothioates in mice. Proc Natl Acad Sci USA 88:7595–7599

Agrawal S, Zhang X, Lu Z, Zhao H, Tamburin JM, Yan J, Cai H, Diasio RG, Habus I, Jiang Z, Iyer RP, Yu D, Zhang R (1995a) Absorption, tissue distribution and in vivo stability in rats of a hybrid antisense oligonucleotide following oral administration. Biochem Pharmacol 50:571–576

Agrawal S, Temsamani J, Galbraith W, Tang J (1995b) Pharmacokinetics of antisense oligonucleotides. Clin Pharmacokinet 28:7–16

Anderson KP, Driver VB, Fox MC, Martin M, Azad RF (1996) Inhibition of cytomegalovirus immediate early gene expression by an antisense oligonucleotide complementary to immediate early RNA. Antimicrob Agents Chemother

Azad RF, Driver VB, Tanaka K, Crooke RM, Anderson KP (1993) Antiviral activity of a phosphorothioate oligonucleotide complementary to RNA of the human cytomegalovirus major immediate-early region. Antimicrob Agents Chemother 37:1945–1954

Azad RF, Driver VB, Buckheit R Jr, Anderson KP (1995) Antiviral activity of a phosphorothioate oligonucleotide complementary to human cytomegalovirus RNA when used in combination with antiviral nucleoside analogs. Antiviral Res 28:101–111

Bennett CF, Chiang M-Y, Chan H, Shoemaker JEE, Mirabelli CK (1992) Cationic lipids enhance cellular uptake and activity of phosphorothioate antisense oligonucleotides. Mol Pharmacol 41:1023–1033

Bennett CF, Chiang M-Y, Wilson-Lingardo L, Wyatt JR (1994) Sequence specific inhibition of human type II phospholipase A2 enzyme activity by phosphorothioate oligonucleotides. Nucleic Acids Res 22:3202–3209

Bennett CF, Condon T, Grimm S, Chan H, Chiang MY (1995) Inhibition of endothelial cell-leukocyte adhesion molecule expression with antisense oligonucleotides. J Immunol 152:3530–3540

Bennett CF, Kornbrust DJ, Henry SP, Stecker K, Howard R, Cooper S, Dutson S, Hall W, Jacoby HI (1997) An ICAM-1 antisense oligonucleotide prevents and reverses dextran sulfate sodium-induced colitis in mice. J Pharmacol Exp Ther 280:988–1000

Bennett RM, Gabor GT, Merritt MM (1985) DNA binding to human leukocytes. Am Soc Clin Invest 76:2182–2190

Bergot BJ, Egan W (1992) Separation of synthetic phosphorothioate oligodeoxynucleotides from their oxygenated (phosphodiester) defect species by strong-anion-exchange high-performance liquid chromatography. J Chromatogr:35–42

Bock LC, Griffin LC, Latham JA, Vermaas EH, Toole JJ (1992) Selection of single stranded DNA molecules that bind and inhibit human thrombin. Nature 355:564–566

Boman NL, Bally M.B, Cullis PR, Mayer LD, Webb MS (1995) Encapsulation of vincristine in liposomes reduces its toxicity and improves its anti-tumor efficacy. J Lipid Res 5:523–541

Boussif O, Lezoualc'h F, Zanta MA, Mergny MD, Scherman D, Demeneix B, Behr J-P (1995) A versatile vector for gene and oligonucleotide transfer into cells in culture and in vivo: Polyethylenimine. Proc Natl Acad Sci USA 92:7297–7301

Branda RF, Moore AL, Mathews L, McCormack JJ, Zon G (1993) Immune stimulation by an antisense oligomer complementary to the rev gene of HIV-1. Biochem Pharmacol 45:2037–2043

Carlson RH, Bozick AH, Fitchett JR (1994) Comparison of the accuracy and precision between capillary gel electrophoresis versus polyacrylamide gel electrophoresis for assay and impurity determinations of pharmsceutical polynucleotide samples. In Sixth International Symposium on Capillary Electrophoresis, San Diego, CA, USA

Chiang MY, Chan H, Zounes MA, Freier SM, Lima WF, Bennett CF (1991) Antisense oligonucleotides inhibit intercellular adhesion molecular 1 expression by two distinct mechanisms. J Biol Chem 266:18162–18171

Chiasson BJ, Hooper ML, Murphy PR, Robertson HA (1992) Antisense oligonucleotide eliminates in vivo expression of c-fos in mammalian brian. Eur J Pharmacol 227:451–453

Cohen AS, Vilenchik M, Dudley JL, Gemborys MW, Bourque AJ (1993) High-performance liquid chromatography and capillary gel electrophoresis as applied to antisense DNA. J Chromatogr:293–301

Cornish KG, Iversen P, Smith L, Arneson M, Bayever E (1993) Cardiovascular effects of a phosporothioate oligonucleotide with sequence antisense to p53 in the conscious rhesus monkey. Pharmacol Commun 3:239–247

Cossum PA, Sasmor H, Dellinger D, Troung L, Cummins L, Owens SR, Markham PM, Shea JP, Crooke S (1993) Disposition of the 14c-labeled phosphorothioate oligonucleotide ISIS 2105 after intravenous administration to rats. J Pharmacol Exp Ther 267:1181–1190

Cossum PA, Troung L, Owens SR, Markham PM, Shea JP, Crooke ST (1994) Pharmacokinetics of a 14c-labeled phosphorothioate oligonucleotide, ISIS 2105, after intradermal administration to rats. J Pharmacol Exp Ther 269:89–94

Crooke ST, Bennett CF (1996a) Progress in antisense oligonucleotide therapeutics. Annu Rev Pharmacol Toxicol 36:107–129

Crooke ST, Graham MJ, Zuckerman JE, Brooks D, Conklin BS, Cummins LL, Greig MJ, Guinosso CJ, Kornbrust D, Manoharan M, Sasmor HM, Schleich T, Tivel KL, Griffey RH (1996b) Pharmacokinetic properties of several novel oligonucleotide analogs in mice. J Pharmacol Exp Ther 277:923–937

Dean NM, McKay R (1994) Inhibition of PKC-alpha expression in mice after systemic administration of phosphorothioate antisense oligodeoxynucleotides. Proc Natl Acad Sci USA 91:11762–11766

Dean NM, McKay R, Condon TP, Bennett CF (1994) Inhibition of protein kinase C-alpha expression in human A549 cells by antisense oligonucleotides inhibits induction of intercellular adhesion molecule 1 (ICAM-1) mRNA by phorbol esters. J Biol Chem 269:16416–16424

Diamantstein T, Blistein-Willinger E (1978) Specific binding of poly (I) – poly (C) to the membrane of murine B lymphocyte subsets. Eur J Immunol 8:896–899

Diamantstein T, Wagner B, Beyse I, Odenwald MV, Schultz G (1971) Stimulation of humoral antibody formation by polyanions. I. The effect of polyacrylic acid on the primary immune response in mice immunized with sheep red blood cells. Eur J Immunol 1:335–340

Feldman MA, Parsons JM, Lee JA, Srivatsa GS (1995) The inhibition effect of phosphorothioate oligonucleotides on the LAL test is reduced through the $CaCl_2$ precipitation of the oligonucleotides. In Sixth International Symposium on Pharmaceutical and Biomedical Analysis, St. Louis, MO, USA

Felgner PL, Kumar R, Basava C, Border RC, Hwang-Felgner J-Y (1995) Cationic lipids for intracellular delivery of biologically active molecules. In United States Patent Office. USA

Fuhrman LC, Ocheltree TW, Godwin DA, Michniak BB, Bennett CF (1995) Evaluation of several liposomal formulations and preparation techniques for the dermal delivery of phosphorothioate antisense oligonucleotides in hairless mouse skin in vitro. In AAPS Annual Meeting. Miami Beach, FL, USA

Galbraith WM, Hobson WC, Giclas PC, Schechter PJ, Agrawal S (1994) Complement activation and hemodynamic changes following intravenous administration of phosphorothioate oligonucleotides in the monkey. Antisense Res Dev 4:201–206

Geary RS, Leeds JM, Fitchett J, Burckin T, Truong L, Spainhour C, Creek M, Levin AA (1997) Pharmacokinetics and metabolism in mice of a phosphorothioate oligonucleotide antisense inhibitor of C-raf-Kinase expression. Drug Metab Dispos 25:1212–1281

Glover JM, Leeds JM, Mant TGK, Amin D, Kisner DL, Zuckerman J, Levin AA, Shanahan WR (1997) Phase 1 safety and pharmacokinetic profile of an ICAM-1 antisense oligodeoxynucleotide (ISIS 2302). J Pharmacol Exp Ther 282:1173–1180

Henry SP, Larkin R, Novotny WF, Kornbrust DJ (1994) Effects of ISIS 2302, a phosphorothioate oligonucleotide, on in vitro and in vivo coagulation parameters. Pharm Res II:S-353

Henry SP, Leeds J, Giclas PC, Gillett NA, Pribble JP, Kornbrust DJ, Levin AA (1996) The toxicity of ISIS 3521, a phosphorothioate oligonucleotide, following intravenous (IV) and subcutaneous (SC) administration in cynomolgus monkeys. Toxicol 30:112

Henry SP, Monteith DK, Levin AA (1997a) Antisense oligonucleotide inhibitors for the treatment of cancer: (2). Toxicological properties of phosphorothioate oligodeoxynucleotides. Anticancer Drug Des 12:395–408

Henry SP, Taylor J, Midgley L, Levin AA, Kornbrust DJ (1997b) Evaluation of the toxicity of ISIS 2302 a phosphorothioate oligonucleotide, in a 4 week study in CS-1 mice. Antisense Nucleic Acid Drug Dev 7:473–481

Hooper ML, Chiasson BJ, Robertson HA (1994) Infusion into the brain of an antisense oligonucleotide to the immediate-early gene c-fos suppresses production of fos and produces a behavioral effect. Neuroscience:917–924

Iversen P (1991) In vivo studies with phosphorothioate oligonucleotides: pharmacokinetics prologue. Anticancer Drug Des 6:531–538

Kambhampati RVB, Chiu Y-Y, Chen CW, Blumenstein JL (1993) Regulatory concerns for the chemistry, manufacturing, and controls of oligonucleotides for use in clinical studies. Antisense Res Dev 3:405–410

Katz SM, Browne B, Pham T, Wang ME, Bennett CF, Stepkowski SM, Kahan BD (1995) Efficacy of ICAM-1 antisense oligonucleotide in pancreatic islet transplantation. Transplant Proc 27:3214

Klinman DM, AE-Kyung Y, Beaucage SL, Conover J, Krieg AM (1996) CpG motifs present in bacterial DNA rapidly induce lymphocytes to secrete interleukin 6, interleukin 12, and interferon γ. Proc Natl Acad Sci USA 93:2879–2883

Krieg AM, Yi A-K, Matson S, Waldschmidt TJ, Bishop GA, Teasdale R, Koretzky GA, Klinman DM (1995) CpG motifs in bacterial DNA trigger direct B-cell activation. Nature 374:546–549

Kumasaka T, Quinlan WM, Doyle NA, Condon TP, Sligh J, Takei F, Beaudet AL, Bennett CF, Doerschuk CM (1996) The role of ICAM-1 in endotoxin-induced pneumonia evaluated using ICAM-1 antisense oligonucleotides, anti-ICAM-1 monoclonal antibodies, and ICAM-1 mutant mice. J Clin Invest

Lasic DD, Needham D (1995) The "stealth" liposome: a prototypical biomaterial. Chem Rev 95:2601–2628

Lee K-D, Oh YK, Portnoy DA, Swanson JA (1996) Delivery of macromolecules into cytosol using liposomes containing hemolysin from listeria monocytogenes. J of Biol Chem 271:7249–7252

Leeds JM, Graham MJ, Truong L, Cummins LL (1996b) Quantitation of phosphorothioate oligonucleotides in human plasma. Anal Biochem 235:36–43

Leeds JM, Williams K, Scherrill S, Levin AA, Bistner S, Henry SP (1996a) Potential for retinal accumulation of a phosphorothioate oligonucleotide (ISIS 2922) after intravitreal injection in cynomolgus monkeys. In Society of Toxicology Annual Meeting. Anaheim, CA, USA

Lewis JG, Lin K-Y, Kothavale A, Flanagan WM, Matteucci MD (1996) A serum-resistant cytofectin for cellular delivery of antisense oligodeoxynucleotides and plasmid DNA. Proc Natl Acad Sci USA 93:3176–3181

Lisziewicz J, Sun D, Metelev V, Zamecnik P, Gallo RC, Agrawal S (1993) Long-term treatment of human immunodeficiency virus-infected cells with antisense oligonucleotide phosphorothioates. Proc Natl Acad Sci USA 90:3860–3864

McIntyre KW, Lombard-Gillooly K, Perez JR, Kunsch C, Sarmiento UM, Larigan JD, Landreth KT, Narayanan R (1993) A sense phosphorothioate oligonucleotide directed to the initiation codon of transcription factor nf-kb p65 causes sequence-specific immune stimulation. Antisense Res Dev 3:309–322

Monia BP, Johnston JF, Ecker DJ, Zounes MA, Lima WF, Freier SM (1992) Selective inhibition of mutant Ha-ras mRNA expression by antisense oligonucleotides. J Biol Chem 267:19954–19962

Nestle FO, Mitra RS, Bennett CF, Chan H, Nickoloff BJ (1994) Cationic lipid is not required for uptake and selective inhibitory activity of ICAM-1 phosphorothioate antisense oligonucleotides in keratinocytes. J Invest Dermatol 103:569–575

Oberbauer R, Schreiner GF, Meyer TW (1995) Renal uptake of an 18-mer phosphorothioate oligonucleotide. Kidney Int 48:1226–1232

Ocheltree TW, Fuhrman LC, Mehta R, Michniak BB, Shah JC (1996) Epidermal and dermal penetration of anionic and zwitterionic liposomally encapsulated antisense oligonucleotides into hairless mouse skin. In Amer Assoc of Pharm Scientists, ed. R. Mehta. Seattle, WA, USA

Plenat F, Klein-Monhoven N, Marie B, Vignaud J-M, Duprez A (1994) Cell and tissue distribution of synthetic oligonucleotides in healthy and tumor-bearing nude mice. Am J Pathol 147:124–135

Rappaport J, Hanss B, Kopp JB, Copeland TD, Bruggeman LA, Coffman TM, Klotman PE (1995) Transport of phosphorothioate oligonucleotides in kidney implications for molecular therapy. Kidney Int 47:1462–1469

Ravikumar VT, Andrade M, Wyrzykiewicz TK, Scozzari A, Cole DL (1995) Large-scale synthesis of oligodeoxyribonucleotide phosphorothioate using controlled-pore glass as support. Nucleosides Nucleotides 14:1219–1226

Rosenberger RD (1989) Biochemistry of heparin antithrombin interactions, and the physiologic role of this natural anticoagulant mechanism. Am J Med 87:2S-3S

Saijo Y, Perlaky L, Wang H, Busch H (1994) Pharmacokinetics, tissue distribution, and stability of antisense oligodeoxynucleotide phosphorothioate ISIS 3466 in mice. Oncol Res 6:243–249

Sands H, Gorey-Feret LJ, Cocuzza AJ, Hobbs FW, Chidester D, Trainor GL (1994) Biodistribution and metabolism of internally 3H-labeled oligonucleotides. I. Comparison of a phosphodiester and phosphorothioate. Mol Pharmacol 45:932–943

Sarmiento UM, Perez JR, Becker JM, Narayanan R (1994) In vivo toxicological effects of rel A antisense phosphorothioates in CD-1 mice. Antisense Res Dev 4:99–107

Sawai K, Miyao T, Takakura Y, Hashida M (1995) Renal disposition characteristics of oligonucleotides modified at terminal linkages in the perfused rat kidney. Antisense Res Dev 5:279–287

Schuette J, Srivatsa GS, Cole DL (1994) Development and validation of a method for routine base composition analysis of phosphorothioate oligonucleotides. J Pharm Biomed Anal 12:1345–1353

Schuette J, Pieles U, Maleknia S, Srivatsa GS, Cole DL, Moser HE, Afeyan NB (1995) Sequence analysis of phosphorothioate oligonucleotides via matrix-assisted laser desorption ionization time-of-flight mass spectrometry. J Pharm Biomed Anal 13:1195–1203

Srinivasan SK, Iversen P (1995) Review of in vivo pharmacokinetics and toxicology of phosphorothioate oligonucleotides. J Clin Lab Anal 9:129–137

Srivatsa GS, Batt M, Schuette J, Carlson R, Fitchett J, Lee C, Cole DL (1994) Quantitative capillary gel electrophoresis (QCGE) assay of phosphorothioate oligonucleotide in pharmaceutical formulations. J Chromatogr 680:469–477

Srivatsa GS, Klopchin P, Batt M, Feldman M, Carlson RH, Cole DL (1997) Selectivity of anion exchange chromatography and capillary gel electrophoresis for the analysis of phosphorothioate oligonucleotides. J Pharm Biomed Anal 16:619–630

Stein CA (1995) Does antisense exist? Nat Med 1:1119–1121

Stein CA, Cheng Y-C (1993) Antisense oligonucleotides as therapeutic agents-Is the bullet really magical? Science 261:1004–1012

Stein CA, Krieg AM (1994) Problems in intrepretation of data derived from in vitro and in vivo use of antisense oligodeoxynucleotides. Antisense Res Dev 4:67–69

Stein CA, Neckers M, Nair BC, Mumbauer S, Hoke G, Pal R (1991) Phosphorothioate oligodeoxycytidine interferes with binding of HIV-1 gp120 to CD4. J Acquir Immune Defic Syndr 4:686–693

Stepkowski SM, Tu Y, Condon TP, Bennett CF (1994) Blocking of heart allograft rejection by intercellular adhesion molecule-1 antisense oligonucleotides alone or in combination with other immunosuppresive modalities. J Immunol 10:5336–5346

Suska A, Grajkowski A, Wilk A, Uznanski B, Blaszczyk J, Wieczorek M, Stec WJ (1993) Antisense oligonucleotides: Stereocontrolled synthesis of phosphorothioate oligonucleotides. Pure Appl Chem 65:707–714

Szklarczyk A, Kaczmarek L (1995) Antisense oligodeoxyribonucleotides: Stability and distribution after intracerebral injection into rat brain. J Neurosci Methods:181–187

Talmadge JE, Adams J, Phillips H, Collins M, Lenz B, Schneider M, Schlick E, Ruffmann R, Wiltrout RH, Chirigos MA (1985) Immunomodulatory effects in mice of polyinosinic-polycytidylic acid complexed with poly-L-lysine and carboxymenthycellulose1. Cancer Res 45:1058–1065

Temsamani J, Tang J-Y, Padmapriya A, Kubert M, Agrawal S (1993) Pharmacokinetics, biodistribution, and stability of capped oligodeoxynucleotide phosphorothioates in mice. Antisense Research and Development 3:277–284

Warren WJ, Vella G (1993) Analysis of synthetic oligodeoxyribonucleotides by capillary gel electrophoresis and anion-exchange HPLC. BioTechniques 14:598–606

Watson PH, Pon RT, Shiu RPC (1992) Inhibition of cell adhesion to plastic substratum by phosphorothioate oligonucleotide. Exp Cell Res 202:391–397

Weiss B, Zhou L-W, Zhang S-P, Qin Z-H (1993) Antisense oligodeoxynucleotide inhibits D2 dopamine receptor-mediated behavior and D2 messanger RNA. Neuroscience 55:607–612

Whitesell L, Geselowitz D, Chavany C, Fahmy B, Walbridge S, Alger JR, Neckers LM (1993) Stability, clearance, and disposition of intraventricularly administered oligodeoxynucleotides: implications for therapeutic application within the central nervous system. Proc Natl Acad Sci USA 90:4665–4669

Wu-Pong S, Weiss TL, Hunt AC (1994) Antisense *c-myc* oligonucleotide cellular uptake and activity. Antisense Res Dev 4:155–163

Wyrzykiewicz TK, Cole DL (1994) Sequencing of oligonucleotide phosphorothioates based on solid-supported desulfurization. Nucleic Acids Res 22:2667–2669

Wyrzykiewicz TK, Cole DL (1995) Stereo-reproducibility of the phosphoramidite method in the synthesis of oligonucleotide phosphorothioates. Bioorg Chem 23:33–41

Yamamoto T, Yamamoto S, Kataoka T, Tokunaga T (1994) Ability of oligonucleotides with certain palindromes to induce interferon production and augment natural killer cell activity is associated with their base length. Antisense Res Dev:119–122

Yazaki T, Ahmad S, Chahlavi A, Zylber-Katz E, Dean NM, Martuza RL, RI G (1996Submitted) Treatment of glioblastoma U-87 by systemic administration of an antisense protein kinase C-alpha phosphorothioate oligodeoxynucleotide. Mol Pharmacol 50:236–242

Zhou L-W, Zhang S-P, Qin Z-H, Weiss B (1994) In vivo administration of an oligodeoxynucleotide antisense to the D2 dopamine receptors in mouse striatum. J Pharmacol Exp Ther 268:1015–1023

CHAPTER 8

Formulation and Delivery of Nucleic Acids

H.E.J. Hofland and L. Huang

A. Introduction

The aim of gene therapy is to correct diseases at their origin by delivery and subsequent expression of exogenous DNA, which encodes for a missing or defective gene product. Two distinct strategies for DNA delivery can be employed. The ex vivo strategy uses the cells of a patient, into which the DNA is introduced in vitro. The genetically modified cells are selected, expanded and, finally, transplanted back into the patient (Berns et al. 1995). This is a relatively efficient method for gene delivery. Most disorders, however, require direct gene transfer, i.e., delivery of the nucleic acids directly into the affected tissues in vivo. One of the major challenges for the direct gene-transfer approach is to develop a vehicle that is able to protect the nucleic acids from degradation, while delivering the genes of interest to specific tissues and target-cell compartments.

Vehicles for gene transfer, which have successfully demonstrated the delivery of exogenous genes in vivo, can be divided into two major groups: the viral and the non-viral vectors, each with their own specific advantages and disadvantages. Viruses are very efficient gene-transfer vehicles; however, significant limitations are inherent to their use. Retroviral vectors, for instance, are very useful for the ex vivo approach, but since they only transduce dividing cells, their use for direct gene transfer is limited. Adenoviruses have been shown to transfect somatic cells efficiently and with a surprising persistence (since their genome is not integrated into the host genome, as is the case with retroviruses and adeno-associated viruses). However, they do have the disadvantage of being immunogenic. Success has been obtained in deleting parts or even the whole wild-type viral genome, such that transduced cells do not express viral proteins. However, proteins carried on the viral envelope are still seen by the immune system, which prevents subsequent reinfections with the same vector. Other less-characterized viruses have been used for direct gene transfer, such as adeno-associated virus, herpes virus, vaccinia virus, polio virus or human immunodeficiency virus (HIV). One of the more common problems associated with viral vectors is obtaining pharmaceutically significant quantities and/or concentrations. Furthermore, the fact that their gene products, even of replication incompetent viruses, may be toxic, and the slight possibility that replication competent viruses could emerge through recombination or contamination of the stock, make these vehicles less attractive for use in a clinical setting (Mulligan 1993).

Considering these limitations, synthetic carriers offer an attractive alternative. Non-viral vectors are being developed under the assumption that they will overcome the problems associated with viral gene delivery. Although non-viral gene delivery vehicles are still relatively inefficient and the resulting gene expression is transient in nature, their non-immunogenic nature allows for repeated treatments. Problems associated with these non-viral systems will be discussed in this chapter, as well as some of the strategies used to overcome them.

B. Formulation of DNA

I. Naked-DNA Injections

In special situations, polynucleotide transfer and expression does not require a delivery system. In fact, it has been well established that DNA delivery systems, such as lipid/DNA complexes, actually inhibit gene uptake and expression by myofibers, compared with direct injection of naked DNA into skeletal muscle (Wolff et al. 1990; Jiao et al. 1992), heart muscle (Ardehali et al. 1995), and liver (Hickman et al. 1994). One of the advantages of direct injection into the muscle is that the expression appears to be more persistent than that found in other tissues. Although the mechanisms of polynucleotide uptake by the muscle are unknown, parameters such as needle type, speed and site of injection, and the type of solute used appear to have a strong impact on the transfection efficiency (Wolff et al. 1991). The levels of expression, however, are still relatively low. Muscle regeneration, induced by the myotoxic local anesthetic bupivacaine, significantly increased gene expression following plasmid injection (Wells 1993). Much of this effect is attributed to the uptake and expression of the plasmid by a greater number of muscle fibers. Possible clinical applications may be found in genetic muscle diseases, such as the expression of the dystrophin gene as a therapy for Duchenne's muscular dystrophy (Acsadi et al. 1991) or heterologous expression of a transgene in diseases in which muscle is not involved, such as blood-clotting factor IX. Most of the efforts, however, are focused on genes encoding antigens to be used for vaccines (Ulmer et al. 1993).

The use of unformulated plasmid DNA, "naked DNA", is only useful in two specialized cases. In general plasmid DNA, which is a large and negatively charged polymer, will not cross the hydrophobic, negatively charged cell membranes without the help of a carrier. In addition, the plasmid DNA needs to be protected from degradation by endonucleases during transport from the site of administration to the site of action. DNA delivery vehicles should be able to do both. In the following section, different DNA formulations and delivery strategies will be discussed.

II. Gene Guns

Gene guns offer a physical method to literally shoot plasmid DNA into cells. Plasmid DNA is coated with gold particles and transferred to a mylar carrier sheet. At gene transfer, the particles on the carrier sheet are accelerated by an electric arc, which is generated by a high-voltage discharge. The discharge will accelerate the DNA-coated gold particles to high velocity, enabling efficient penetration of target organs in vivo or single-cell layers in vitro (YANG et al. 1990; ANDREE et al. 1994; SUN et al. 1995). Important parameters that determine the efficiency of DNA transfer by this technique include DNA to gold ratio, the amount of DNA/gold particles per carrier sheet, and the voltage used at discharge. Although the DNA transfer is relatively efficient, the major drawback of this system is that, for in vivo use, the target tissues (except for the skin) have to be surgically exposed.

III. Polymer-Based Formulations

There is a wide variety of cationic polymers that have been shown to complex to DNA and facilitate gene transfer. The structure of the polymer does not appear to have a major effect on the transfection efficiency, as is demonstrated by the variety of structures that have been used, for example: (1) linear polymers, such as poly(L-lysine), (2) branched polymers, such as polyethylenimine (PEI), and (3) spheroidal polymers, such as starburst dendrimers (Fig. 1). In the following sections, the use of each of these types of polymers for gene transfer will be discussed in more detail.

Poly(L-lysine)/DNA complexes have been widely used for DNA transfections. However, this polymer alone only binds to the DNA, but does not promote its cellular uptake. Gene transfer is achieved by conjugating the poly(L-lysine) to other molecules that do not bind to DNA by themselves, but do enhance transport across membranes, such as the lipid *N*-glutarylphosphatidyl-ethanolamine (ZHOU et al. 1991) or by conjugating the poly(L-lysine) to adenoviral envelopes (KUPFER et al. 1994; GAO et al. 1993). Another way to enhance the transfection efficiency of poly(L-lysine)/DNA complexes is by conjugating the polymer to a targeting ligand (Table 2).

PEI is a very efficient and very affordable means of gene delivery. Every third atom of PEI is a protonable amino nitrogen atom, which allows for DNA binding and offers a substantial buffering capacity below physiological pH (see Sect. E). PEI/DNA complexes, at only slight excess positive charge, led to efficient luciferase gene expression both in vitro in a variety of cell types and in vivo by direct intracranial injections into newborn mice (BOUSSIF et al. 1995).

Starburst dendrimers are spheroidal polycations that are synthesized in a well-defined manner, allowing for a very predictable size, shape and surface charge of the polymer, which makes them an interesting tool for DNA delivery

Poly(L-lysine)

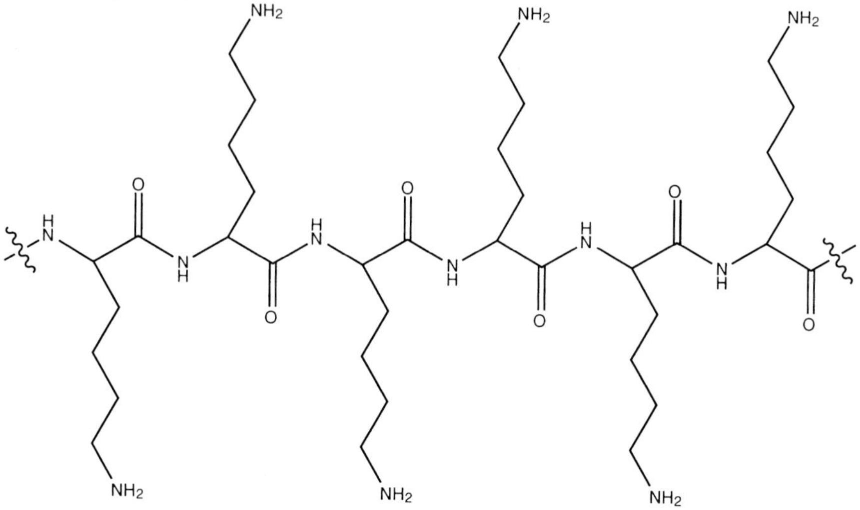

a

PEI

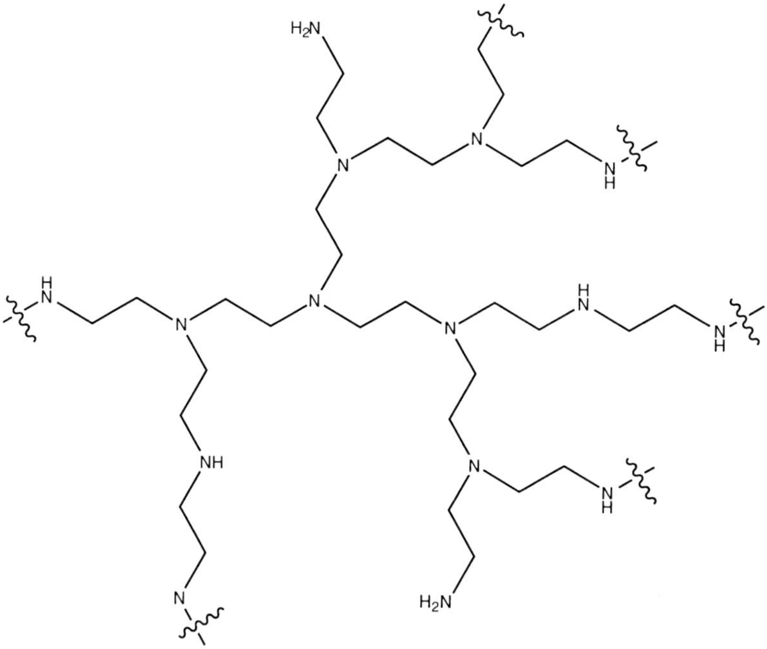

b

Fig. 1a–c. Three examples of morphologically different polymers that are used for nucleotide delivery. A linear polymer, poly(L-lysine) (**a**); a branched polymer, polyethylenimine (PEI, **b**); and a spheroidal polymer, polyamidoamine (PAMAM) starburst dendrimer (**c**)

Starburst Dendrimer

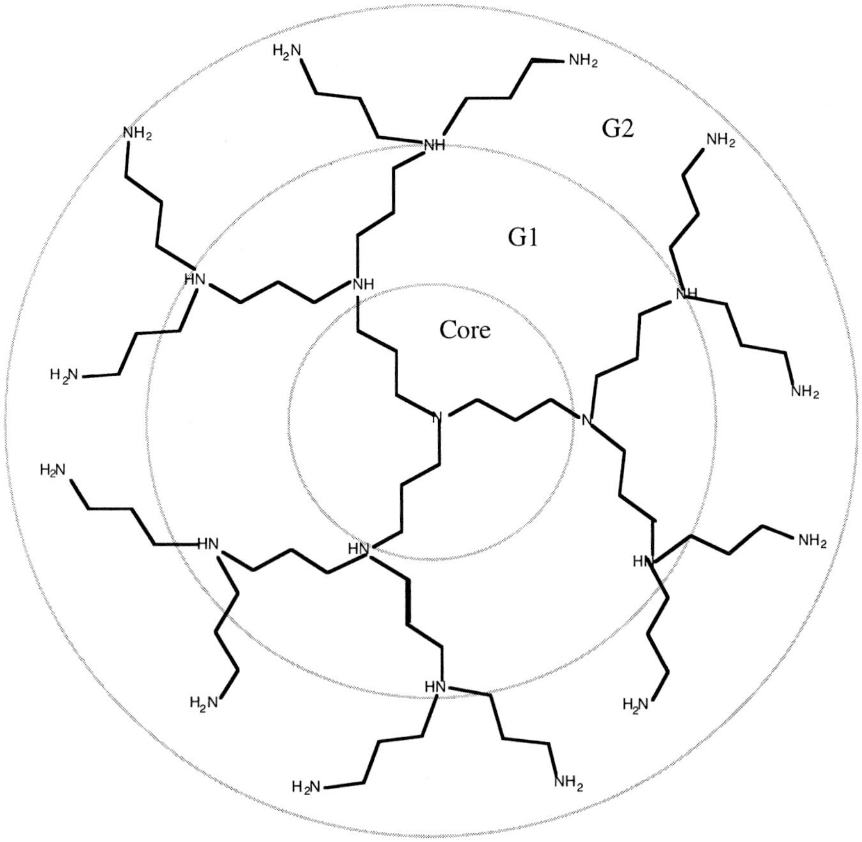

c

Fig. 1a–c. *Continued*

(Haensler and Szoka 1993a; Kukowska-Latallo et al. 1996). The capability of dendrimers to transfect cells depends on the size and number of primary amino groups on the surface, and the shape of the polymer. These characteristics are determined by the core molecule and by the number of polymerization rounds, i.e., the number of generations of the dendrimer. Increasing transfection efficiency was found with increasing generation from G5 to G10, with a plateau at G8. No transfection activity was found for generations smaller than 5. The restriction of transfection efficiency to G5 and higher is most likely due to the charge density and the spherical shape of these dendrimers. However, larger dendrimers with even higher density of surface charge will form large aggregates when complexed to the DNA that are no longer able to transfect cells in vitro. Kukowska-Latallo et al. (1996) observed that DEAE-Dextran added to the media during transfection significantly enhances the transfection efficiency of the dendrimer. One advantage of

Table 1. Studies employing "passive targeting" by local administration of the DNA formulation in vivo

Route of administration	Vehicle	Reference
Injection into skeletal muscle	Naked DNA	Wolff et al. 1990, 1991; Acsadi et al. 1991; Jiao et al. 1992; Wells 1993; Ulmer et al. 1993
Injection into heart muscle	Naked DNA	Ardehali et al. 1995
Injection into liver	Naked DNA	Hickman et al. 1994
Epidermis	Naked DNA (gene gun)	Andree et al. 1994
Liver, skin, muscle, mammary tissue	Naked DNA (gene gun)	Yang et al. 1990
Intra-tumor	Naked DNA (gene gun)	Sun et al. 1995
Intra-tumor	Cationic liposomes	Nabel 1992; Nabel et al. 1993; Son and Huang 1994; Missol et al. 1995
Intra-tracheal installation	Cationic liposomes	Yoshimura et al. 1992; Hyde et al. 1993
Intra-tracheal aerosol administration	Cationic liposomes	Stribling et al. 1992; Brigham and Schreier 1993
Intra-nasal	Cationic liposomes	Caplen et al. 1995
Intraperitoneal	Cationic liposomes	Yu et al. 1995; Aoki et al. 1995
Balloon isolated regions of vasculature	Cationic liposomes	Plautz et al. 1993; Rajawalia et al. 1995
Bile duct	Cationic liposomes	Takehara et al. 1995
Intracranial	Cationic liposomes	Schwartz et al. 1995; Cao et al. 1995
Portal vein	Cationic liposomes	Hara et al. 1995b
Intra-liver	HVJ-anionic liposomes	Kato et al. 1993
Infusion into coronary artery	HVJ-anionic liposomes	Sawa et al. 1995
Direct into liver	HVJ-anionic liposomes	Tomita et al. 1993; Kato et al. 1993
Renal artery	HVJ-anionic liposomes	Tomita et al. 1992
Portal vein injection	HVJ-anionic liposomes	Kaneda et al. 1989
Intracranial	PEI polymer	Boussif et al. 1995
Caudal vena cava	Poly(L-lysine)	Perales et al. 1994
Tail vein	Poly(L-lysine)	Stankovics et al. 1994
Tail vein	Poly(L-lysine)	Trubetskoy et al. 1992a,b
Caudal vena cava	Poly(L-lysine)	Ferkol et al. 1995
Intra-tracheal instillation	Poly(L-lysine)	Gao et al. 1993

HVJ, Hemagglutinating Virus of Japan; PEI, polyethylenimine

using dendrimers for DNA transfer is their relatively low toxicity. Another major advantage is the fact that the complexes are highly soluble in aqueous solutions, allowing up to 1 mg DNA to be complexed to 65 mg polymer, leading to very active polymer/DNA complexes. Dendrimer/DNA complexes are stable in solution for weeks and retain their transfection efficiency for many cell types in vitro over this period of time.

IV. Lipid-Based Formulations

1. Liposome Encapsulation

Polynucleotides can be encapsulated inside liposomes and delivered to cells in vivo (OSTRO et al. 1977; NICOLAU et al. 1983; SORIANO et al. 1983; ALINO et al. 1996). Different preparation methods, such as the freeze–thaw method (ZHOU et al. 1992c) or the reverse-phase evaporation method (LEIBIGER et al. 1991) will lead to the formation of large unilamellar liposomes with a high encapsulation volume, capable of encapsulating up to 40% of the DNA. The encapsulated DNA is protected from the environment, which is essential for intravenous administration. A long history of liposome research showed that liposomes are readily taken up by the reticulo endothelial system (RES) and destroyed. Both size and surface characteristics of the liposomes are very important for an increased circulation half-life of the liposomes. Reducing the size of the liposomes avoids nonspecific entrapment of the liposomes in the capillary beds, and subsequent uptake and elimination by macrophages. Another strategy that can be followed is coating the liposomes with molecules that provide a steric barrier for opsonisation by serum proteins, which leads to escape from RES uptake. Although conventional liposomes are able to package DNA, their DNA transfer and subsequent expression of the genes is very low. This is most likely due to the fact that these liposomes do not escape the lysosomal degradation once taken up by the cells. A very elegant way to escape the endosome is by utilizing pH-sensitive liposomes. The pH drop in the late endosomes will cause these liposomes to fuse with the endosome membrane, thus offering an escape for the entrapped DNA (ZHOU et al. 1992c). Another exciting application for liposomal delivery systems lies in the fact that they can be targeted to specific sites in the body by using ligands, such as immunoglobulins (WANG and HUANG 1987a,b, 1989; ZHOU and HUANG 1992b).

A combination of the use of negatively charged pH-sensitive liposomes containing condensed folate PEG and DNA/poly(L-lysine) complex, associated as a targeting ligand, was shown to be capable of delivering the DNA specifically to cells expressing the folate receptor (LEE and HUANG 1996).

2. Cationic Lipid/Nucleotide Complex

FELGNER et al. (1987) introduced the use of positively charged lipids in the form of liposomes for the introduction of DNA into mammalian cells. They showed that these cationic liposome/DNA complexes were very active in transfecting cells and are also very easy to use. In a direct comparison of the transfection efficiency of cationic liposome/DNA complexes to pH-sensitive and non-pH-sensitive liposomes, it was found that the transfection efficiency of pH-sensitive liposomes was only 1–30% that of cationic liposome/DNA complexes, and no transfection efficiency could be observed for non-pH-sensitive liposomes (LEGENDRE and SZOKA 1992). This is likely due to the very high, nonspecific cellular uptake of DNA complexed with cationic lipids. Ever since this initial report, the field of synthetic gene delivery vehicles has grown immensely. The idea of using synthetic, i.e., non-immunogenic and non-toxic vehicles for in vivo gene transfer is very appealing. Cationic lipid-mediated gene transfer has become so important that a large portion of this chapter is devoted to it. Lipid-mediated DNA (BARTHEL et al. 1993) or RNA (LU et al. 1994) transfer is dependent on a number of parameters. Important are the type of lipid used, the amount and type of helper lipid and the lipid to DNA ratio. DNA quality and dose concentration has a major impact on the formation of the complex, cationic liposome size, temperature during assembly of the complex, cell type in vitro, and in vivo route of administration. In order to understand the mechanisms behind DNA binding, condensation and, ultimately, transport into the cells, a number of studies have been reported on structure of the cationic lipid in relation to its ability to transfect cells. Each cationic lipid can be divided into four functional units that have an impact on DNA binding, transfection efficiency and toxicity of the lipid. These units are: (1) a DNA-binding moiety, (2) a hydrophobic moiety (membrane anchor), (3) a spacer between the hydrophilic headgroup and the hydrophobic alkyl chains, and (4) the chemical bond linking the headgroup to the lipid (GAO and HUANG 1995). Another very important component of the cationic lipid formulation is the helper lipid. In the following sections, each of these moieties will be discussed in more detail.

3. DNA-Binding Moiety

It appears that regular DNA-binding molecules, such as spermine (BEHR et al. 1989) or poly(L-lysine) (ZHOU et al. 1991), can be made into efficient gene delivery vehicles, simply by making these molecules more hydrophobic. Protonated amines are generally used as the DNA-binding domains. Polycationic lipids, such as DOGS (Fig. 2) or DOSPA (Fig. 3), contain both primary and secondary amines. Monocationic lipids, such as DOTMA (Fig. 4), usually have

Fig. 2a–c. Structure–function relationships of lipids were investigated (REMY et al. 1994). The DNA binding motif was systematically varied by changing the number of positive charges in the headgroup of the lipid, e.g., spermine (3+) (**a**), ornithine (2+) (**b**), glycine (1+) (**c**). The following relative transfection efficiencies were found: spermine/ornithine/glycine = 20/1/0

spermine (DOGS)

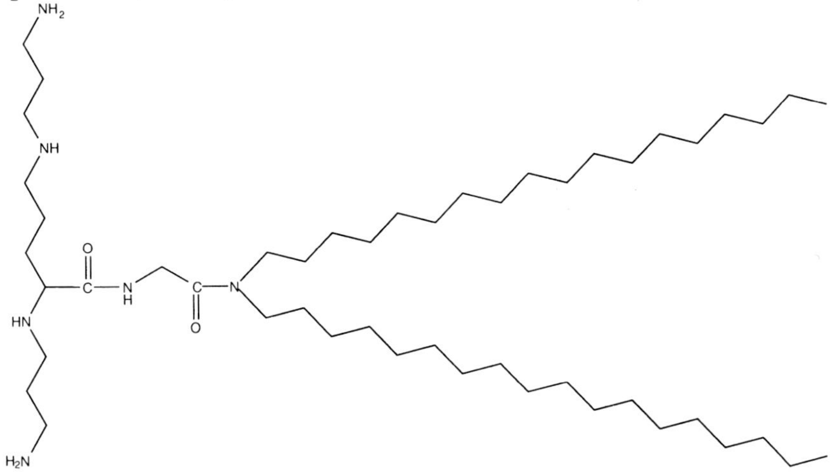

a

ornithine

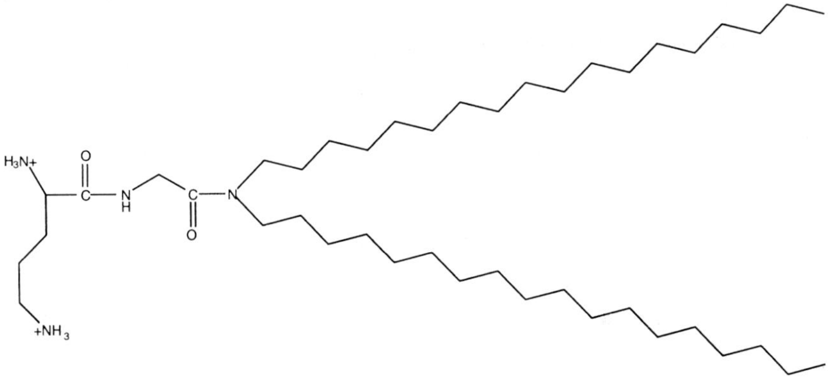

b

glycine

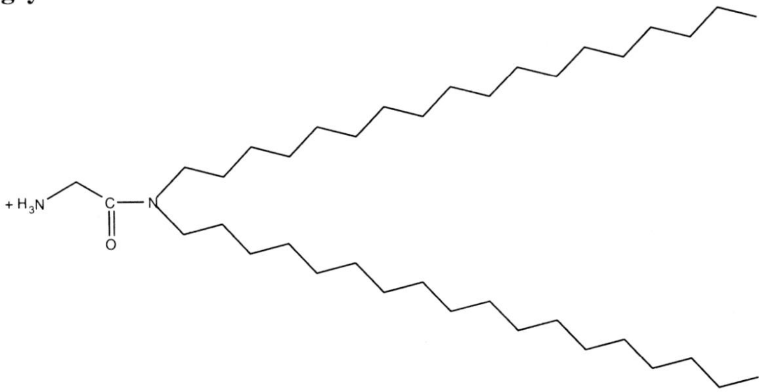

c

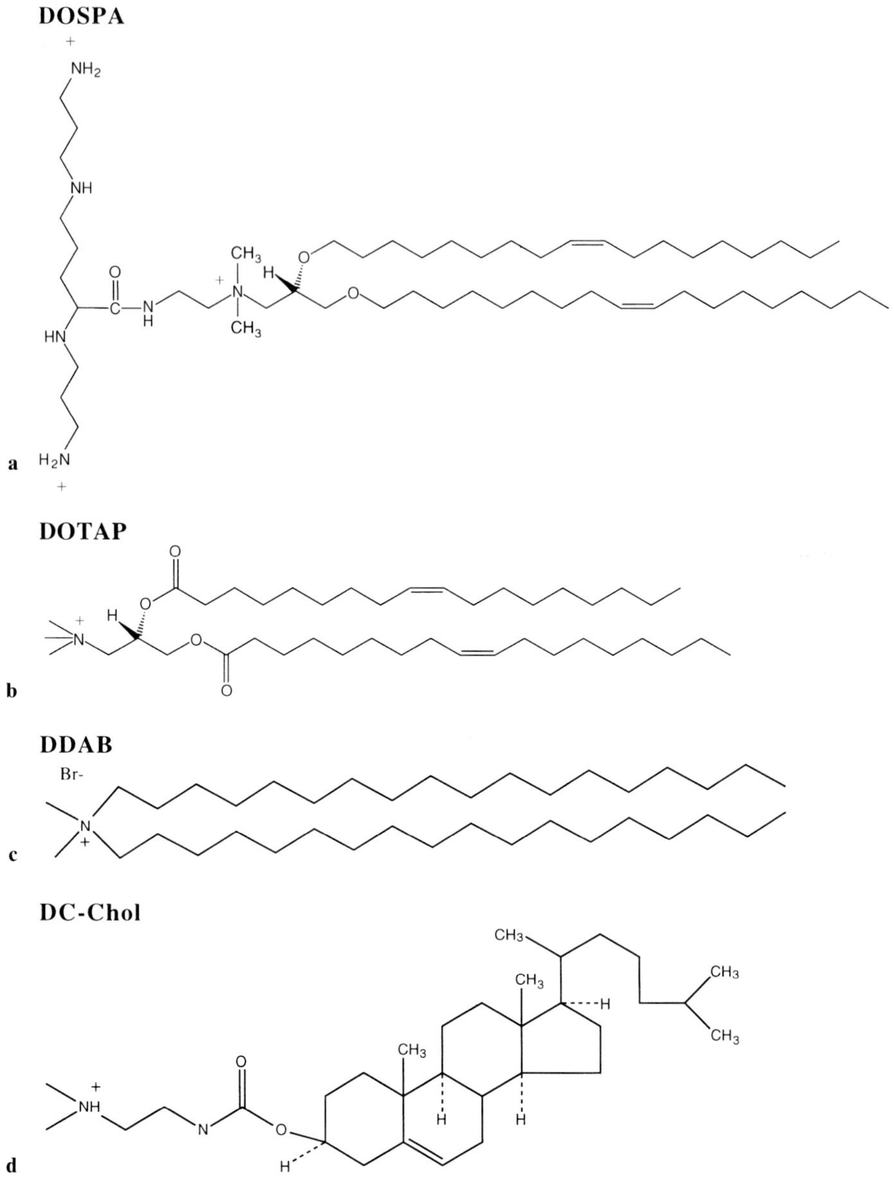

Fig. 3a–d. Four examples of cationic lipids used for nucleotide delivery. Each cationic lipid can be divided into four functional units: the DNA-binding moiety, a hydrophobic moiety (membrane anchor), a spacer between the hydrophilic headgroup and the hydrophobic alkyl chains, and the chemical bond linking the headgroup to the lipid. **a** DOSPA; **b** DOTAP; **c** DDAB; **d** DC-Chol

Cationic Lipid Headgroup Variation

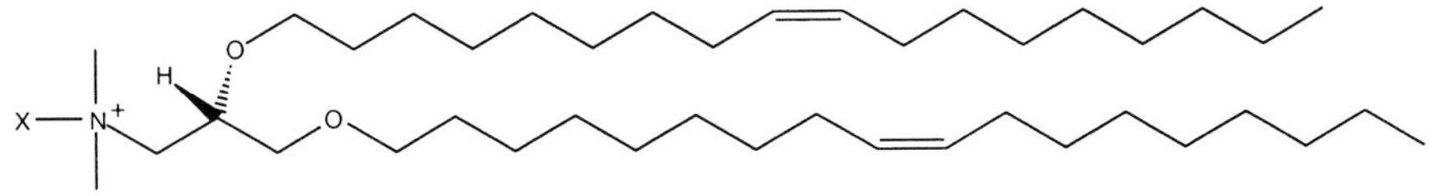

X = CH_3	DOTMA
X = $HO(CH_2)_2$	DORIE
X = $HO(CH_2)_3$	DORIE-HP
X = $HO(CH_2)_4$	DORIE-HB
X = $HO(CH_2)_5$	DORIE-HPe

a

Cationic Lipid Alkyl Chain Variation

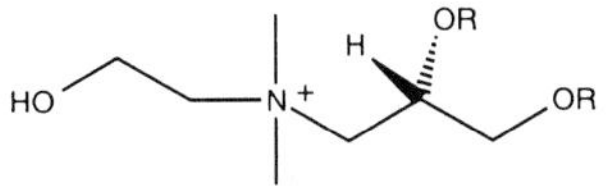

R = $CO(CH_2)_7CH{=}CH(CH_2)_7CH_3$	DORI
R = $(CH_2)_8CH{=}CH(CH_2)_7CH_3$	DORIE
R = $(CH_2)_{13}CH_3$	DMRIE
R = $(CH_2)_{15}CH_3$	DPRIE
R = $(CH_2)_{17}CH_3$	DSRIE

b

Fig. 4a,b. Structure–function relationships of cationic lipids were investigated by systematically varying the headgroup (*X*) or the lipid tails (*R*). The following relative transfection efficiencies were found: **a** DORIE/DORIE-HP/DORIE-HB/DORIE-HPe/DOTMA = 2.2/1.9/1.4/1.2/1.0. **b** DMRIE/DORIE/DORI/DPRIE/DSRIE = 5.2/2.2/2.1/2.0/0.8. (FELGNER et al. 1994)

a quaternary amine. However, FARHOOD et al. (1992) found that DC-Chol with a tertiary amine is more active and less toxic than the same molecule with a quarternary amine. FELGNER et al. (1994) reported a systematic study on the effects of changes in the chemical structure of the cationic lipid. A general cationic lipid structure was used in the cationic headgroup (H) and was varied systematically (Fig. 4). Each formulation contained 50 mol% DOPE. The following enhancements on the transfection efficiency were found (the transfection efficiency of DOTMA was defined as 1): DORIE/DORIE-HP/DORIE-HB/DORIE-HPe/DOTMA = 2.2/1.9/1.4/1.2/1.0. The substitution of one of the methyls by a hydroxyethyl group in DOTMA led to an enhanced transfection efficiency. It is suggested that the hydroxyl group increases the polarity of the headgroups, which in turn enhances the hydration and stabilization of the membranes. Cationic liposomes with a polyvalent cationic headgroup show better transfection efficiency than the monovalent ones (BEHR et al. 1989; HAWLEY-NELSON et al. 1993). One explanation is that polyvalent cationic liposomes tend to form more compact complexes with the DNA, resulting in an enhanced delivery of the DNA. Another explanation is the buffering capacity of the polyamines, which may lead to rupture of the endosomes (Sect. E). REMY et al. (1994) also studied structure–function relationships of a wide range of lipids that were systematically varied, both in the lipid-binding motif with various positive charges, such as polyamines (spermine[3+], ornithine[2+]), quaternary ammonium-bearing lipids [glycine (+) (Fig. 2)], or lipids conjugated to DNA-binding molecules, such as acridine, ornithine, netropsin, and in the lipid moiety of the molecules. DNA was complexed to these compounds at various ratios. The lipid with the spermine headgroup was 20-fold more active than the ornithine lipid, and the single positively charged lipid showed no detectable expression. Polyamines have been shown to hydrogen bond to the floor of the DNA minor groove. The nonionic acridine will intercalate in between the base pairs, and netropsin derivatives will fill the minor groove. Similar results were found for polymer/DNA complexes. When the DNA-binding moiety consisted of intercalating compounds, no gene transfer could be observed (HAENSLER and SZOKA 1993b; WAGNER et al. 1991a). An explanation for the transfection efficiency of polyamine-lipids lies in the fact that they are not only able to bind to the DNA but are also able to condense it. In addition, polyamines serve as a proton buffer, which may enhance escape from the endosomal compartment (Sect. E). The same explanation holds true for differences found in transfection efficiencies for various polyamine lipids containing different numbers of positive charge. Binding affinity of these molecules for the DNA is only part of the explanation. The other part remains unclear, since a number of monocationic lipids have been shown to be at least as active as the ornithine lipid (DOTMA or DDAB).

4. Hydrophobic Moiety

Liposome membranes can be made more fluid by decreasing the number of carbon atoms in the alkyl chain, or by introducing *cis* unsaturated bonds. This results in a reduction of the phase-transition temperature, which allows for a higher membrane–membrane interaction and may lead to enhanced fusion (lipid mixing between membranes) of the lipid/DNA complex with the endosome. FELGNER et al. (1994) found a relationship between lipid chain length and transfection efficiency. The following transfection enhancements were found upon changing the lipid tail (Fig. 4): DMRIE/DORIE/DORI/DPRIE/DSRIE = 5.2/2.2/2.1/2.0/0.8. Again, these numbers are relative to DOTMA (defined as having a transfection efficiency of 1), which has a slightly different headgroup. These relative transfection efficiencies were obtained using equal transfection conditions with respect to DNA and lipid concentrations. However, considering the fact that each of these formulations have different optimum transfection conditions, these transfection enhancements are less pronounced if the optimal formulation was chosen in each case, e.g., in a matrix experiment systematically altering 64 conditions, the optimal DMRIE formulation is only 1.4-fold better than the optimal DOTMA formulation. For the lipospermines this effect is even less pronounced (REMY et al. 1994). Although the spermine headgroup appeared to be essential for transfection efficiency, the lipid moiety appeared not to be. Several spermine derivatives were synthesized (diacyl chain stearyl (C18)2-Spermine vs oleyl (C9=9)2-Spermine vs palmitoyl (C16)2-Spermine vs mono acyl chain (C18)-Spermine) with no significant change in transfection efficiency. Using a cholesterol as the lipid moiety instead of alkyl chains offers several advantages. Cholesterol derivatives have been shown to be very efficient for gene transfer (GAO and HUANG 1991; FARHOOD et al. 1992; GUO et al. 1993), with relatively less toxicity.

5. Spacer

The spacer that is used between the positively charged headgroup and the alkyl chain should allow for uncompromised binding to the DNA. The effect of the length of the spacer arm between the cationic headgroup and the alkyl chains on the transfection activity varies. Some lipids are active with no or only one carbon spacer, such as DDAB (Fig. 3) or DOTMA (Fig. 4). Increasing their spacer up to 5 C atoms did not improve their transfection efficiency (FELGNER et al. 1994). For cholesterol derivatives (Fig. 3), however, the length of the spacer appears to play an important role. A spacer length between 3 C atoms and 6 C atoms gives optimal transfection efficiency, as reported by FARHOOD et al. (1992). Similar results were reported by ITO et al. (1990) for didodecyl- and dimyristyl-*N*-[*p*-(2-trimethylammonioethyloxy)(CH2)*n*]-(L)-glutamate bromide, where the optimal spacer length (*n*) was between 2 C atoms and 6 C atoms. The transfection efficiency was substantially reduced upon increasing the spacer arm to 8–11 C atoms. Lipids containing polyamine headgroups are also sensitive to the spacer length. Removal of the glycine spacer of DOGS (Fig. 2) totally removes the transfection activity (REMY et al. 1994). It is suggested that the spacer allows the spermine headgroup to bind in the minor groove of the DNA.

6. Linker

The linker between the lipid tails and the cationic headgroup determines stability of the lipid. However, if the lipids are more stable, they are also potentially more toxic. Therefore, it is preferable to use biodegradable linkers that provide enough stability as well as reduce toxicity (ITO et al. 1990; FARHOOD et al. 1992; GAO and HUANG 1991; REMY et al. 1994). Biodegradable amide bonds or carbamoyl bonds may be used; however, ester bonds may be too unstable.

7. Helper Lipid

Most cationic lipid formulations require the use of a second, neutral "helper" lipid, such as DOPE. DOPE is a non-bilayer-forming lipid, which can only be incorporated in membranes in the presence of another highly hydratable lipid. The DOPE molecule has a cone-shaped configuration, which gives the compound the ability to form hexagonal-II phases (Litzinger and Huang 1992). Hexagonal structures are membrane-fusion intermediates. Therefore, altering the shape of the helper lipid changes its ability to form these hexagonal structures and, thus, inhibits its ability to fuse membranes. Membrane fusion seems to be essential for the escape of the DNA from the endosomal compartment. The importance of DOPE cone shape was illustrated by Felgner et al. (1994). Upon increasing the size of the hydrophilic headgroup, the cone shape was transformed into a cylindrical shape. The result was a reduction in the transfection efficiency. The following relative transfection enhancements were found: 1.0/0.3/0.25/0.15/0.05/0.01 for DOPE/PMME/DPE/PDME DOPC/CPE, respectively (Fig. 5a). Similarly, the cone shape of DOPE was transformed into a cylinder shape by either decreasing the number of C atoms of the alkyl chains or by removing their unsaturated bonds. This led to a significant reduction in the transfection efficiency of the formulation, i.e., DOPE was approximately fivefold more active than DMPE, DPPE, or DSPE (Fig. 5b). The use of single-alkyl chain neutral lipids had a very poor transfection efficiency of 1% compared with DOPE (Fig. 5c). Another interesting observation was that larger vesicles (MLVs diameter 300–700 nm) appeared to have a higher transfection efficiency than that obtained with smaller vesicles (SUVs diameter 50–100 nm) in vitro. Optimal transfection conditions in vivo with respect to particles' size, lipid/DNA ratio, final charge of the complex, etc. are likely to differ substantially from optimal conditions found in vitro.

One of the major problems encountered in formulating DNA delivery systems is the high degree of variability. The behavior of DNA–polycation complexes is very much dependent on the method for complexing. Reproducibility, especially of in vivo results is difficult (Perales et al. 1994). The inherent disadvantage of the use of liposomes is that they are relatively large and rigid structures that provide a scaffold for the DNA to wrap around. Over time, the liposome/DNA complexes will aggregate, forming complexes that continuously increase in size (Sternberg et al. 1994). Eventually these large complexes are no longer able to transfect cells. Similar problems are encountered when working with polymer-based systems. The instability of the complex dictates that the lipids and DNA are mixed at the bedside of the patient and used immediately. A single-vial, stable pharmaceutical formulation is preferred. Recently, a publication by Hofland et al. (1996) offered a solution to this instability problem. The rationale was to use the cationic lipids in the form of micelles rather than liposomes. Micelles are much smaller than liposomes and, therefore, alleviate the three-dimensional packing constraints due to the morphology of the liposomes. This allows for a much more efficient

Neutral Lipid Headgroup Variation

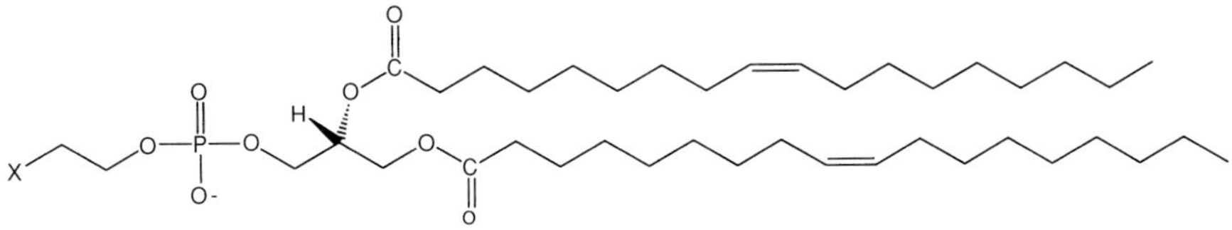

$X = H_3N^+$ DOPE
$X = H_3CH_2N^+$ PMME
$X = (CH_3)_2HN^+$ PDME
$X = (CH_3)_3N^+$ DOPC
$X = H_3N^+\ (CH_2)_6$ CPE
$X = H_3N^+\ (CH_2)_{12}$ DPE

a

Neutral Lipid Double Alkyl Chain Variation

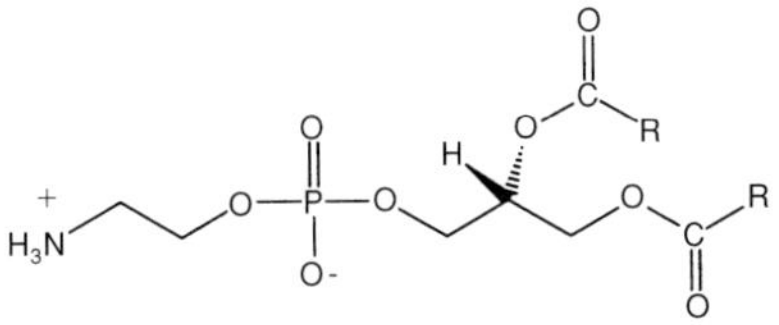

$R = (CH_2)_8CH{=}CH(CH_2)_7CH_3$ DOPE
$R = (CH_2)_{13}CH_3$ DMPE
$R = (CH_2)_{15}CH_3$ DPPE
$R = (CH_2)_{17}CH_3$ DSPE

b

Neutral Lipid Single Alkyl Chain Variation

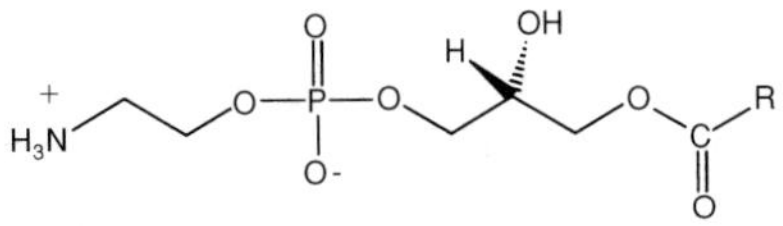

$R = (CH_2)_8CH{=}CH(CH_2)_7CH_3$ lysoOPE
$R = (CH_2)_{13}CH_3$ lysoMPE
$R = (CH_2)_{15}CH_3$ lysoPPE
$R = (CH_2)_{17}CH_3$ lysoSPE

c

Fig. 5a–c. Structure–function relationships of various helper lipids were investigated by systematically varying the headgroup (*X*) or the lipid tails (*R*). The following relative transfection efficiencies were found: **a** DOPE/PMME/DPE/PDME/DOPC/CPE = 1.0/0.3/0.25/0.15/0.05/0.01. **b** DOPE is approximately fivefold more active than either DMPE, DPPE, or DSPE. **c** Using single alkyl chain neutral lipids had very poor transfection efficiency of 1% as compared to DOPE coating of the DNA by the lipids. After removal of the detergent (octyl glucoside) by dialysis, lipid/DNA complexes were obtained that maintained their transfection efficiency over at least 3 months when stored at 4°C or at least 9 months when frozen at –20°C. Also, a marked increase in stability of the complex in the presence of serum was found, which has an important impact on the in vivo application of these stable complexes. Finally, it was shown that after complexation, the free uncomplexed lipid could be removed from the lipid/DNA complex. Thus, removing the toxicity of the formulation, which is associated with the free cationic lipids, allowed for much higher lipid/DNA doses to be delivered and much higher gene expression.

C. Delivery to Target Cells

Nonviral gene delivery systems rely on normal cellular uptake mechanisms. A schematic view of the processes involved in uptake and subsequent expression of the trans gene is shown in Fig. 6. Before the nucleotide delivery system is able to reach the cell surface, several hurdles need to be taken. For example, after intravenous administration, serum inactivation and DNA degradation must be avoided or, in the case of cystic-fibrosis gene therapy (using intratracheal administration), the delivery system must be able to penetrate a thick layer of mucus covering the target cells. The most simple way of helping the delivery system reach the target cells is by "passive targeting" or local administration, i.e., physically placing the formulation in the proximity of the cells that need to be transfected. A number of studies have been published using a variety of administration sites (Table 1). The systemic biodistribution of gene expression is also dependent on the route of administration. Tierry et al. (1995) showed that after intravenous administration of (DOGS/DOPE)/DNA complexes, the highest luciferase expression was found in the lung. Relative to this expression level, 78% was found in the spleen and 63% in the liver. After intraperitoneal injection, however, the highest expression was found in the spleen, and only 25% and 18% of this level was found in the liver and lung, respectively. Expression in spleen was also the highest after subcutaneous injection. Respective liver and lung expression were 28% and 72%.

Knowing the mechanisms that are involved in biodistribution and cellular uptake is the first step in developing targeted gene delivery vehicles for intravenous administration. Mahato et al. (1995a) showed that following intravenous injection of [^{32}P]-labeled DNA, the radioactivity was rapidly eliminated from the plasma. Approximately 70% of the dose was taken up by the liver within the first minute after injection, and no gene expression was found. Although Liu et al. (1995) did observe some gene expression after intravenous injection of naked DNA, they showed that expression levels were significantly increased upon using lipid/DNA complexes. The blood clearance was found to be equally rapid when DNA was complexed to cationic lipids (DDAB/DOPE). The percentage of total radioactivity that accumulated in

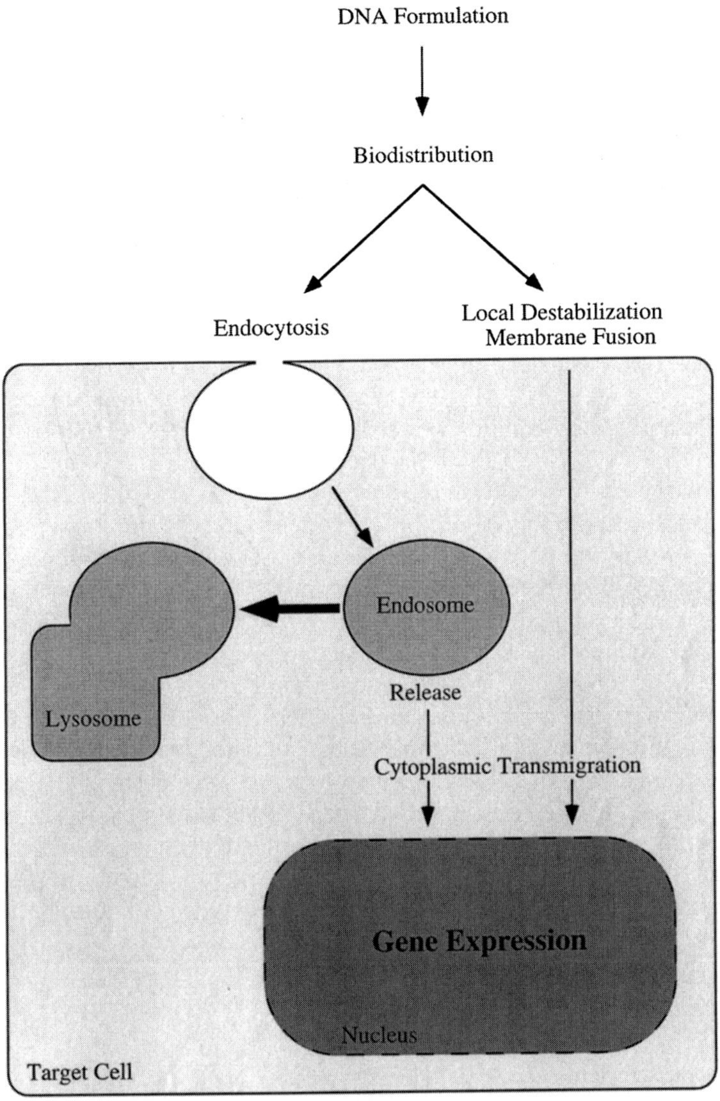

Fig. 6. A schematic overview of processes involved in DNA delivery in vivo

the lung and in the liver within the first minute after injection was 35% and 40%, respectively. Similarly, for (DOTMA/DOPE)/DNA complexes, the accumulation in the liver after 1 min was 55% and in the lung, 25%. Hepatic accumulation occurred preferentially in the non-parenchymal (Kupffer) cells.

Mahato et al. (1995b) demonstrated that the overall charge of the lipid/DNA complex is an important factor determining its biodistribution. Changing the Zeta potential of the complex from negative to positive results in a sharp decrease in lung accumulation, accompanied by an increase in spleen accumulation. However, the effect of charge on biodistribution can be clouded by the effect of the size of the complex on the biodistribution. Large particles (>400 nm) will simply embolize in the capillary beds, which explains the accumulation in the lung. Furthermore, large particles are known to be rapidly cleared from the circulation by the reticuloendothelial system.

D. Cell Entry

When the DNA reaches the cells, the cellular uptake is most likely not the cause of the low transfection efficiency of non-viral vectors. In an in vitro setting, it was demonstrated that 6 h after adding lipid/DNA complex at a dose of 2 μg DNA to 2×10^5 COS cells, each cell contained an average of 3×10^5 copies of the plasmid. However, only less than 50% of these cells actually expressed the transgene. Therefore, the inefficiency of non-viral vectors appears not to be due to cell entry (ZABNER et al. 1995). This suggests that the success of non-viral delivery of DNA is merely due to a mass-action effect. Electron micrographs showed that both lipid/DNA complexes (ZABNER et al. 1995) and also lipopoly(L-lysine)/DNA complexes (ZHOU and HUANG 1992a) are predominantly taken up by endocytosis. After endocytosis, the DNA-containing particles are retained in perinuclear vesicles. Escape from these vesicles is thought to be a major barrier for transfection. These studies are in contradiction to earlier suggestions that the main mechanism of cell entry is by membrane fusion (FELGNER and RINGOLD 1989; LEGENDRE and SZOKA 1993). Other studies find that the effect of lysosomotropic agents is dependent on both the formulation and cell type, suggesting that both membrane fusion and endocytosis may occur, depending on the cell type and formulation used (LEGENDRE and SZOKA 1992).

I. Receptor-Mediated Uptake

Uptake of non-viral DNA delivery systems is a rather nonspecific process. Due to the excess positive charge of the complex, it will bind to the negatively charged cell membrane and can subsequently be taken up by either endocytosis or membrane fusion. The specificity of gene expression can be increased and possible toxicity can be reduced by adding targeting ligands to the surface of the DNA delivery vehicle. Various targeting ligands have been described in literature for a variety of DNA delivery systems; an overview is given in Table 2. Using proteins as targeting ligands, such as asialoglycoprotein or antibodies, is inherently associated with the problem of evoking an immune response (STANKOVICS et al. 1994). However, reducing the size of antibodies to Fab fragments, or smaller, may prevent immune responses against these molecules.

Table 2. Studies employing "active targeting" by the use of targeting ligands

Targeting ligand	Targeted receptor/tissue	Vehicle	Reference
In vitro			
Cyclic RGD peptide	Integrin	Poly(L-lysine)	HART et al. 1995
Fab fragment	Polymeric immunoglobulin receptor on (lung) epithelial cells	Poly(L-lysine)	FERKOL et al. 1993
Adenovirus		Poly(L-lysine)	KUPFER et al. 1994
IgG	CD-3 T-cell receptor	Poly(L-lysine)	BUSCHLE et al. 1995
Surfactant Protein A	Airway epithelial cells	Poly(L-lysine)	ROSS et al. 1995
Transferrin	Tf-receptor	Poly(L-lysine)	WAGNER et al. 1991b, 1992
Folate	Folate receptor on tumor cells	Poly(L-lysine)/ anionic liposome	LEE and HUANG 1996
Asialofetuin	Asialoglycoprotein receptor of hepatocytes	Cationic liposome	HARA et al. 1995a; REMY et al. 1995
Galactose	Asialoglycoprotein receptor of hepatocytes	Bisacridine	HAENSLER and SZOKA 1993b
In vivo			
Asialofetuin	Asialoglycoprotein receptor of hepatocytes	Cationic liposome	HARA et al. 1995b
Asialoorosomucoid	Asialoglycoprotein receptor of hepatocytes	Poly(L-lysine)	STANKOVICS et al. 1994
Galactose	Asialoglycoprotein receptor of hepatocytes	Poly(L-lysine)	PERALES et al. 1994
IgG	Thrombomodulin of lung endothelial cells	Poly(L-lysine)	TRUBETSKOY et al. 1992a; TRUBETSKOY et al. 1992b
Fab fragment	Polymeric immunoglobulin receptor on (lung) epithelial cells	Poly(L-lysine)	FERKOL et al. 1995

Strategies, as discussed above, may be combined. TRUBETSKOY et al. (1992a) showed that the transfection efficiency of DNA complexed with N-terminal modified poly(L-lysine) conjugated with anti-thrombomodulin antibody, could be enhanced 10- to 20-fold by addition of cationic liposomes.

E. Endosomal Release

Although the mechanisms behind cellular uptake of DNA with or without carrier are still unclear, the current belief is that, at least in an in vivo situation,

DNA enters the cell by endocytosis (LEGENDRE and SZOKA 1992; ZHOU and HUANG 1994; ZABNER et al. 1995). This route of entry is (still) very inefficient, since most of the DNA will either be retained in perinuclear vesicles or end up in the lysosomes, where it will be degraded. Therefore, a number of strategies have been explored to enhance endosomal release. An overview of the strategies is given in Table 3. Endosomal release implies rupture of the endosome compartment. Two distinctly different strategies are followed: first, by using fusogenic lipids or peptides to disrupt membranes, which may form hexagonal structures or pores in the membranes; second by using DNA delivery systems with high buffering capacity, which will avoid acidification of the endosome and subsequently lead to rupture of the endosomal membrane (REMY et al. 1994). The rationale that polymer/DNA complexes are taken up by cells via endocytosis, and that endosomes can be ruptured if the pH drop of the late endosomes is inhibited by the buffering capacity of the formulation, led to the use of PEI as a DNA delivery polymer (BOUSSIF et al. 1995). Liposome-mediated gene delivery can be greatly enhanced when a pH-sensitive formulation is used (ZHOU et al. 1992c). Enhanced expression of genes transferred by pH-sensitive liposomes is subscribed to the instability of the formulation at a lower pH. The pH-sensitive liposomes will destabilize the endosome, thus allowing escape into the cytoplasm before lysosomal degradation can take place. Other groups also describe the use of fusogenic delivery systems. Plasmid DNA was encapsulated into liposomes containing the nuclear protein high-mobility group I and fused with inactivated Hemagglutinating Virus of Japan (HVJ). Expression of various genes could be obtained with this fusion hybrid of liposome and virus (KATO et al. 1993; TOMITA et al. 1992, 1993; KANEDA et al. 1989; SAWA et al. 1995). In vitro levels of transfection found with this system were similar or slightly better than lipofectin-mediated transfection, depending on the cell type (MORISHITA et al. 1993). One problem for

Table 3. Studies employing delivery systems containing endosomolytic agents

Endosomolytic agent	Vehicle	Reference
In vitro		
Adenovirus	Poly(L-lysine)	KUPFER et al. 1994; CURIEL et al. 1992; WAGNER et al. 1992
Fusogenic peptide from influenza virus	Poly(L-lysine) Cationic liposomes	MIDOUX et al. 1993; LIANG et al. 1996; MORADPOUR et al. 1996
GALA	Polyamidoamine dendrimers	HAENSLER and SZOKA 1993a
In vivo		
Sendai Virus	Anionic liposomes	TOMITA et al. 1992, 1993; KATO et al. 1993; KANEDA et al. 1989
Adenovirus	Poly(L-lysine)	GAO et al. 1993
Adenovirus	Cationic lipids	RAJAWALIA et al. 1995

intravenous delivery of these fusogenic liposomes is their size, which is well over 1 μm (KANEDA et al. 1987). These large liposomes are readily taken up by the reticuloendothelial system. The circulation half-life could be improved by size reduction, but this in turn means an extreme reduction of the encapsulation efficiency of the DNA. Cationic lipid/DNA complexes generally use the fusogenic helper lipid, DOPE (FARHOOD et al. 1995; LITZINGER and HUANG 1992). Upon adhesion of the cationic lipids to the negatively charged lipids in the membranes, it is believed that phase separation may occur. This initiates the presence of DOPE-rich regions inducing hexagonal phases and membrane destabilization.

F. Nuclear Localization

XU and SZOKA (1996) showed that the DNA in a cationic liposome/DNA complex can be displaced from the complex by adding anionic liposomes. This DNA release is suggested to be due to the multivalent nature of the anionic lipid surface, and the collaborative effects of electrostatic interactions and hydrophilic–hydrophobic interactions of the lipids. Lipid mixing results in neutralization of the charges, which allows for diffusion of the cationic lipids away from the DNA. Thus, the anionic lipids in the liposomes will compete with the DNA for binding to cationic lipids. Likewise, it is suggested that the anionic lipids present in the endosome membrane can displace the DNA from cationic lipid/DNA complex. This study shows that DNA release by competition is a very efficient process. Equal moles of anionic lipid added to the cationic lipid/DNA complex leads to 80% release of the DNA. This suggests that the rate-limiting step in transfection is neither the release from the endosome nor the uncoating of the DNA, but the transport of the DNA from the cytoplasm into the nucleus. Microinjection studies also revealed that this transport is an inefficient process (CAPECCHI 1980). DNA injected into the nucleus led to expression of the transgene, where microinjection of the same DNA into the cytoplasm did not. Microinjection studies also revealed that the DNA has to be uncoated before it enters the nucleus, because lipid/DNA complexes that were directly injected into the nucleus did not lead to gene expression. This is in agreement with the studies of XU and SZOKA (1996), who observed that even at a 100-fold charge, excess DNA, tRNA, ATP, poly(glutamic acid), spermine, spermidine and histones were not able to displace DNA from the cationic lipid/DNA complex.

One strategy to circumvent the requirement for nuclear transport is given by a powerful cytoplasmic expression system utilizing the T7 promoter. It was shown that the purified T7 RNA polymerase could be co-delivered with the DNA, using DC-Chol liposomes (GAO and HUANG 1993; GAO et al. 1994). This system is particularly suitable for expression of short RNA molecules, such as anti-sense or ribozymes.

G. Gene Expression

After conquering all the hurdles described above, the DNA finally reaches the nucleus, where the transgene can be expressed. ZABNER et al. (1995) showed that upon microinjection of the monocationic lipid DMRIE/DNA complex into the nucleus, the transfection efficiency was dependent on the lipid/DNA ratio. At ratios at which the lipid/DNA was optimal for transfection, the expression was less when the complex was microinjected, compared with microinjection of naked DNA. At suboptimal lipid/DNA ratios, however, the expression was increased. This suggests that gene expression may be inhibited due to lack of dissociation of the lipid from the DNA. REMY et al. (1994) suggested that an explanation for the higher transfection efficiencies of polyamine-containing delivery systems lies in the high affinity of polyamines for DNA and leads to competitive uncoating of the plasmid DNA by the chromosomal DNA.

Many studies focus on improving the transfection efficiency from the level of transcription, i.e., gene expression. SON and HUANG (1994) reported an interesting finding. Injection of cisplatin in the intraperitoneal cavity, 1 week before intra-tumor injection of the DC-Chol/DNA complex, leads to an enhancement of 10- to 20-fold that of the transfection efficiency. This phenomenon is specific for cisplatin. Other anti-cancer drugs, such as transplatin, carboplatin and other unrelated agents, did not have an effect. Although the exact mechanism behind this enhanced transfection is not fully understood, it is likely to be associated with cisplatin's induced damage to DNA and proteins, which triggers increased repair mechanisms, anti-oxidant levels and PKC activation, which leads to increased gene expression. This offers new possibilities for gene-therapy treatment of cancer patients that fail to respond to conventional chemotherapy. The expression of the transgene can directly be enhanced by incorporating protein kinase C activator, i.e., phobol ester, into the formulation (ZHOU et al. 1992c).

From a molecular point of view, it is clear that one of the major factors in gene expression is the promoter that drives the transcription of the transgene. LI et al. (1992) showed that the use of cytomegalovirus (CMV) promoter yielded higher transfection efficiencies than the use of rous sarcoma virus (RSV), Simian virus-40 (SV-40) or human alpha-1 antitrypsin (hAAT) promoters.

Although some groups have reported some longer persistence of transgene expression (ZHU et al. 1993; THIERRY et al. 1995), a typical time course of expression following non-viral DNA delivery in vivo is 100% for the first 3 days, followed by a rapid decrease, such that after 1 week only 1–10% of the initial expression is found. The expression of the transgene can be prolonged for up to 3 months by including the human papovirus BKV sequence of the viral early region and origin of replication, so that the plasmid is maintained episomally (THIERRY et al. 1995; LIU et al. 1995). Another way to improve the duration of expression is by using vectors that are integrated into the host genome. Adenovirus associated viral (AAV) vectors have an advantage over retroviruses, as the integration is site specific in a non-coding region of chromosome 19. Therefore, the chance of oncogenesis should be reduced and the transgene should be present for the entire life span of the host cell and its progeny (Muzyczka 1992).

When gene delivery is no longer the limiting factor in reaching therapeutic levels of the transgene, the problem of how to regulate gene expression must be addressed. Tissue-specific promoters, or even inducible promoters, would be an elegant way to keep the expression of the desired transgene within the therapeutic window. One example is given by Wang and Huang (1989), who used pH-sensitive anionic immunoliposomes for the delivery of the Herpes tyrosine-kinase gene under the rat promoter for the phosphoenolpyruvate carboxykinase gene. This promoter can be regulated by cAMP drugs (Wynshaw-Boris et al. 1984). Another example of inducible gene expression is via the tetracycline (tet)-regulated transactivation system (Shockett and Schatz 1996). This system is based on the transcriptional transactivators that interact specifically with bacterial *cis*-regulatory elements, and tetracycline that modulates the binding of the transactivators. The gene will be switched on by giving the patient low, non-toxic doses of the antibiotic.

References

Acsadi G, Dickson G, Love DR, Jani A, Walsh FS, Gurusinghe A, Wolff JA, Davies KE (1991) Human dystrophin expression in mdx mice after intramuscular injection of DNA constructs. Nature 352:815–818

Alino SF, Bobadilla M, Crespo J, Lejarreta M (1996) Human alpha1-antitrypsin gene transfer to in vivo mouse hepatocytes. Hum Gene Ther 7:531–536

Andree C, Swain WF, Page CP, Macklin MD, Slama J, Hatzis D, Eriksson E (1994) In vivo transfer and expression of a human epidermal growth factor gene accelerates wound repair. Proc Natl Acad Sci USA 91:12188–12192

Aoki K, Yoshida T, Sugimura T, Terada M (1995) Liposome-mediated in vivo gene transfer of antisense K-ras construct inhibits pancreatic tumor dissemination in the murine peritoneal cavity. Cancer Res 55:3810–3816

Ardehali A, Fyfe A, Laks H, Drinkwater DC, Qiao JH, Lusis AJ (1995) Direct gene transfer into donor hearts at the time of harvest. J Thorac Cardiovasc Surg 109: 716–720

Barthel F, Remy JS, Loeffler JP, Behr JP (1993) Gene transfer optimization with lipospermine-coated DNA. DNA Cell Biol 12:553–560

Behr JP, Demeneix B, Loeffler JP, Perez-Mutul J (1989) Efficient gene transfer into mammalian primary endocrine cells with lipopolyamine-coated DNA. Proc Natl Acad Sci USA 86:6982–6986

Berns AJM, Clift S, Cohen LK, Donehower RC, Dranoff G, Hauda KM, Jaffee EM, Lazenby AJ, Levitsky HI, Marshall FF, Mulligan RC, Nelson WG, Owens AH, Pardoll DM, Parry G, Partin AH, Piantadosi S, Simons JW, Zabora JR (1995) Clinical Protocol: PhaseI study of non-peplicating autologous tumor cell injections using cells prepared with or without GM-CSF gene transduction in patients with metastatic renal cell carcinoma. Hum Gene Ther 6:347–368

Boussif O, Lezoualc'h F, Zanta MA, Mergny MD, Scherman D, Demeneix B, Behr JP (1995) A versatile vector for gene and oligonucleotide transfer into cells in culture and in vivo: polyethylenimine. Proc Natl Acad Sci USA 92:7297–7301

Brigham KL, Schreier H (1993) Cationic liposomes and DNA delivery. J Liposome Res 3:31–49

Buschle M, Cotten M, Kirlappos H, Mechtler K, Schaffner G, Zauner W, Birnstiel ML, Wagner E (1995) Receptor-mediated gene transfer into human T lymphocytes via binding of DNA/CD3 antibody particles to the CD3 T cell receptor complex. Hum Gene Ther 6:753–761

Cao L, Zheng ZC, Zhao YC, Jiang ZH, Liu ZG, Chen SD, Zhou CF, Liu XY (1995) Gene therapy of parkinson disease model rat by direct injection of plasmid DNA-lipofectin complex. Hum Gene Ther 6:1497–1501

Capecchi MR (1980) High efficiency transformation by direct microinjection of DNA into cultured mammalian cells. Cell 22:479–488

Caplen NJ, Alton EWFW, Middleton PG, Dorin JR, Stevenson BJ, Gao X, Durham SR, Jeffery PK, Hodson ME, Coutelle C, Huang L, Porteous DJ, Williamson R, Geddes DM (1995) Liposome-mediated CFTR gene transfer to the nasal epithelium of patients with cystic fibrosis. Nat Med 1:39–46

Curiel DT, Wagner E, Cotten M, Birnstiel ML, Agarwal S, Li C (1992) High-efficiency gene transfer mediated by adenovirus coupled to DNA-polylysine complexes. Hum Gene Ther 3:147–154

Farhood H, Bottega R, Epand RM, Huang L (1992) Effect of cationic cholesterol derivatives on gene transfer and protein kinase c activity. Biochim Biophys Acta submitted

Farhood H, Serbina N, Huang L (1995) The role of dioleoyl phosphatidylethanolamine in cationic liposome mediated gene transfer. BBA-Biomembranes 1235:289–295

Felgner PL, Gadek TR, Holm M, Roman R, Chan HW, Wenz M, Northrop JP, Ringold GM, Danielsen M (1987) Lipofection: a highly efficient, lipid mediated DNA-transfection procedure. Proc Natl Acad Sci USA 84:7413–7417

Felgner PL, Ringold GM (1989) Cationic liposome-mediated transfection. Nature 337: 387–388

Felgner JH, Kumar R, Sridhar CN, Wheeler CJ, Tsai YJ, Border R, Ramsey P, Martin M, Felgner PL (1994) Enhanced gene delivery and mechanism studies with novel series of cationic lipid formulations. J Biol Chem 269:2550–2561

Ferkol R, Kaetzel CS, Davis PB (1993) Gene transfer into respiratory epithelial cells by targeting the polymeric immunoglobulin receptor. J Clin Invest 92:2394–2400

Ferkol T, Perales JC, Eckman E, Kaetzel CS, Hanson RW, Davis PB (1995) Gene transfer into the airway epithelium of animals by targeting the polymeric immunoglobulin receptor. J Clin Invest 95:493–502

Gao X, Huang L (1991) A novel cationic liposome reagent for efficient transfection of mammalian cells. Biochem Biophys Res Commun 179:280–285

Gao X, Huang L (1993) Cytoplasmic expression of a reporter gene by co-delivery of T7 RNA polymerase and T7 promoter sequence with cationic liposomes. Nucleic Acids Res 21:2867–2872

Gao L, Wagner E, Cotten M, Agarwal S, Harris C, Romer M, Miller L, Hu P, Curiel D (1993) Direct in vivo gene transfer to airway epithelium employing adenovirus-polylysine-DNA complexes. Hum Gene Ther 4:17–24

Gao X, Jaffurs D, Robbins PD, Huang L (1994) A sustained, cytoplasmic transgene expression system delivered by cationic liposomes. Biochem Biophys Res Commun 200:1201–1206

Gao X, Huang L (1995) Cationic liposome-mediated gene transfer. Gene Ther 2:710–722

Guo LSS, Radhakrishnan R, Redemann CT, Brunette EN, Debs RJ (1993) Cationic liposomes containing noncytotoxic phospholipid and cholesterol derivatives. J Liposome Res 3:51–70

Haensler J, Szoka FC (1993a) Polyamidoamine cascade polymers mediate efficient transfection of cells in culture. Bioconjugate Chem 4:372–379

Haensler J, Szoka FC (1993b) Synthesis and characterization of a trigalactosylated bisacridine compound to target DNA to hepatocytes. Bioconjugate Chem 4:85–93

Hara T, Aramaki Y, Takada S, Koike K, Tsuchiya S (1995a) Receptor-mediated transfer of pSV2CAT DNA to a human hepatoblastoma cell line HepG2 using asialofetuin-labeled cationic liposomes. Gene 159:167–174

Hara T, Aramaki Y, Takada S, Koike K, Tsuchiya S (1995b) Receptor-mediated transfer of pSV2CAT DNA to mouse liver cells using asialofetuin-labeled liposomes. Gene Ther 2:784–788

Hart SL, Harbottle RP, Cooper R, Miller A, Williamson R, Coutelle C (1995) Gene delivery and expression mediated by an integrin-binding peptide. Gene Ther 2:552–554

Hawley-Nelson P, Ciccarone V, Gebeyehu G, Jessee J, Felgner PL (1993) LipofectAMINE reagent: a new, higher efficiency polycationic liposome transfection reagent. Focus 15:80

Hickman MA, Malone RW, Lehmann-Buinsma K, Sih TR, Knoell D, Szoka FC, Walzem R, Carlson DM, Powell JS (1994) Gene expression following direct injection of DNA into liver. Hum Gene Ther 5:1477–1483

Hofland HEJ, Shephard L, Sullivan SM (1996) Formation of stable cationic lipid/DNA complexes for gene transfer. Proc Natl Acad Sci USA (in press)

Hyde S, Gill D, Higgins C, Trezise A, MacVinish LJ, Cuthbert AW, Ratcliff R, Evans MJ, Colledge WH (1993) Correction of the ion transport defect in cystic fibrosis transgenic mice by gene therapy. Nature 362:250–255

Ito A, Miyazoe R, Mitoma J, Akao T, Osaki T, Kunitake T (1990) Synthetic cationic amphiphiles for liposome-mediated DNA transfection. Biochem Int 22: 235–241

Jiao S, Williams P, Berg RK, Hodgeman BA, Liu L, Repetto G, Wolff JA (1992) Direct gene transfer into nonhuman primate myofibers in vivo. Hum Gene Ther 3:21–33

Kaneda Y, Uchida T, Kim J, Ishiura M, Okada Y (1987) The improved efficient method for introducing macromolecules into cells using HVJ (Sendai virus) liposomes with gangliosides. Exp Cell Res 173:56–69

Kaneda Y, Iwai K, Uchida T (1989) Introduction and expression of the human insulin gene in adult rat liver. J Biol Chem 264:12126–12129

Kato K, Dohi Y, Yoneda Y, Yamamura K, Okada Y, Nakanishi M (1993) Use of the hemagglutinating virus of Japan (HVJ)-liposome method for analysis of infiltrating lymphocytes induced by hepatitis B virus gene expression in liver tissue. Biochim Biophys Acta 1182:283–290

Kukowska-Latallo JF, Bielinska AU, Johnson J, Spindler R, Tomalia DA, Baker JR (1996) Efficient transfer of genetic material into mammalian cells using starburst polyamidoamine dendrimers. Proc Natl Acad Sci USA 93:4897–4902

Kupfer JM, Ruan XM, Liu G, Matloff J, Forrester J, Chaux A (1994) High-efficiency gene transfer to autologous rabbit jugular vein grafts using adenovirus-transferrin/polylysine-DNA complexes. Hum Gene Ther 5:1437–1443

Lee RJ, Huang L (1996) Folate-targeted, anionic liposome-entrapped polylysine-condensed DNA for tumor cell-specific gene transfer. J Biol Chem 271:8481–8487

Legendre J-Y, Szoka FC (1992) Delivery of plasmid DNA into mammalian cell lines using pH-sensitive liposomes: comparison with cationic liposomes. Pharm Res 9:1235–1242

Legendre J-Y, Szoka F (1993) Cyclic amphipathic peptide-DNA complexes mediate high-efficiency transfection of adherent mammalian cells. Proc 90:893–897

Leibiger B, Leibiger I, Sarrach D, Zuhlke H (1991) Expression of exogenous DNA in rat liver cells after liposome-mediated transfection in vivo. Biochem Biophys Res Commun 174:1223–1231

Li AP, Myers CA, Kaminski DL (1992) Gene transfer in primary cultures of human hepatocytes in vitro. Cell 28A:373–375

Liang WW, Shi X, Deshpande D, Malanga CJ, Rojanasakul Y (1996) Oligonucleotide targeting to alveolar macrophages by mannose receptor-mediated endocytosis. Biochim Biophys Acta 1279:227–234

Litzinger DC, Huang L (1992) Phosphatidylethanolamine liposomes: drug delivery, gene transfer and immunodiagnostic applications. Biochim Biophys Acta 1113: 201–227
Liu Y, Liggitt D, Zhong W, Tu G, Gaensler K, Debs R (1995) Cationic liposome-mediated intravenous gene delivery. J Biochem 270:24864–24870
Lu D, Benjamin R, Kim M, Conry RM, Curiel DT (1994) Optimization of methods to achieve mRNA-mediated transfection of tumor cells in vitro and in vivo employing cationic liposome vectors. Cancer Gene Ther 1:245–252
Mahato RI, Kawabata K, Takakura Y, Hashida M (1995a) Control of in vivo disposition of plasmid DNA using cationic liposomes. J Drug Target 3:149–157
Mahato RI, Kawabata K, Nomura T, Takakura Y, Hashida M (1995b) Physicochemical and pharmacokinetic characteristics of plasmid DNA/cationic liposome complexes. J Pharm Sci 84:1267–1271
Midoux P, Mendes C, Legrand A, Raimond J, Mayer R, Monsigny M, Roche AC (1993) Specific gene transfer mediated by lactosylated poly-L-lysine into hepatoma cells. Nucleic Acids Res 21:871–878
Missol E, Sochanik A, Szala S (1995) Introduction of murine Il-4 gene into B16(F10) melanoma tumors by direct gene transfer with DNA-liposome complexes. Cancer Lett 97:189–193
Moradpour D, Schauer JI, Zurawski VR, Wands JR, Boutin RH (1996) Efficient gene transfer into mammalian cells with cholesteryl-spermidine. Biochem Biophys Res Commun 221:82–88
Morishita R, Gibbons GH, Kaneda Y, Ogihara T, Dzau VJ (1993) Novel in vitro gene transfer method for study of local modulators in vascular smooth muscle cells. Hypertension 21:894–899
Mulligan R (1993) The basic science of gene therapy. Science 260:926–932
Muzyczka N (1992) The use of adeno-associated virus as a general transduction vector for mammalian cells. Curr Top Microbiol Immunol 158:97–129
Nabel G, Chang A, Nabel EG, Plautz G, Fox BA, Huang L, Shu S (1992) Immunotherapy of malignancy by in vivo gene transfer into tumors. Hum Gene Ther 3:399–410
Nabel G, Nabel E, Yang Z, Fox B, Plautz G, Gao X, Huang L, Shu S, Gordon D, Chang A (1993) Direct gene transfer with DNA-Liposome complexes in melanoma: expression, biologic activity, and lack of toxicity in humans. Proc 90:11307–11311
Nicolau C, Le Pape A, Soriano P, Fargette F, Juhel MF (1983) In vivo expression of rat insulin after intravenous administration of the liposome-entrapped gene for rat insulin I. Proc Natl Acad Sci USA 80:1068–1072
Ostro M, Giacomoni D, Dray S (1977) Incorporation of high molecular weight RNA into large artificial lipid vesicles. Biochem Biophys Res Commun 76:836–842
Perales JC, Ferkol T, Beegen H, Ratnoff OD, Hanson RW (1994) Gene transfer in vivo: Sustained expression and regultaion of genes introduced into the liver by receptor-targeted uptake. Proc Natl Acad Sci USA 91:4086–4090
Plautz G, Nabel E, Nabel G (1993) Liposome Mediated Gene Transfer Into Vascular Cells. J Liposome Res 3:179–199
Rajawalia R, Webber J, Naftilan J, Chapman GD, Naftilan AJ (1995) Enhancement of liposome-mediated gene transfer into vascular tissue by replication deficient adenovirus. Gene Ther 2:521–530
Remy J-S, Sirlin C, Vierling P, Behr JP (1994) Gene transfer with a series of lipophilic DNA-binding molecules. Bioconjugate Chem 5:647–654
Remy JS, Kichler A, Mordvinov V, Schuber F, Behr JP (1995) Targeted gene transfer into hepatoma cells with lipopolyamine-condensed DNA particles presenting galactose ligands: a stage toward artificial viruses. Proc Natl Acad Sci USA 92:1744–1748
Ross GF, Morris RE, Ciraolo G, Huelsman K, Bruno M, Whitsett JA, Baatz JE, Korfhagen TR (1995) Surfactant protein A-polylysine conjugates for delivery of DNA to airway cells in culture. Hum Gene Ther 6:31–40

Sawa Y, Suzuki K, Bai HZ, Shirakura R, Morishita R, Kaneda Y, Matsuda H (1995) Efficiency of in vivo gene transfection into transplanted rat heart by coronary infusion of HVJ liposome. Circulation 92:479–482
Schwartz B, Benoist C, Abdallah B, Scherman D, Behr JP, Demeneix B (1995) Lipospermine-based gene transfer into the newborn mouse brain is optimized by a low lipospermine/DNA charge ratio. Hum Gene Ther 6:1515–1524
Shockett PE, Schatz DG (1996) Diverse strategies for tetracycline-regulated inducible gene expression. Proc Natl Acad Sci USA 93:5173–5176
Son K, Huang L (1994) Exposure ofhuman ovarian carcinoma to cisplatin transiently sensitizes the tumor cells for liposome-mediated gene transfer. Proc Natl Acad Sci U S A 91:12669–12672
Soriano P, Legrand A, Spanjer H, Londos-Gagliardi D, Roerdink F, Scherphof G, Nicolau C (1983) Targeted and non-targeted liposomes for in vivo transfer to rat liver cells of a plasmid containing preproinsulin I gene. Proc Natl Acad Sci USA 80:11277–11281
Stankovics J, Crane AM, Andrews E, Wu CH, Wu GY, Ledley FD (1994) Overexpression of human methylmalonyl CoA mutase in mice after in vivo gene transfer with asialoglycoprotein/polylysine/DNA complexes. Hum Gene Ther 5:1095–1104
Sternberg B, Sorgi FL, Huang L (1994) New structures in complex formation between DNA and cationic liposomes visualized by freeze-fracture electron microscopy. FEBS Lett 356:361–366
Stribling R, Brunette E, Liggitt D, Gaensler K, Debs R (1992) Aerosol gene delivery in vivo. Proc 89:11277–11281
Sun WH, Burkholder JK, Sun J, Culp J, Lu XG, Pugh TD, Ershler WB, Yang NS (1995) In vivo cytokine gene transfer by gene gun reduces tumor growth in mice. Proc Natl Acad Sci USA 92:2889–2893
Takehara T, Hayashi N, Miyamoto Y, Yamamoto M, Mita E, Fusamoto H, Kamada T (1995) Expression of the hepatitis C virus genome in rat liver after cationic liposome-mediated In vivo gene transfer. Hepatology 21:746–751
Thierry AR, Lunardi-Isakandar Y, Bryant JL, Rabinovich P, Gallo RC, Mahan LC (1995) Systemic gene therapy: biodistribution and long-term expression of a transgene in mice. Proc Natl Acad Sci U S A 92:9742–9746
Tomita N, Higaki J, Morishita R, Kato K, Mikami H, Kaneda Y, Ogihara T (1992) Direct in vivo gene introduction into rat kidney. Biochem Biophys Res Commun 186:129–134
Tomita N, Higaki J, Kaneda Y, Yu H, Morishita R, Mikami H, Ogihara T (1993) Hypertensive rats produced by in vivo introduction of the human renin gene. Circ Res 73:898–905
Trubetskoy VS, Torchilin VP, Kennel S, Huang L (1992a) Cationic liposomes enhance targeted delivery and expression of exogenous DNA mediated by N-terminal modified poly(L-lysine)-antibody conjugate in mouse lung endothelial cells. Biochim Biophys Acta 1131:311–313
Trubetskoy VS, Torchilin VP, Kennel SJ, Huang L (1992b) Use of N-terminal modified poly(L-lysine)-antibody conjugate as a carrier for targeted gene delivery in mouse lung endothelial cells. Bioconjugate Chem 3:323–327
Ulmer JB, Donnelly JJ, Parker SE, Rhodes GH, Felgner PL, Dwarki VJ, Gromkowski SH, Deck RR, De Witt CM, Friedman A (1993) Heterologous protection against influenza by injection of DNA encoding a viral protein. Science 259:1745–1749
Wagner E, Cotten M, Mechtler K, Kirlappos H, Birnstiel ML (1991a) DNA binding transferrin conjugates as funtional gene-delivery agents: synthesis by linkage of polylysine or ethidium homodimer to the transferrin carbohydrate moiety. Bioconjugate Chem 2:226–231
Wagner E, Cotten M, Foisner R, Birnstiel ML (1991b) Transferrin-polycation-DNA complexes: The effect of poly-cations on the structure of the complex and DNA delivery to cells. Proc Natl Acad Sci USA 88:4255–4259

Wagner E, Zatloukal K, Cotten M, Kirlappos H, Mechtler K, Curiel DT, Birnstiel ML (1992) Coupling of adenovirus to transferrin-polylysine-DNA complexes greatly enhances receptor-mediated gene delivery and expression of transfected genes. Proc Natl Acad Sci USA 89:6099–6103

Wang C-Y, Huang L (1987a) Plasmid DNA adsorbed to pH-sensitive liposomes efficiently transforms the target cells. Biochem Biophys Res Commun 147:980–985

Wang C-Y, Huang L (1987b) pH-sensitive immunoliposomes mediate target-cell-specific delivery and controlled expression of a foreign gene in mouse. Proc Natl Acad Sci USA 84:7851–7855

Wang C-Y, Huang L (1989) Highly efficient DNA delivery mediated by pH-sensitive immunoliposomes. Biochemistry 28:9508–9514

Wells D (1993) Improved gene transfer by direct plasmid injection associated with regeneration in mouse skeletal muscle. FEBS Lett 332:179–182

Wolff JA, Malone RW, Williams P, Chong W, Acsadi G, Jani A, Felgner PL (1990) Direct gene transfer into mouse muscle in vivo. Science 247:1465–1468

Wolff JA, Williams P, Acsadi G, Jiao S, Jani A, Chong W (1991) Conditions affecting direct gene transfer into rodent muscle in vivo. BioTechniques 11:474–485

Wynshaw-Boris A, Lugo TG, Short JM, Fournier RE, Hanson RW (1984) Identification of a cAMP regulatory region in the gene for rat cytosolic phosphoenolpyruvate carboxykinase (GTP). Use of chimeric genes transfected into hepatoma cells. J Biol Chem 259:12161–12169

Xu Y, Szoka FC (1996) Mechanism of DNA release from cationic liposome/DNA complexes used in cell transfection. Biochemistry 35:5616–5623

Yang N-S, Burkholder J, Roberts B, Martinell B, McCabe D (1990) In vivo and in vitro gene transfer to mammalian somatic cells by particle bombardment. Proc Natl Acad Sci USA 87:9568–9572

Yoshimura K, Rosenfeld MA, Nakamura H, Scherer EM, Pavirani A, Lecocq JP, Crystal RG (1992) Expression of the human cystic fibrosis transmembrane conductance regulator gene in the mouse lung after in vivo intratracheal plasmid-mediated gene transfer. Nucleic Acids Res 20:3233–3240

Yu DH, Matin A, Xia WY, Sorgi F, Huang L, Hung MC (1995) Liposome-mediated in vivo E1 A gene transfer suppressed dissemination of ovarian cancer cells that overexpress HER-2/neu. Oncogene 11:1383–1388

Zabner J, Fasbender AJ, Moninger T, Poellinger KA, Welsh MJ (1995) Cellular and molecular barriers to gene transfer by a cationic lipid. J Biol Chem 270:18997–19007

Zhou X, Klibanov A, Huang L (1991) Lipophilic polylysines mediate efficient DNA transfection in mammalian cells. Biochim Biophys Acta 1065:8–14

Zhou X, Huang L (1992a) Liposomes containing lipopolylysine as a vehicle for gene transfer II: mechanism of action. J Biol Chem submitted

Zhou X, Huang L (1992b) Targeted delivery of DNA by liposomes and polymers. J Control Rel 19:269–274

Zhou X, Klibanov A, Huang L (1992c) Improved encapsulation of DNA in pH-sensitive liposomes for transfection. J Liposome Res 2:125–139

Zhou X, Huang L (1994) DNA transfection mediated by cationic liposomes containing lipopolylysine: characterization and mechanism of action. Biochim Biophys Acta 1189:195–203

Zhu N, Liggit D, Liu Y, Debs R (1993) Systemic gene expression after intravenous DNA delivery into adult mice. Science 261:209–211

CHAPTER 9

Safe, Efficient Production of Retroviral Vectors

H. KOTANI and G.J. MCGARRITY

A. Introduction

The first clinical trials in gene therapy were held in the United States in 1990 at the National Institutes of Health (NIH). The first trial was for gene marking. Lymphocytes were removed from a melanoma biopsy, expanded in vitro and transduced with a retroviral vector carrying a gene that conferred resistance to the neomycin analogue, G-418. The clinical objective was to determine the distribution and longevity of these marked lymphocytes. Shortly thereafter, on 14 September 1990, the first clinical trial in gene therapy was initiated. A child with a form of severe combined immune deficiency (SCID), due to a defect in the adenosine deaminase (ADA) gene, was re-infused with her own lymphocytes which had been propagated ex vivo and transduced with the retroviral vector (LASN) that contained the wild-type ADA gene. Results of this and a related ADA trial, which involved the introduction of the gene into $CD34^+$ cells from umbilical-cord blood obtained from neonates, have been published (BLAESE et al. 1995; KOHN et al. 1995).

The initial clinical studies in gene therapy were for trials involving a small number of patients, usually less than 15. Typically, this meant that volume of vector was not a deciding factor. More often, the deciding factor was defining and achieving good manufacturing practices (GMP) as they applied to the new field of gene therapy. Also critical was the development of quality-control (QC) assays that were essential for lot release. Many of the required QC assays for gene-therapy products were similar to other products manufactured from mammalian cell lines (SMITH et al. 1996).

In the United States, the Food and Drug Administration (FDA) is the appropriate regulatory agency that has oversight of human trials in gene therapy. At one point, the Recombinant DNA Advisory Committee (RAC) of the NIH held public meetings to review and approve clinical proposals for clinical trials in this field, although the FDA always had to give final approval. Because of the role of the RAC, a public record exists of the disease targets and the type of gene delivery system, or vectors, that are in use and development. Eventually, RAC review of proposed clinical trials ended; the FDA is now the sole reviewer of clinical trials in the United States. Regulatory authorities in many European countries, Japan and the Peoples Republic of China have also approved clinical trials in the specialty. The journal, *Human*

Gene Therapy, publishes an update of United States and international clinical protocols in the field (ANONYMOUS 1997).

The purpose of this chapter is to review safe and effective methods that have been employed to manufacture retroviral vectors in large scale. This technology is still at an early stage. At the time of writing, no product employing retroviral or other vectors for gene therapy has been approved for commercial use. Nevertheless, it is believed that material presented here should contribute to an understanding of the issues needed to produce these vectors on a commercial scale.

B. Vectors

A variety of methods have been utilized to deliver genes to cells. The instrument of gene delivery is defined as a vector. Two major types of vectors have been employed; viral and non-viral, or synthetic. Viral vectors have been developed from different viruses. To date, vectors derived from retrovirus, adenovirus, herpes virus, adeno-associated virus (AAV) and a canary pox virus have been used in human trials. Each viral-vector type has specific characteristics that may make it more appropriate for a specific cell target or a specific clinical condition. To develop a viral vector from its parent virus, the vector is rendered replication incompetent and a significant portion of the viral genome is removed to accommodate the foreign genes to be inserted. The vector will generally contain only sequences of the transgene(s), plus any promoter, enhancer or related sequences that aim to maximize or control gene expression, as well as sequences needed to assemble or package the vector, such as *psi* for retroviruses. The structural genes, e.g., capsid of adenovirus or envelope of retrovirus, are encoded in *trans* in packaging elements in a cell referred to as a packaging cell (without vector sequences) or a producer cell for retroviral vectors (with packaging sequences).

The major viral vectors employed in gene-therapy trials have been retroviral, adenoviral and AAV systems; these have been reviewed (MCGARRITY and CHIANG 1997; BLAESE 1997). Non-viral vector systems have also been described, including the use of plasmid DNA, as well as DNA packaged within liposomes and other carriers.

I. Retroviral Vectors

Retroviral vectors have been the most frequently used gene-transfer system. Retroviral vectors have been developed from the Moloney strain of mouse leukemia virus (Mo-MuLv). Replication-incompetent retroviral vectors have been widely used in early gene marking and phase-I/-II clinical trials (MCGARRITY and CHIANG 1997). Among the more than 170 RAC-approved human gene-therapy protocols reported at the end of 1997, approximately 70% have employed retroviral vectors. The Mo-MuLV has a simple genome, consisting of three genes: *gag*, *pol* and *env*, which are contained within long terminal repeats (LTR) at the 5′ and 3′ ends of the genome. The parental virus is diploid, meaning that two RNA molecules are present. The RNA genome is contained within a viral envelope, which is partially obtained from the plasma membrane of the host cell. Following incorporation of the virus (or vector) into the host cell via specific receptors, the RNA is converted to DNA by viral reverse transcriptase. The proviral DNA is integrated randomly into the host genome. Viral proteins are transcribed from the provirus, and RNA is packaged into virions utilizing viral nucleotide sequences known as *psi*, or encapsidation factor. The virions are released by budding into the extracellular fluid. The infection does not lyse the host cells, and viruses are continuously generated. The difference between the life cycle of the wild-type virus and the vector is that the entire set of molecular components necessary to generate vectors are found only in the producer cells, not in the cells transduced by the vectors. The components necessary to generate vectors are provided by the producer cells on at least two DNA fragments. If the vector is used to transduce a target cell, the proviral DNA is integrated into the host genome and the transgene DNA is expressed. However, progeny vectors are not produced.

Current retroviral vectors can incorporate approximately 8 kb of nucleotides. To generate replication-incompetent retroviral vectors, packaging cell-line systems have been developed from a variety of permanent cell cultures (Table 1). To generate amphotropic packaging cell lines, viral structural genes (*pol*, *gag* and *env*) were transfected into mouse or human cells. Vector sequences were then transfected into the packaging cell containing the transgene(s), along with the *psi* sequence and appropriate internal promoters for the second transgene (if any) into the packaging cells. Vectors are elaborated from vector producer cells (VPC) into the culture supernatant. These supernatants are then collected and processed to yield the vector product. A description of retroviral packaging and producer cells is shown in Fig. 1. Retroviral vectors budding from the plasma membrane of producer cells are shown in Fig. 2.

Retroviral vectors have been demonstrated to efficiently transduce many different types of human cells in vitro and have been extensively employed for

Table 1. Retroviral packaging cell lines

Host	Packaging cells	Parent cells	Helper genome	References
Mouse	PA317	NIH3T3	Single helper genome	Miller and Buttimore 1986
Mouse	ΨCRIP	NIH3T3	Split genome	Danos and Mulligan 1988
Mouse	GP + envAM 12	NIH3T3	Split genome	Markowitz et al. 1988
Dog	DA	D17	Split genome	Mento 1994
Human	ProPak	293	Split genome	Rigg et al. 1996

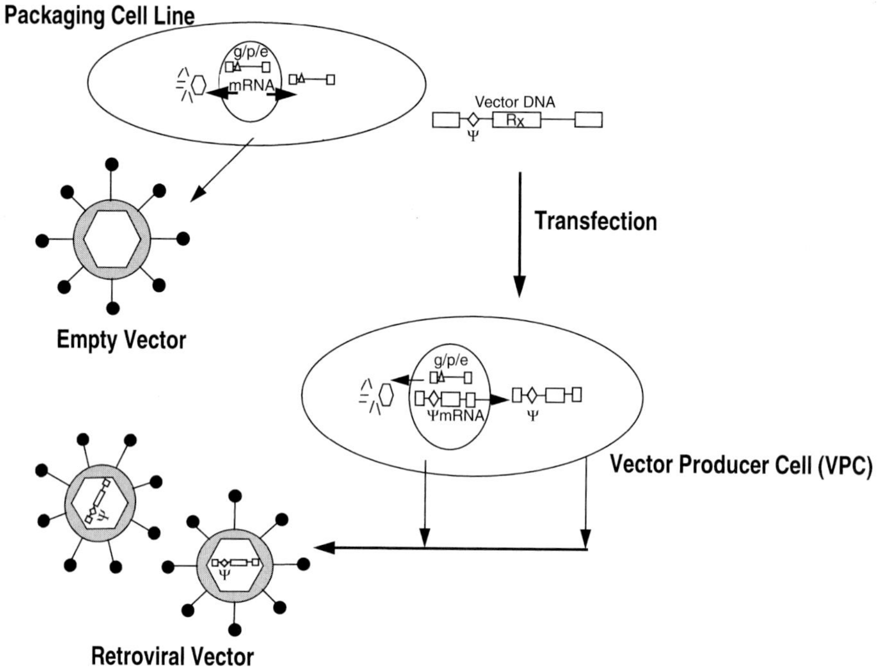

Fig. 1. Production of retroviral vector producer cells (VPCs). The retroviral genes *gag*, *pol* and *env* are depicted as *g*, *p*, and *e*, respectively. *Psi* (4) refers to the retroviral packaging or encapsulation sequences. *Rx* refers to the transgene

ex-vivo gene-therapy protocols, including hematopoietic cells, fibroblasts and tumor cells. Retroviral-mediated gene-therapy protocols have been used for a variety of different diseases, including genetic diseases, acquired diseases, especially acquired immunodeficiency syndrome (AIDS), and cancer. Current retroviral-vector systems have limitations since they can only transduce replicating cells and are rapidly inactivated by human serum. The efficiency of retroviral vectors to transduce certain human cells, such as peripheral lymphocytes and hematopoietic stem cells, is low. Nevertheless, these vectors have found broad utilization in early clinical trials, especially in in vitro protocols, where cells are removed from the patient, propagated in the laboratory, transduced with the retroviral vector and re-infused into the patient.

C. Production of Retroviral Vectors

A variety of methods have been developed for growing mammalian cells in large scale. These have been reviewed (FINTER et al. 1990; LUBINIECKI 1990). Most of these methods are appropriate for gene therapy in general, and retroviral vectors specifically. However, mouse retroviral VPCs are typically

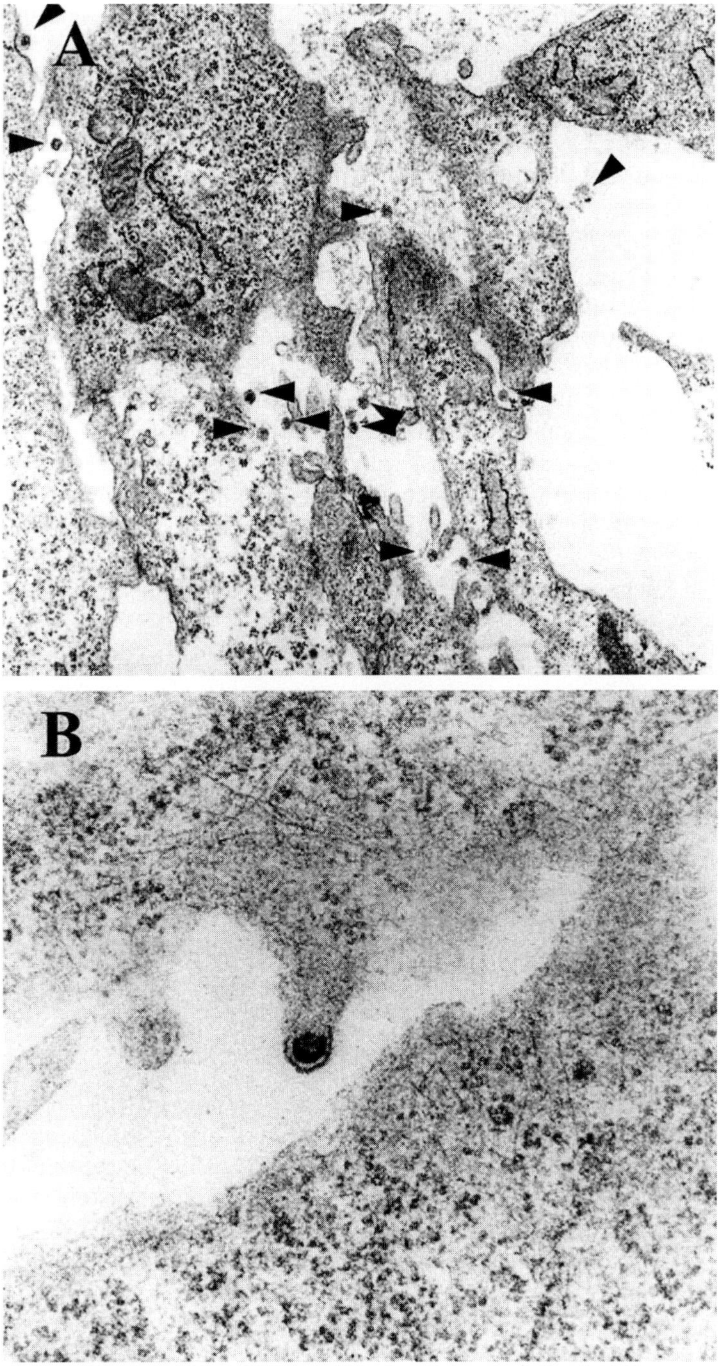

Fig. 2A,B. Electron micrographs of G1Tk1SvNa vector (**A** ×12,500, **B** ×50,000)

anchorage dependent. Efficient retroviral production has not been reported in suspension cultures. For vector production, the cells are cultured as monolayers. In early manufacturing, tissue-culture flasks and roller bottles have been widely used. More recently, technology for larger scale production of retroviral vectors has been developed, and several articles have been published. A significant production concern in gene therapy is the potential of the system to generate replication-competent retroviruses (RCR). This can occur in the producer cell through the recombination of sequences from vector and helper sequences of packaging cells. Different packaging cell lines have been generated to minimize the potential of RCR generation. This is achieved by minimizing the homology between vector and helper sequences and by placing the viral structural genes on two or more molecules instead of one, to generate a split packaging cell. Theoretically, recombination can also occur between the vector sequences and other endogenous retroviral sequences in either the packaging or the transduced host cells. Different producer cells have been generated from murine cells, which contain a significant amount of retroviral sequences. In addition, it is also theoretically possible for a recombination event to occur between the retroviral vector and endogneous-retroviral sequences in the human target cell, although this has never been observed in an in vitro system.

I. Production Methods

1. Batch Systems

In our laboratories, VPCs derived from the amphotropic retroviral-vector packaging cells PA317, which were derived from NIH 3T3 cells, have been extensively used. PA317 has been described (MILLER and BUTTIMORE 1986). Vectors bud from VPCs into the culture supernatant without causing cell lysis. Multiple supernatant harvests are collected at intervals following addition of fresh medium to confluent cells. These VPCs have been propagated on flat surfaces in a variety of culture vessels.

2. Roller Bottles

Roller-bottle technology remains a widely used method for viral-vaccine production, despite being labor intensive and lacking control systems such as pH- and dissolved-oxygen monitors used in bioreactors. Robotic roller-bottle systems offer potential to increase process reliability and consistency. Robotic systems can automate virtually all roller-bottle processes: inoculation, medium change, decantation of supernatant, cell washing, trypsinization and cell harvesting in a standardized, efficient manner. The surface areas of roller bottles yield cell numbers that are commercially relevant, using either 850 cm^2 or 1700 cm^2 roller bottles.

For the production of retroviral-vector supernatant, we reported results of vector titers from roller bottles under different conditions, including different cultivation temperatures, surface areas of roller bottle and harvest times of supernatant using many different producer cell lines, all derived from PA317 packaging cells (KOTANI et al. 1994). A non-GMP production of G1Na.40 vector production in 850 cm^2 roller bottles at 37°C is shown in Fig. 3. The medium was changed every day, and daily samples were taken after cell confluence was reached. The vector titer was 5×10^6 colony-forming units (CFU)/ml at day 5 and remained at approximately that level until day 11, when it declined slowly. These results suggested that vectors can be collected continuously for at least for 2 weeks, although clonal variability can influence this characteristic. These collections were free of RCR. The maximum number of harvests is dependent on the type and clone of the producer cell line. The vector titer is partially dependent on the number of cells because vector yield per cell is limited. Some cell lines and clones tend to detach from the surface of the roller bottle after reaching confluence. We reported selection of one clone due to its increased titer, rapid growth and effectiveness in vitro and in vivo (LYONS et al. 1995). When different cultivation temperatures of producer cell lines were compared, highest titers were always at 32°C, an approximate 1-log increase at 32°C, compared with at 37°C. Collection of supernatants every

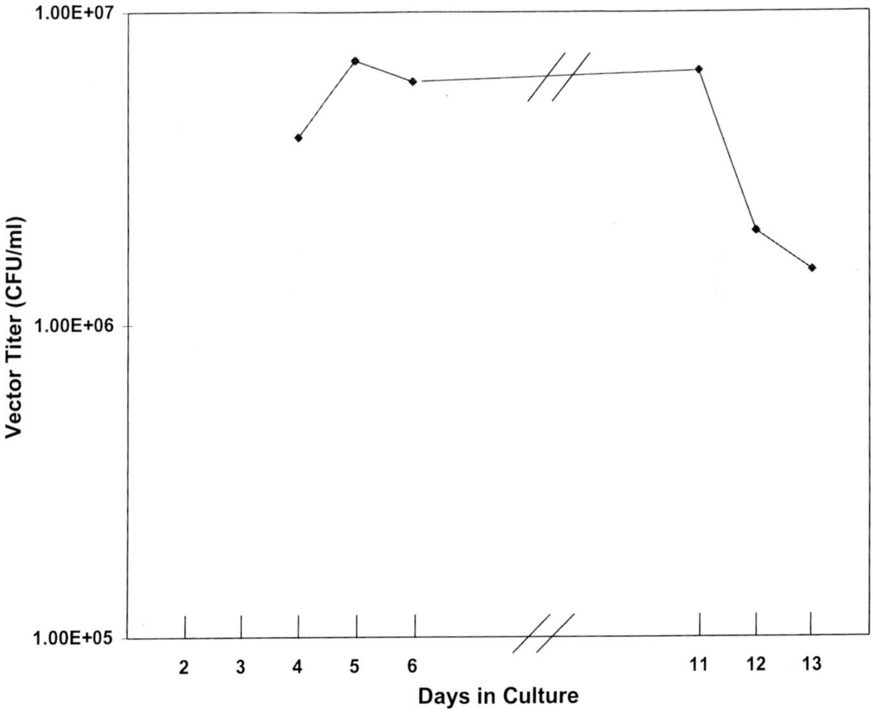

Fig. 3. Production of G1Na.4 vectors in roller bottles

48 h can further increase the number of retroviral vector particles collected. This is influenced by the half-life of retroviral vectors at 37°C, estimated to be 4–6 h (KOTANI et al. 1994). Roller-bottle production can be a useful method for small-scale production with or without robotics.

Additional vector-transduction efficiency can be achieved by centrifugation of the vector onto the target cells (KOTANI et al. 1994) or having the vector flow over the target cells (CHUCK and PAULSSON 1996).

3. Multilayered Propagator

Various efforts have been made to increase the surface area for cell growth by using multiple surfaces with multiplates mounted horizontally in a vessel such as Cell Factory (NUNC). Retroviral VPC lines have been grown at 32°C in this system. Medium and cells are added manually. Supernatant is collected by gravity. New medium can be added into the same system for further harvests as multiple collection at least three or five times. In limited studies in this laboratory, titers were similar to that obtained in roller bottles.

II. Bioreactors

1. CellCube Bioreactor

The CellCube bioreactor is a perfused system with controlled monitors that contain multiple plates in a densely stacked module for anchorage-dependent cells. This is scaleable to a maximum of 16 cube units (340,000 cm^2). The system has been used to expand a variety of anchorage-dependent cells as an alternate method to roller bottles. We investigated the possibility of large-scale retroviral-vector production by means of the CellCube system. Three different producer cell lines were cultured with Dulbecco's modified Eagle's medium (DMEM) containing 10% fetal bovine serum (FBS) in 1 cube unit (21,250 cm^2) at 32°C using a batch collection. There were 2 l supernatants collected every 48 h. High-titer vectors ($>7 \times 10^6$ CFU/ml) were generated for more than 3 weeks after the study was ended without the presence of RCR. The vector titers obtained from this system were significantly higher than that of the roller bottle.

The results of a development non-GMP, run in 16 cube units, with PA317/G1NaSvAd.24-producer cell line that produce an ADA vector, are shown in Fig. 4. VPCs were cultivated at 32°C with perfusion for 23 days. The vector reached a titer of 6×10^6 CFU/ml on day 3 with 20 l of harvest/day and yielded the highest titer on day 8 with 30 l harvest/day. Titer declined gradually thereafter, possibly due to the rapid decline of glucose concentration because of uncontrolled dissolved oxygen. Cell metabolism increased as evidenced by glucose fermentation by the large number of cells (approximately 1.6×10^{11} cells) in the system. The oxygen limitation in the long culture period may limit scale up. However, a newer generation oxygenerator is now available, which may alleviate this problem.

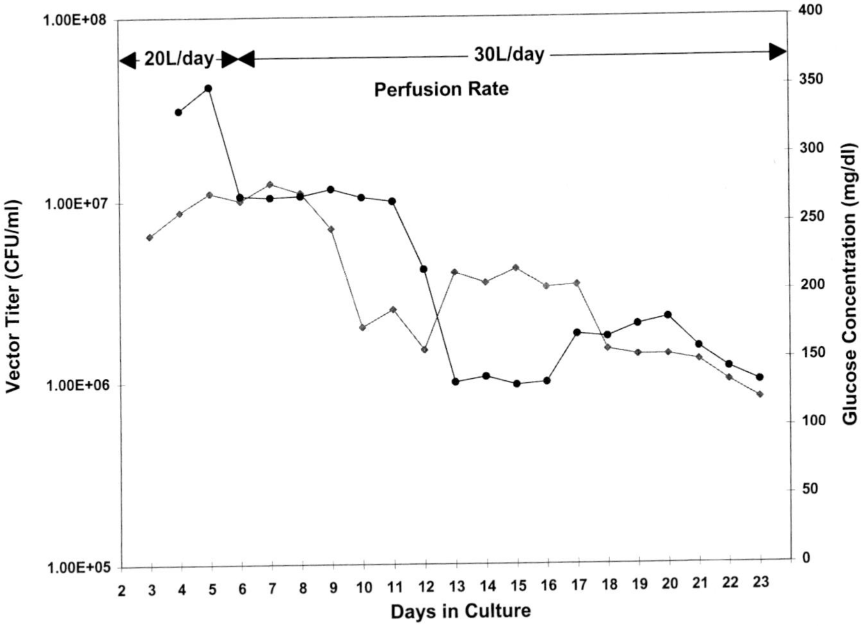

Fig. 4. Production of G1NaSvAd.24 vector in CellCube bioreactor. Vector titer (*squares*), glucose concentration (*circles*)

The summary of gene-therapy protocols from other laboratories list 1 CellCube unit and 2 CellCube units that have been successfully employed, and have produced clinical-grade vector supernatants for phase-I/-II cancer gene-therapy protocols (Smiley et al. 1997). The supernatant was produced consistently for 3 weeks without detectable RCR. Due to a high concentration of cell density and a reduced volume, vector titer became higher than that of standard monolayer culture systems.

2. Hollow-Fiber Bioreactor

Hollow-fiber perfusion bioreactors have been used to cultivate mammalian cells for large-scale production of secreted proteins and monoclonal antibodies. This technology was also used for production of viruses, including retroviruses and viral proteins for vaccines or diagnostic purposes. In another laboratory, the PA317/LNL-6-producer cell line (carrying the gene conferring resistance to the neomycin analogue G418) was grown in the Cellco Cellmax type-B artificial capillary cartridge with DMEM+10% FBS at 34°C under monitoring of lactate production and glucose-consumption rates (Blaese, personal communication). The medium was replaced with AIM-V serum-free medium, and daily supernatant harvests were collected. The vector titer from the cartridge was significantly higher than that from flasks in DMEM containing 10% FBS (1.2×10^7 CFU/ml vs 8.6×10^5 CFU/ml), due to high cell density. Further scale-up capability of this system as a single unit has not been reported.

3. Microcarrier Beads in Bioreactor

Anchorage-dependent cells require a surface. Microcarriers have been developed for large-scale vaccine production and therapeutic and diagnostic proteins (Griffiths 1991). Several types of microcarriers are available. Collagen microcarriers allow cells to grow, not only the surface of beads, but the inside of beads as well, resulting in higher cell density. Developments in this field resulted in effective microcarriers for fluidized beds, but were less successful for stirred suspensions. Microcarrier beads in stirred-tank fermentors with batch or semi-batch systems and perfusion systems have been widely employed for production of cells and secreted proteins from engineered mammalian cells in industry. Fluidization produces less mechanical stress to microcarriers. A report has been published on the production of retroviral vectors (Mento 1994). Producer cells were grown on Sephadex beads (Pharmacia Cytodex) in a CelliGen bioreactor (New Brunswick Scientific, New Brunswick, NJ) with a continuous perfusion system. The supernatant containing vector was processed for downstream processing, including concentration, purification by size-exclusion column, and lyophilization for a direct injection of vectors into patients, who were human immunodeficiency virus (HIV) positive, as gene-therapy protocol (Martineau et al. 1997).

4. Packed-Bed Air-Lift Bioreactor

Several methods have been employed for the larger scale production of retroviral vectors. All systems have advantages and disadvantages. In general, we believe that methods used in mammalian cell culture in biotechnology should work well for retroviral-vector production. In gene transfer, high titer of non-replicating vectors is one of the critical issues for efficient gene transfer into target cells. The number of vector particles in culture supernatants depends, in part, on the number of cells in the culture system per defined time period. High-cell-density culture systems generate higher titer in the supernatant than that in low-cell-density culture systems. However, the highest attainable titers achieved from current retroviral producer cells are typically 10^6–10^7 CFU/ml.

To increase the vector concentration and stability and remove chemical contaminants and inhibitors, further downstream processing is necessary. This includes clarification, concentration, purification and lyophilization. Retroviral vectors are relatively fragile. The mouse leukemia-virus envelope contains gp70, which is sensitive to shear. Supernatants that are viscous with many foreign proteins from medium and mammalian cells are more difficult to clarify, concentrate and purify. Despite this, we have been able to generate retroviral vector having titer in excess of 5×10^8 CFU/ml following downstream processing (KOTANI unpublished results). We reported the lyophilization of retroviral vectors (KOTANI et al. 1994).

We have been investigating various technologies to produce retroviral vectors. High-cell-density culture systems using macroporous carriers in packed-bed bioreactors have been used in the biotechnology field for production of monoclonal antibodies and secreted proteins. Those bioreactors, however, usually have an external oxgenator and recirculation loop for medium exchange. The external-medium recirculation system becomes the limiting factor for scalability of bioreactors.

We have designed a novel bioreactor called a packed-bed air-lift bioreactor (GTI-AL2) to overcome the above problems for long-term, continuous production of retroviral vectors (U.S. Patent 5,563,068). The AL series of bioreactor is a packed-bed bioreactor containing macroporous glass beads using an internal air-lift mechanism for both mixing and oxygenation of medium. High-titer vectors have been successfully produced in the bioreactor, in the order of 10^7–10^8 CFU/ml. Producer cells grow on macroporous glass beads in a packed bed. Medium mixing and oxygenation in the bioreactor were easily provided by an air-lift sparging mechanism. The configuration of the bioreactor is shown in Fig. 5. Macroporous glass beads were packed in a second column that is situated in the center of the bioreactor. The packed-bed column was made of stainless-steel mesh. Oxygenation by sparging was performed in the medium between the reactor column and the column where the beads are held. Oxygen supply to cells grown on macroporous beads in the packed-bed is very effective by medium circulation through the bead bed. This sized AL reactor contains a packed-bed volume consisting of 700 ml porous borosilicate glass beads 3–5 mm in diameter. During production, 5% CO_2 in air was constantly sparged into the medium. Air flow was controlled. Dissolved oxygen in the bioreactor was monitored; an oxygen monitor in the culture medium was maintained above 50% of air saturation by regulating the air flow rate. The pH in the reactor sustained in the range of 7.1–7.3. PA317/G1Tk1SvNa.7 VPCs grown on borosilicate beads are shown in Fig. 6.

5. Serum-Containing Production

PA317/G1Na.40 producer cell line (Neo^R gene) was cultured in an 850-cm^2 roller bottle in DMEM+10% FBS and inoculated (2×10^8 cells) into the AL2 bioreactor and incubated at 37°C. The medium used for perfusion was DMEM+10% heat-inactivated FBS, 0.1% pluronic F-68 and 10 ppm antifoam C. The perfusion rate was controlled by the glucose and lactate concentrations in the supernatant by a peristaltic pump. The vector titer, vector yield, glucose/lactate concentrations and medium perfusion rates for a representative 34-day run in the AL2 bioreactor is shown in Fig. 7. Vector titer increased logarithmically during the first week of culture from 5×10^5 CFU/ml to 9×10^7 CFU/ml. Total vector yield per day reached more than $2–3\times10^{11}$ CFU by increasing

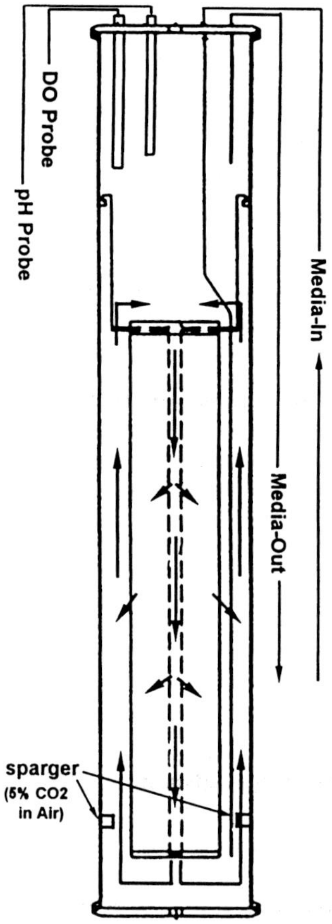

Fig. 5. Schematic configuration of the packed-bed air-lift bioreactor. *Arrows* indicate patterns of medium flow

medium perfusion from day 1 through day 15 and then maintaining a constant perfusion rate until the end of the run. The titer gradually declined to 10^6 CFU/ml. The highest vector yield was 9×10^{11} CFU/day. The reason for the decline in vector titer after 3 weeks of culture is unknown. This phenomenon has been observed in different producer cells using different cultivation procedures in both small and large scale. The decline in the amount of vector produced may be due to cessation of an essential vector component, for example, *gag*. Two other producer cell lines were also cultured for more than 1 month in the bioreactor, with similar results for titer and total vector yield. A total of 12 l/day of vector supernatant was harvested continuously. Samples from different stages of cultures in the bioreactor were free of detectable RCR, as determined by a biological S+/L− assay.

Cell growth in the bioreactor was estimated by the cell-specific metabolic rates of glucose and lactose. The total number of cells in this size bioreactor was estimated to be 4×10^{10}. Cells grew exponentially during 10 days in culture

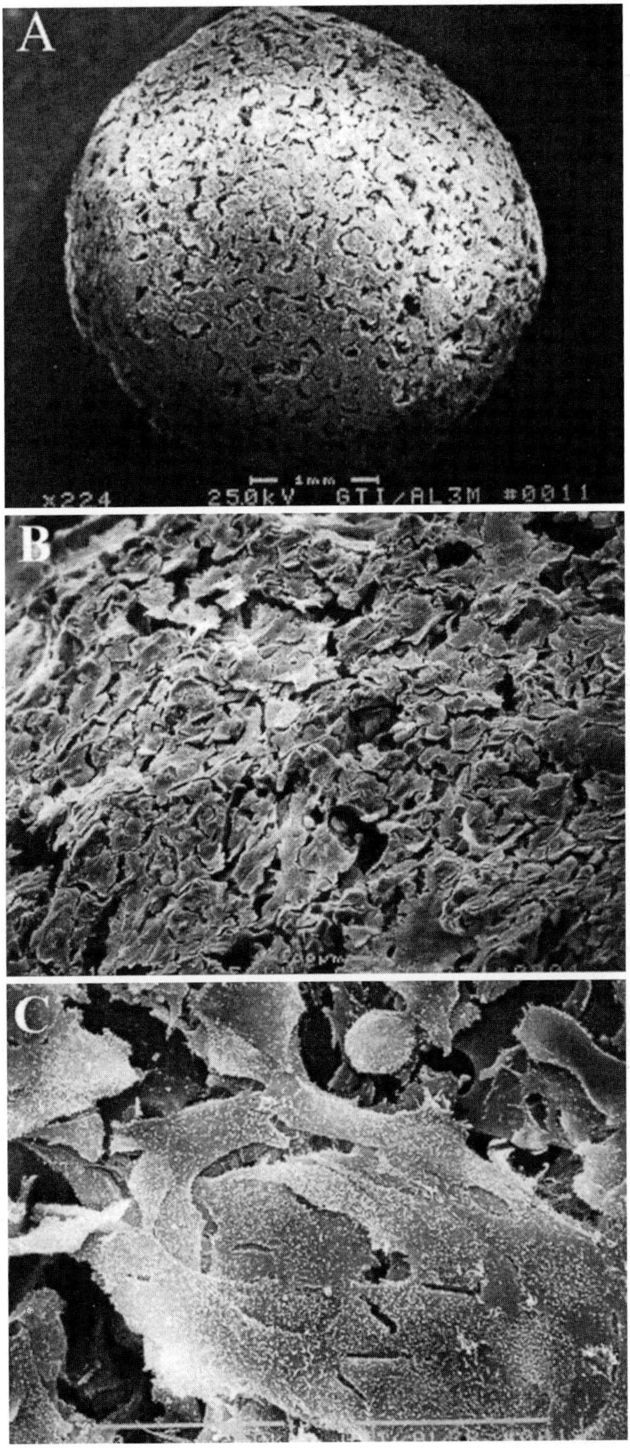

Fig. 6A–C. Electron micrographs of PA317/G1Tk1SvNa.7 vector producer cells grown on a borosilicate bead. (**A**, ×22.4; **B**, ×210; **C**, ×1820)

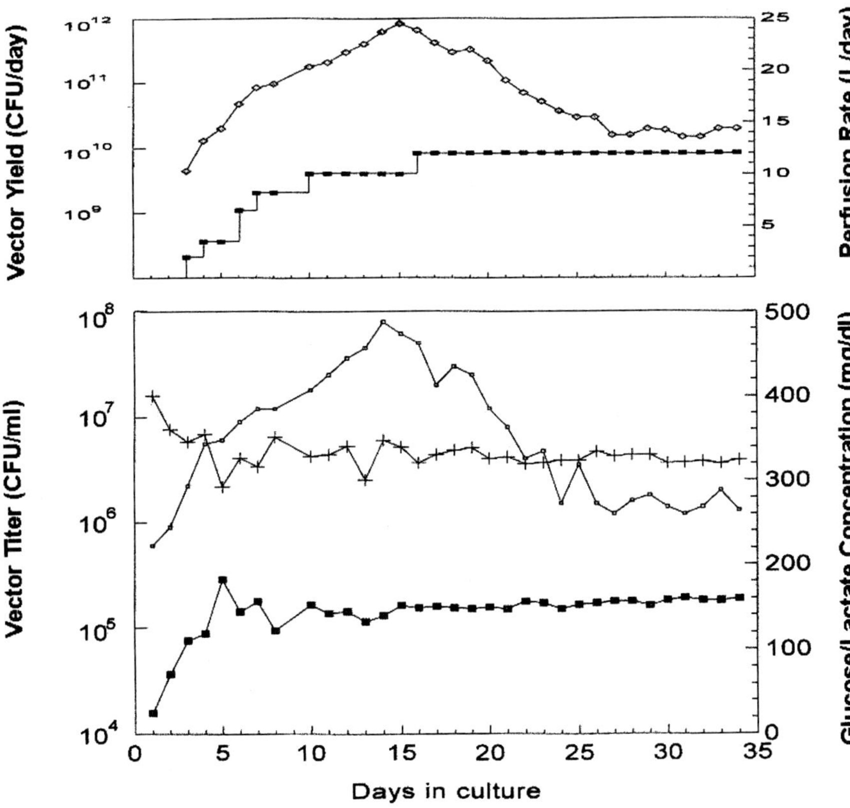

Fig. 7. Production of G1Na.40 vector in AL2 bioreactor. Vector titer (*open squares*), vector yield (*open diamonds*), glucose concentration (+), lactate concentration (*filled squares*)

and remained at plateau levels until the end of culture, demonstrating a 200-fold increase from the initial cell inoculum (2×10^8 cells). Consequently, the cell numbers needed to inoculate the bioreactor were relatively small, only 2% of final cell concentration. Cell-doubling time in the bioreactor was approximately 27 h, similar to 28 h measured in a 175-cm^2 flask. Cell concentration in the packed-bead bed was estimated as 1×10^7 cells/bead.

The AL2 bioreactor demonstrated the capability to maintain high cell density and continuous high-titer production of retroviral vectors, without detectable RCR events for at least 1 month of culture run with 20 different runs. The downstream procedures involve clarification, concentration, purification and eventual lyophilization. Based on the result from these pilot studies, the packed-bed air-lift bioreactor is scaleable for a large production of retroviral vectors. Larger versions of the AL2 bioreactor described here have been constructed (AL3, 2.4L and AL4, 11.5L bead volume) and perform similarly to the AL2 regarding vector titer.

Table 2. Comparison of different technologies for retroviral vector production

	Roller bottle ($850 cm^2$)	CellCube	AL2 Bioreactor
No. of units	100	4	1
Total no. of cells	4×10^{10}	4×10^{10}	4×10^{10}
Harvest/day (l)	15	10	13
Vector titer (CFU/ml)	2×10^7	$2–3 \times 10^7$	$2–5 \times 10^7$
Production mode	Batch	Continuous	Continuous
Run time (day)	3	>21	>28
Total harvest per run (l)	80	210	280
Labor	Intensive	Moderate	Minimum
RCR	Neg	Neg	Neg

RCR, replication competent retrovirus; CFU, colony forming units; Neg, negative

Most large-scale production techniques developed for mammalian cell cultures can also be utilized to produce retroviral vectors. In the large-scale mammalian cell culture, factors such as hydrodynamic shear, oxygen-transfer limitation, surface-area limitation and lactate build up must be addressed. Table 2 shows the comparison of three different technologies: roller bottles, CellCube and AL2 bioreactor. These comparisons were performed on the same numbers of PA317/G1Tk1SvNa.7 producer cells in each system (4×10^{10}). The average vector titer ($2–3 \times 10^7$ CFU/ml) was similar among the three methods. However, the duration of operation time, labor intensity and the harvested volume are significantly different. The roller-bottle production was harvested for at least 72 h after 100% confluence; the harvest was 80 l. However, CellCube and AL2 bioreactors can operate at least for 21 days and 28 days, respectively, before the significant decline of vector titer is observed, typically after 3 weeks of culture. Due to the continuous perfusion system, the harvest of supernatants in the CellCube and AL2 were 210 l/run and 280 l/run, respectively. The most significant difference is labor for upstream operation among the three methods. The AL2 bioreactor required minimum labor, suggesting the potential for future large-scale production. No RCRs were detected by biological assay, even following a 1-month long production period.

D. Downstream Processing

Currently, the majority of human trials in gene therapy utilizing retroviral gene transfer are ex-vivo, meaning that patient cells, e.g., lymphocytes, bone marrow, tumor cells, skin fibroblast, etc., were transduced with the vector during logarithmic growth of the cells in vitro (or ex vivo) and then administered to the patient. The majority of clinical trials in the field are phase-I or phase-I/-II clinical trials. Therefore, supernatants as sources of vectors have been used for transduction of target cells in vitro. Many ex vivo gene-transfer protocols are limited, due to the target cell or low vector titers. Low

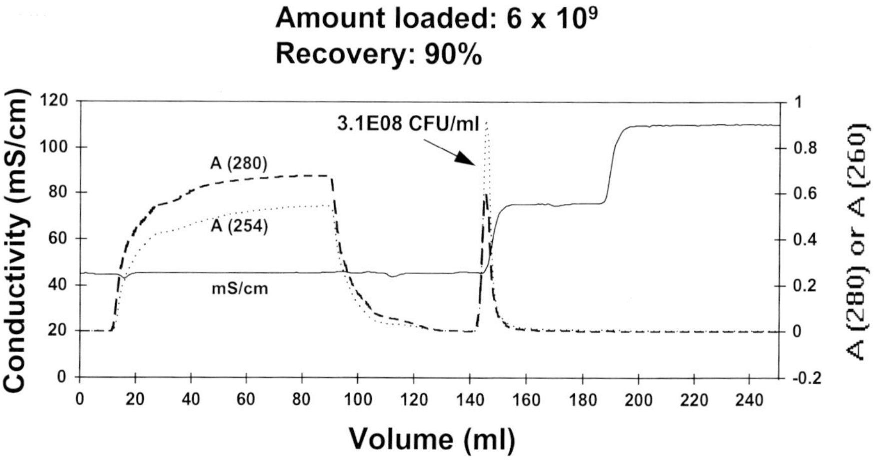

Fig. 8. Column chromatograph of G1Tk1SvNa.7 vector

transduction efficiency may also be due to the presence of inhibitors or contaminants in the harvested supernatant. For example, vector-free gp70, empty vectors and defective vectors can be present in these preparations and may compete with infectious vector for the retroviral receptor on the target cell. Retroviral vectors are relatively stable against phycochemical treatments, such as pH, ionic strength and shear forces. To increase titer, therefore, vector supernatants can be concentrated by several methods, including polyethylene glycol (PEG) precipitation, sucrose-gradient ultracentrifugation or tangential-flow ultrafiltration. To achieve a high efficiency of gene transfer to target cells, such as hematopoietic cells, myoblasts or endotherial cells, vectors must demonstrate not only high titer but also a high degree of purity.

Purification techniques have been employed for live viruses, including vaccines in the laboratory. Column-chromatographic purification technology has been incorporated for gene-therapy vectors. An ion-exchange column demonstrated purification capability of retroviral vectors due to strong anionic-charge properties of viral vectors. Size-exclusion chromatography has also been employed (Mccormack et al. 1997). Recovery of a high yield of retroviral vector from a G1Tk1SvNa.7 vector supernatant by column chromatography is demonstrated in Fig. 8.

E. GMP Production of Retroviral Vectors

The present state of gene-therapy product development is comparable to the early stages of recombinant proteins and monoclonal antibodies. Many clinical protocols have been approved by the FDA and other regulatory agencies throughout the world. However, as stated, most are phase I and often are for diseases that have no successful alternate therapies. Many patients have a

Table 3. Quality control (QC) assays

QC Assays for master cell bank
1. Sterility
2. Mycoplasma
3. Transmission electron microscopy for viruses
4. Assays for adventitious viruses
   - MAP test
   - In vitro virus assay
   - In vivo virus assay
5. Identity
   - Isoenzyme and cytogenetics analysis
   - Southern analysis
   - Functional test (if necessary)
6. Tumorigenicity
7. Potency
   - Biological vector titer
8. Replication competent retrovirus
   - Mus dunni cocultivation followed by S+/L− assay
   - Mus dunni amplification followed by S+/L− assay

QC assay for product lots
1. Sterility
2. Mycoplasma
3. Pyrogenicity
4. Identity
   - Southern analysis and/or function test
5. Potency
   - Vector titer
6. RCA assay by cocultivation and amplification
   - Supernatant 5% of total supernatant
   - Producer cells 1% of total pooled cells or 10^8 cells

short life expectancy. Typically, companies and academic institutions manufacture vectors for phase-I or phase-I/-II clinical trials involving a small number of patients. For making clinical-grade vectors, production and QC must be carried according to GMP requirements. Currently, the retroviral vectors have been primarily produced in batch systems. Scale-up in most cases was achieved by increasing vessel size or number of vessels, including multilayered vessels and roller bottles. However, those technologies have limitations.

I. Cell Banking

Clinical-grade retroviral vectors are produced by the cells from cell banks [Master Cell Bank (MCB) and Working Cell Bank (WCB)] under GMP. Cell lines used in MCB for gene therapy must meet the same requirements as those used for production of recombinant proteins and monoclonocal antibodies (Wiebe and May 1990), as well as those criteria specific for gene therapy, such as analysis for the transgene(s) by Southern analysis and/or a functional bioassay as well as assays for RCR. Each cell bank must demonstrate genetic stability, and be free of adventitious agents, including bacteria, fungi, mycoplasma and viruses. Tests are shown in Table 3. Both culture supernatant and

VPC must be assayed for RCR according to FDA recommendations, which is 1% for VPC, up to a maximum of 1×10^8, and 5% of the supernatant volume for RCR assays.

Validation of vector production under serum-containing medium or serum-free (SF) medium is important for safety and standardization concerns. It is believed that most clinical-grade vectors for phase I or phase I/II used supernatant produced from VPCs in FBS-containing medium. The use of FBS raises several issues, including FBS regulatory concerns over adventitious agents. To overcome these regulatory hurdles, we developed SF cell banking and production systems. A complete SF process of retroviral-vector production is achievable. The VPCs from an original MCB is adapted in a SF medium and cryopreserved; development of freezing medium is essential for SF cell banks.

The characterization of cell banks for use in gene therapy are similar to that described for cell banks for use in production of recombinant proteins (WEIBE and MAY, 1990). The principles for gene therapy are a modification of those described by WIEBE and MAY:

1. The origin and general characterization of the cells should be known. This means the origin and passage history of the cell lines. The origin should include definition of the species, sex, organ and/or tissue of the cells. Deliberate alterations to the cell line should be described, such as transfection of plasmids to make the packaging cell, as well as transfection of vector DNA to make the retroviral VPC. Other characterization should also be noted, including morphology, cloning data, vector titer and confirmation of animal species.
2. The cell bank itself and cells derived from the bank must be demonstrated to be free of adventitious agents by sensitive, validated assays.
3. Particular attention must be made to demonstrate absence of RCR, by means of sensitive, validated assays. In addition, the product identity must be demonstrated by inoculation of the vector made from the cell banks into appropriate target cells to show delivery and expression of the specific transgene. This can be accomplished by molecular probes in a Southern assay and an appropriate bioassay.
4. A maximum limit of population doublings for continuous cell culture during production should be defined, and the cells characterized at appropriate passages at, prior to, and beyond that passage level. The population doublings from the WCB through the end of production and the post-production period should be calculated. Any production that goes beyond this number of population doublings must be re-characterized to minimize the potential of spurious findings in QC assays. Parameters to be considered include cell growth, morphology, vector titer, absence of RCR, overall product quality and transducibility of target cells with subsequent expression of the transgene. These characteristics should be defined in development runs, and this information should be used to generate a certificate of analysis for final product lots.

5. Cell-culture-derived contaminants must be defined and appropriate sensitive assays developed. Appropriate assays are needed to validate removal of these materials during purification.

Upon controlled freezing and storage in liquid nitrogen, the cell banks maintain viability for long periods. High viability of cell banks after 20 years and 27 years of storage has been described (Weibe and May 1990). We have had similar experiences with a large number of animal and human cells, often spanning 25–30 years (unpublished). Many of our early packaging cells and producer cells for retroviral vectors have been maintained for 7–9 years in liquid nitrogen, and no change in viability as been detected.

For cryopreservation of cells grown in SF medium, the freeze medium must contain 7 mg/ml of protein plus cryopreservatives.

II. Serum-Free Upstream Processing

Retroviral vectors have been produced by cells grown in SF medium. The AL3 bioreactor is a larger version of the AL2 described earlier and has a 9.5-l vessel for GMP production. The VPCs from one frozen SF-WCB vial are recovered and expanded in SF medium in flasks and roller bottles and then inoculated into an AL3 bioreactor. Cells attach and grow on the glass surface of the borosilicate beads and appropriate oxygen transport and medium circulation is obtained by air sparging via air lift. The AL3 bioreactor is a perfused and continuous harvesting system. Comparative results of G1Tk1SvNa7 vector production in the AL3 bioreactor with DMEM+2% FBS and SF medium are shown in Fig. 9. In the SF system, an inoculum five times larger than that required in serum-containing medium is necessary. The vector titer reached the maximum level on day 3 in the SF system and remained relatively stable throughout the 21 days of operation. The perfusion rate was modulated to maintain a constant glucose-consumption rate. The average titer was 1.5×10^7 CFU/ml and the vector yield was 3×10^{11} CFU/day, demonstrating the comparability of the SF system to the one employing the serum system. We are validating this SF system for GMP production.

III. Serum-Free Downstream Processing

In most early ex vivo clinical trials, culture supernatants containing vectors and prepared in serum-containing medium without purification were used to transduce target lymphocytes, bone marrow or tumor cells. The negative aspect of this approach is low transduction efficiency due to low vector titer and/or inhibitors in the supernatant. Representative titers were 1–10×10^5 CFU/ml. The supernatant contains many undefined components from cell-culture-conditioned medium. We have developed a supernatant production by the AL3 bioreactor under SF conditions, followed by hollow-fiber concentration and the column chromatographic-purification process. The

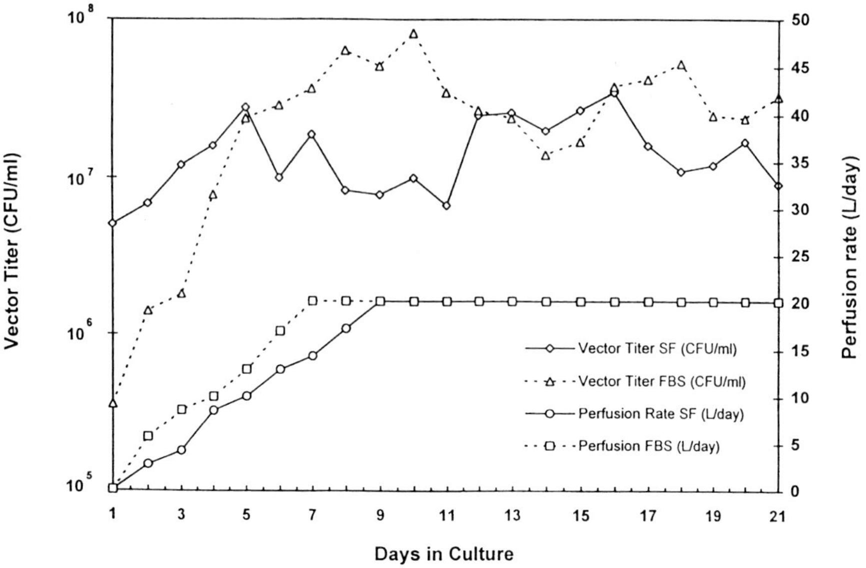

Fig. 9. Comparative studies of G1Tk1SvNa.7 vector production in AL3 bioreactor under serum-containing (FBS) medium and serum-free (SF) medium

stages of this process are outlined in Fig. 10. All harvests are collected and processed as batches. The daily unprocessed vector supernatant (20 l/day) was harvested at 4°C. A collection of 2-day harvests (40 l) was clarified through a 0.22-fm filter to remove cellular debris and concentrated up to 20 times using 300 k molecular weight cut-off hollow fiber. No significant loss of vector occurred with this process. This process effected a partial removal of cellular and medium proteins as well as fragmented vector proteins. The concentrated materials were treated with Benzonase (DNase), resulting in a significant reduction (2–3 log) of host DNA and frozen at –80°C as a batch. Batches were then pooled after thawing to form a lot and then applied to an anion-exchange column chromatography. The retroviral vectors bound to resins and eluted with a high-salt solution. The chromatogram shows a clear vector fraction was obtained by a step gradient. Representative results are shown in Fig. 11. The vector fraction was diafiltrated and concentrated by a hollow fiber. The final vector with the titer of 10^8 CFU/ml was formulated, filled in vials as liquid formulation, and frozen to –80°C. Long-term stability studies at –80°C are in progress. The final vector demonstrated low proteins (10–50 pg/ml) and DNA levels (50–500 pg/patient dose).

F. In-Process Assays

A number of assays have been developed to characterize the upstream- and downstream-production processes. Table 4 shows the list of assays for each

Upstream

MCB-SF
↓
MWCB-SF
↓
Cell Expansion — Tissue Culture Flask
↓
Roller Bottle
↓
AL-3 Bioreactor/SF
↓

Downstream

Harvest
↓
Clarification
↓
Concentration
↓
DNase
↓
Diafiltration
↓
Column Chromatography
↓
Concentration/Diafiltration
↓
Formulation
↓
Final Filtration
↓
Filling
↓
Storage at -80°C

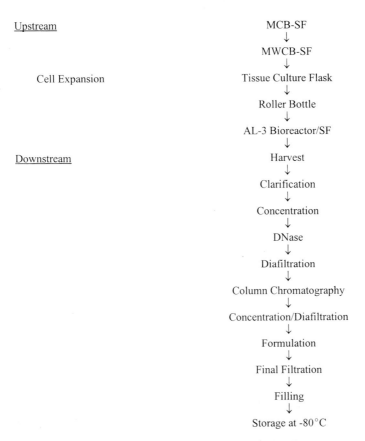

Fig. 10. Flow chart of serum-free (*SF*) retroviral vector production

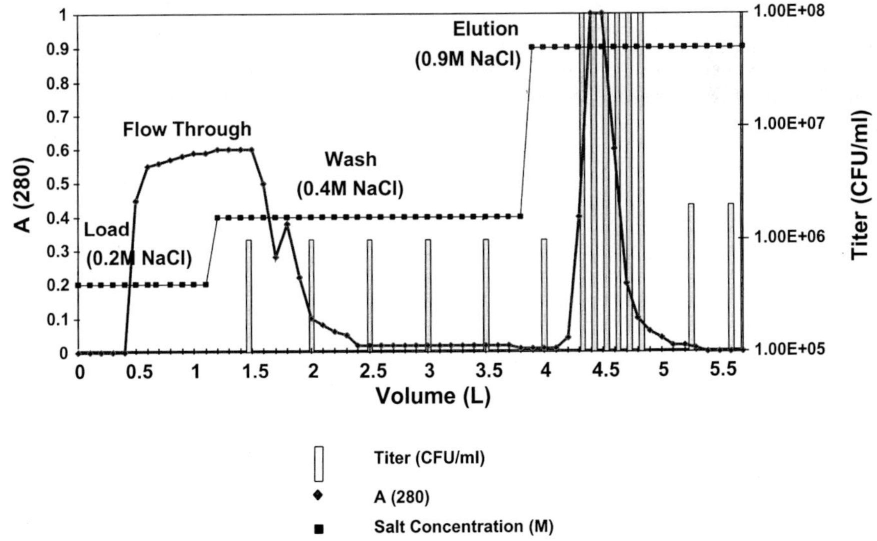

Fig. 11. Column chromatograph of G1Tk1SvNa.7 vector using ion-exchange column 1.1 l

stage of process. For upstream processes, glucose and lactate analyses of supernatant were analyzed to assess the metabolic status of cell growth. Lactate dehydrogenase (LDH) measurement monitored cell viability. The biological vector titer assay was used to monitor vector yield. For the downstream process, total protein at each stage was characterized by sodium dodecyl sulfate polyacrylamide gel electrophoresis (SDS-PAGE)/silver staining, and cellular and vector proteins were analyzed by Western blotting. Cellular DNA was assessed by slot-blot analysis. Adventitious agents, RCR and general safety tests, including endotoxin and microbiological tests, were also assayed, as listed in Table 4. The final vector titer was assayed as a potency parameter, and vector identity was determined by Southern analysis.

To evaluate purity and consistency, final column-purified vectors were compared with the original vector supernatant before any processing. The purity was demonstrated to be approximately 99% and 99.9%, based on protein and DNA concentration, respectively (Table 5).

Table 4. List of in-process assays

Upstream	Downstream	Final
LDH, glucose, lactate	Viral titer	Viral titer
Viral titer	Total protein	Transduction of target cells
Total protein	Specific activity (relative purity)	Specific activity (relative purity)
Southern (cell copy number)	Western blots	SDS-PAGE/silver staining
Western blots	DNA assay	DNA assay
RNA titer	Host-cell proteins	Southern (identity)
RCR	Media-derived proteins	General safety
Adventitious agents	Adventitious agents	Microbiological assay
Microbiological assays	Microbiological assays	RCR

LDH lactate dehydrogenase; *SDS-PAGE* sodium dodecyl sulphate polyacrylamide gel electrophoresis

Table 5. Purity of vectors

	Bioreactor supernatant	Sucrose-gradient purified	Column-chromatography purified
Vector titer (CFU/ml)	$1–2 \times 10^7$	$2–4 \times 10^8$	$1–3 \times 10^8$
Percentage recovery	ND	7 ± 3	65 ± 30
Specific activity (CFU/mg)	$5–10 \times 10^6$	$4–8 \times 10^8$	$2–4 \times 10^9$
DNA/dose (ng)	400–10000	50–500	0.05–0.5

CFU, colony forming units

G. Quality Control

The QC and quality assurance (QA) procedures and principles used for other biotechnology products are applicable to retroviral gene-therapy products. As in these areas, the assays in gene therapy depend on the sensitivity and reproducibility of qualitative and quantitative procedures. Assessment of vector safety must be provided under good laboratory practices (GLP), and the products for patient administration must be manufactured under compliance of GMP. Reviews are available for cell-banking procedures, including QA/QC (Wiebe and May 1990), and for gene therapy from an international perspective (Smith et al. 1996). As noted by Smith et al. (1996), regulatory concerns of these products focus on environmental issues, potential pathogenicity of the agent, and full characterization and QA of the cell banks and the final product (Table 3). These common assays include those for adventitious agents, animal safety and pyrogenicity, among others. Specific assays that are critical for retroviral (and other) vectors are a clear demonstration that the transgene is present and intact in the vector and appropriate target cell(s), and that the final product lot as well as the cell banks are free of detectable RCR. The sensitivity of the Southern and RCR assays must be specified, based on validation studies. It is critical to know at what sensitivity an aberrant form of the vector can be detected. In addition to documenting that the transgene is present and intact, it is critical to demonstrate that the transgene is expressing a protein that is functioning. This can be achieved by a variety of methods, including Northern analysis, immunological detection of the expressed gene product, a biological assay or other procedures.

Sensitive, validated assays are required to detect RCR. At the time of writing, the FDA recommends that 1% of the VPC, up to a maximum of 10^8 cells and 5% of the supernatant be assayed for RCR by means of a biological assay. Both S+/L− and vector-rescue assays are widely used, employing a variety of indicator cell lines, such as mink or feline. The inocula are made into a line of Mus dunni, propagated for 1–2 weeks and then inoculated into the indicator cell line and endpoints, such as focus formation (S+/L−), color change or antibiotic resistance (vector rescue) are measured. Vector rescue is used in this laboratory since we have demonstrated increased sensitivity. Mus dunni cells are available from the American Type Culture Collection.

Final products must be characterized for parameters such as presence of extraneous protein and DNA. Impurities that are of interest include host-cell protein, endotoxins, DNA and other potential contaminants, such as ligands that could interfere with efficient transduction.

H. Safety

In the United States, biosafety levels of human etiologic agents are prescribed by the Center for Disease Control (CDC)/NIH Publication, Biosafety in Microbiological and Biomedical Laboratories (1993). The Moloney amphotropic strain of the murine leukemia virus is the origin of most retroviral vectors. The NIH Guidelines (1997) classify this strain of the virus to be risk-group 2 (RG2) and recommends a containment level appropriate for RG2 human etiologic agents, BL-2. Appendix B-V-1 of the NIH Guidelines states that murine retroviral vectors that contain less than 50% of their respective parental-viral genome and shown to be free of detectable RCR can be maintained, handled and administered under BL1 containment.

For manufacture of replication-incompetent retroviral vectors, BL-2 containment is necessary, since freedom from RCR cannot be assured throughout the manufacturing period. This requirement has added significance in light of the NIH Guidelines requirement for special practices involving volumes larger than 10 l (Appendix K of the NIH Guidelines). In 1997, the NIH RAC modified the large-scale guidelines to allow opening of the bioreactor and removal of the retroviral product prior to sterilization of the vessel. The use of closed systems is necessary for production and downstream processing of the product.

Some animal studies have been performed to evaluate the pathogenic potential of RCR in non-human primates. CORNETTA et al. reported the lack of detectable symptoms in three normal and one immunosuppressed rhesus monkey (CORNETTA et al. 1990). Between 2.5×10^{7} CFU and 1.9×10^{8} CFU of RCR were administered intravenously. The authors noted a rapid clearance from the bloodstream. No virus was detectable in the serum throughout the study period, which was as long as 725 days. A contributing factor to the rapid clearance was the likely rapid inactivation of murine retrovirus by sera of humans and old-world monkeys. In addition to the failure to detect virus, the authors also noted the animals were free of clinical symptoms.

In another study, the development of malignant lymphoma in three monkeys, which were severely immunosuppressed and inoculated with apparently high levels of RCR, drastically altered risk assessment of RCR and retroviral vectors (DONAHUE et al. 1992). The cancers developed from genomic integration of an RCR (not vector) and the malignancy was detected following 6–7 months of retroviremia. Following this publication, FDA's Center for Biologics Evaluation and Research (CBER) issued guidelines for the monitoring of patients receiving retroviral vectors. To date, no RCR has been reported in two studies (MARTINEAU et al. 1997; LONG et al. 1998). At the present time, CBER requires assays to detect presence of antibodies to vector or vector components, proviral DNA and RCR in lymphocytes. In addition, biological assays, such as S+L− or vector rescue, should be performed whenever the above molecular or immunological assays are positive.

GUNTER et al. (1993) summarized testing requirements for the testing of cell banks and product lots for RCR. Even with molecular design of vectors to minimize molecular-recombination events leading to RCR, breakouts still occur. One RCR detected has been characterized (OTTO et al. 1994). This RCR was shown to occur through partial homology between the vector and packaging line, with eight of ten nucleotides in the RCR being homologous. RCR has also been detected in split-packaging cell lines (CHONG and VILE 1996).

The potential effects of RCR in humans is unknown. The only available data come from extrapolation of the results of reporting lymphoma in three of eight severely immunosuppressed rhesus monkeys exposed to apparently large numbers of RCR (DONAHUE et al. 1992). However, clinical correlates of this degree of immunosuppression exist, especially in allogeneic tissue and organ transplantation.

Limited data from patient monitoring exist; no RCR has been detected in these studies (MARTINEAU et al. 1997; LONG et al. 1998). The latter study is particularly informative. This publication described results of patient monitoring for clinical trials that involved the direct injection of VPCs (not vectors) into brain tumors by either stereotactic implantation or injection into residual tumor and peritumoral areas following resection. The VPC obviously had passed all QC assays before the product lot was released. However, it is possible that RCR could be generated in situ from the viable VPC. The failure to detect RCR in this study may be due to: (1) no RCR having been generated, (2) inactivation of any RCR by human serum, or (3) relative insensitivity of the polymerase chain reaction (PCR) and biological assays.

The FDA recently published a paper describing some considerations regarding safety monitoring of patients receiving retroviral vectors. This may lead to alternate recommendations for assays for RCR in product lots, transduced cells used in in vitro clinical protocols prior to administration to patients, and patient monitoring (WILSON et al. 1997). A review is available on the potential mechanism for RCR recombinations and the efficiency to detect RCR in various assays (ANDERSON et al. 1993).

A long history exists in research work on the parental Moloney murine leukemia virus. We are unaware of any untoward effects of working with this agent among laboratory workers or animal caretakers. A general review of safety in the biotechnology industry has been published (MCGARRITY and HOERNER 1995).

I. Summary and Conclusions

The data and recommendations for production of retroviral vectors contained in the review are similar to other uses of mammalian cells for biotechnology products. In gene therapy, as in other cell-based products, a critical characteristic is the need for properly designed production systems consisting of validated processes, facilities and equipment. In addition, QC and manufacturing processes must be strictly followed to generate a safe, consistent and well-characterized product.

Retroviral vector gene-therapy products differ from other cell-based products in two respects. First, since retroviral vectors integrate into the host genome, little or no control is available to remove the product. This differs from standard pharmaceutical products, recombinant proteins and monoclonal antibodies. Each of these have relatively short half-lives in vivo. Retroviral vectors can, theoretically, continue to express their gene products for prolonged periods. While the use of suicide genes, such as thymidine kinase from Herpes simplex virus, has been used for therapeutic purposes, it is not routinely used to destroy cells containing the proviral DNA.

Another difference in gene therapy is the availability of PCR, which offers exquisite sensitivity to detect DNA and RNA. Procedures are routinely used that can detect a single DNA molecule in 500,000 or more cells. With this technology, it is possible to measure drug (vector) distribution in vivo as never before. This has been utilized in toxicology studies, using animal models as well as patient monitoring to assay for either vector or RCR in lymphocytes.

While no gene-therapy product has been approved for commercial use, we are confident that the principles of GLP and GMP can be applied to this emerging technology. These would, by following the precedents successfully set, by our predecessors in the fields of monoclonal antibodies and recombinant proteins.

References

Anderson WF, McGarrity GJ, Moen RC (1993) Report to the NIH recombinant DNA advisory committee on murine replication-competent retrovirus assays. Hum Gene Ther 4:311–321

Anonymous (1997) Human gene marker/therapy clinical protocols. Hum Gene Ther 8: 2301–2338

Blaese RM, Culver KW, Miller AD, Carter C et al. (1995) T lymphocyte-directed gene therapy for ADA-SCID: initial trial results after 4 years. Science 270:475–480

Blaese RM (1997) Gene therapy for cancer. Sci Am 276:111–115

Center for Disease Control/National Instututes of Health (1993) Biological safety on microbiological and biomedical laboratories. US Health and Human Services Publication No (CDC) 93–8395. US PHS Washington, DC

Chong H, Vile RG (1996) Replication-competent retrovirus produced by a "split-function" third generation amphotropic packaging cell line. Gene Ther 3:623–629

Chuck AS, Paulsson BO (1996) Consistent and high rates of gene transfer can be obtained using flow through transduction over a wide range of retroviral vectors. Hum Gene Ther 7:743–750

Cornetta K, Moen RC, Culver K, Morgan RA, McLachlin JR, Sturm S, Selegue J, London W, Blaese RM, Anderson WF (1990) Amphotropic murine leukemia retrovirus is not an acute pathogen for primates. Hum Gene Ther 1:13–26

Danos O, Mulligan RC (1988) Safe and efficient generation of recombinant retroviruses with amphotropic and ecotropic host ranges. Proc Natl Acad Sci USA 85:6460–6464

Donahue RE, Kessler SW, Bodine D, McDonagh K, Byrne E, Raffield M, Moen R, Bacher J, Zsebo KM, Nienhuis AW (1992) Helper virus induced T cell lymphoma in nonhuman primates after retroviral mediated gene transfer. J Exp Med 176:1125–1135

Finter NB, Garland AJM, Telling RC (1990) Large scale mammalian cell culture: a perspective. In: Lubiniecki AS (ed) Large Scale Mammalaian Cell Culture Technology, Marcel Dekker New York, pp 1–14

Griffiths B (1991) Cultural revolutions. Chem Ind, pp 682–684

Gunter K, Khan AS, Noguchi PD (1993) The safety of retroviral vectors. Hum Gene Ther 4:643–645

Kohn DB, Weinberg KI, Nolta JA, Heiss LN et al. (1995) Engraftment of gene modified umbilical cord blood cells in neonates with adenosine deaminase deficiency. Nat Med 10:1017–1023

Kotani H, Newton PB, Zhang S, Chiang YL, Otto E, Weaver L, Blaese RM, Anderson WF, McGarrity GJ (1994) Improved methods of retroviral vector transduction and production for gene therapy. Hum Gene Ther 5:19–28

Long Z, Li L-P, Grooms T, Lockey C, Nader K, Mychkovsky M, Mueller S, Burimski, Ryan P, Kikuchi G, Ennist D, Marcus S, Otto E, McGarrity G (1998) Biosafety monitoring of patients receiving intracerebral injections of retroviral vector producer cells. Human Gene Ther 9:1165–1172

Lubiniecki AS (1990) Continuous cell substrate considerations. In: Lubiniecki AS (ed) Large Scale Mammalian Cell Culture Technology. Marcel Dekker, New York, pp 495–513

Lyons R, Forry-Schaudies S, Otto E et al. (1995) An improved vector encoding the herpes simplex thymidine kinase gene increases antitumor efficacy in vivo. Cancer Gene Ther 2:273–280

Markowitz D, Goff S, Bank A (1988) A safe packaging line for gene transfer: separating viral genes on two different plasmids. J Virol 62:1120–1124

Martineau D, Klump WM, McCormack JE, DePolo NJ, Kamantigue E, Petrowski M, Hanlon J, Jolly D, Mento SJ, Sajjadi N (1997) Evaluation of PCR and ELISA assays for screening clinical trial subjects for replication competent retrovirus. Hum Gene Ther 8:1231–1241

McCormack JE, Martineau D, DePolo N, Maifert S,Akbarian L, Townsend K, Lee W, Irwin M, Sajjadi N, Jolly DJ, Warner J (1997) Anti-vector immunoglobulin induced by retroviral vectors. Hum Gene Ther 8:1263–1273

McGarrity GJ, Hoerner CL (1995) Biological safety in the biotechnoogy industry. In: Fleming DO, Richardson JH, Tulis JJ, Vesley D (eds) Laboratory safety, principles and practices, 2nd edn American Society for Microbiology, Washington DC, pp 119–132

McGarrity GJ, Chiang YL (1997) Gene therapy of local tumors. In: MT Lotze and JT Rubin (eds) Regional therapy of advanced cancer, Lippincott-Raven, Philadelphia, pp 375–390

Mento SJ (1994) Status report on the development of retro vector gene transfer products. Process Development Issues in viral product manufacturing, The Williamsburg Bioprocessing Conference

Miller AD, Buttimore C (1986) Redesign of retrovirus packaging cell lines to avoid recombination leading to helper virus production. Mol Cell Biol 6:2895–2902

National Institutes of Health (1997) Guidelines for research involving recombinant DNA molecules (NIH Guidelines) National Institutes of Health, Bethesda MD Federal Register January 31, 1997 (62 FR 4782)

Otto E, Jones-Trower A, Vanin EF, Stambaugh K, Mueller SN, Anderson WF, McGarrity GJ (1994) Characterization of a replication competent retrovirus resulting from recombination of packaging and vector sequences. Hum Gene Ther 5:567–575

Rigg RJ, Chen J, Dando JS, Forestell SP, Plavec I, Bohnlein E (1996) A novel human amphotropic packaging cell line: high titer, complement resistance, and improved safety. Virology 218:290–295

Smiley WR, Lambert B, Howard BD, Ibang C, Fong TC, Summers WS, Burrows FJ (1997) Establishment of parameters for optimal transduction efficiency and antitumor effects with purified high titer HSV-thymidine kinase retroviral vector in established solid tumors. Hum Gene Ther. 20:965–977

Smith KT, Shepherd AJ, Boyd JE, Lees GM (1996) Gene delivery systems for use in gene therapy: an overview of quality assurance and safety issues. Gene Ther 3:190–200

Wiebe ME, May LH (1990) Cell banking. In: Lubiniecki AS (ed) Large scale mammalian cell culture technology, Marcel Dekker New York, pp 147–160

Wilson CA, Ng T-H, Miller AE (1997). Evaluation of recommendations for replication-competent retrovirus testing associated with use of retroviral vectors. Hum Gene Ther 8:869–874

CHAPTER 10

Clinical Systems for the Production of Cells and Tissues for Human Therapy

R.D. ARMSTRONG, M.R. KOLLER, J. MALUTA, and W.C. OGIER

A. Introduction

Many medical conditions can now be effectively resolved with the use of transplanted cells, tissues or organs. Numerous other conditions are the targets of ongoing preclinical and clinical transplantation research. Despite the remarkable success of transplantation therapies, a number of issues hamper their use in widespread clinical practice (EDGINGTON 1992a,b; KOLLER and PALSSON 1993; LANGER and VACANTI 1993; EMERSON 1996). First, the demand for organs and tissues from suitable donors to be used in allogeneic (non-self) transplantation has far outweighed their availability. Consequently, investigators have increasingly turned to the use of autologous (the patient's own) cells and tissues, or to the use of xenogeneic (non-human) materials for transplantation therapies. Second, transplantation procedures continue to be cost prohibitive for otherwise medically warranted applications. Third, the procurement of suitable tissues is often prohibitively invasive for the donor, in many cases requiring cadaveric material to be used. As a result, the full potential of transplantation therapies has not yet been realized.

The development of ex vivo cell-production processes promises to change this situation, by enabling a large therapeutic number of cells, or even functional tissue, to be produced from a small starting volume of clinical material. Biological science has now progressed such that, for many human tissues, methodology has been developed to grow key cell populations ex vivo. With this achievement comes the opportunity for the increasing and more diverse use of transplantation therapies.

Two key components are necessary in order to grow therapeutic cells and tissues ex vivo: (1) a process for replication and/or differentiation of cells of self or donor origin to form functional tissue or produce large volumes of progenitor cells; and (2) an ex vivo system comprised of biocompatible materials controlled by automated hardware that provides the essential physiological conditions (appropriate cell-growth surface character, medium exchange rate and continuous oxygenation) for cell growth, and enables the complete viable recovery of the cells or tissue for therapy.

Once an effective biological process is developed for the production of therapeutic cells or tissue, there typically exists a corresponding critical need for medical-device technology to implement the process, and to do so on a

Biological Process + **Device Implementation** + **GMP Cell Production** → **Patient Treatment**

Fig. 1. Components of a successful cell-therapy product. A cell-therapy product is first based on a biological process, which generates a cell population that is therapeutically useful. This process must then be implemented in a device, so that the process can be carried out at the clinical scale in a reproducible and sterile manner. Finally, the device must be incorporated into an automated good manufacturing procedures (GMP) process, allowing the procedure to be performed in accordance with regulatory guidelines and in a cost-effective manner. Only then will the ultimate goal of worldwide patient therapy be realized

clinical scale. A further extension of this requirement is the need for automation technology, to bring cost effectiveness and a high degree of reliability to the ex vivo production of cells for therapeutic use. With these two additional technologies – medical devices and process automation of the medical device – the routine production of human cells and tissues for patient therapy is enabled without the use of "artsy" manual techniques requiring specialized research laboratories (Fig. 1). The development of device and automation technologies, thereby, facilitates access by patients and physicians worldwide to life-saving and life-enhancing biological advancements in cell therapy and tissue engineering.

B. Cell Therapy and Tissue Engineering

In response to clinical needs in transplantation and as the result of an expanding understanding of the mechanisms of immunology and cell biology, two related fields of medical expertise have evolved which have been termed "cell therapy" and "tissue engineering." Cell therapy is the older concept and generally refers to the use of living cells, rather than drugs, to treat a clinical disorder or disease (EMERSON 1996). Tissue engineering is the newer field of endeavor, which generally refers to the application of multiple engineering, physiology, and cell-biology disciplines to develop a living (or partly living) tissue, which is then used for a transplantation therapy (EDGINGTON 1992a,b; KOLLER and PALSSON 1993; LANGER and VACANTI 1993). The tissue-engineered product may be different in some key aspect from a directly explanted donor or patient tissue, or be enhanced in quantity while maintaining native or enhanced tissue function. Often, the progenitor cells for a tissue may be produced for the cell therapy and, once infused, will form or repair the targeted tissue in situ. For this review, such approach will be termed "cell-expansion therapy". Numerous university laboratories and companies are engaged in such projects to develop and produce human tissues for transplantation, and tissue engineering has become a recognized academic discipline.

I. Ex Vivo Gene Therapy

In parallel with the advancements in cell-expansion therapy and tissue engineering, and largely dependent upon their success, are the numerous gene-therapy approaches being advanced to initial clinical trials involving the ex vivo genetic manipulation of cells and tissue (MULLIGAN 1993). For gene therapy, cells are isolated ex vivo and a new gene, typically carried by a viral vector, is introduced into the genome of cells. The objective is to achieve correction of a defective gene, regulation of a disease condition or production of a beneficial therapeutic molecule. Once the genetic modification occurs, a therapeutic dose of the cells is needed, requiring production of greater quantities of the "new cells". Accordingly, ex vivo gene therapy requires several critical manipulations of human cells and tissues ex vivo in order to produce the cells needed for therapy. In fact, it has been limitations in these ex vivo procedures that has slowed recent clinical progress, rather than problems with the molecular biological components (such as the genes or vectors).

II. Stem-Cell Therapy

Hematopoietic stem-cell transplantation (such as bone-marrow transplantation) in patients who have received systemic cancer chemotherapy or radiation therapy, resulting in hematopoietic toxicity, is one of the most widely practiced forms of cell therapy today (KOLLER and PALSSON 1993; EMERSON 1996). Hematopoietic stem-cell transplantation involves the reinfusion of early-stage cells that originate in the bone marrow, so that these cells may reconstitute the blood and immune system of a patient along with other components of bone-marrow tissue. Through this cell-therapy process, the hematological toxicity that is a common side effect (or in the case of hematological malignancies, an intended consequence) of aggressive cancer treatment is remedied. This form of transplantation has been practiced for a number of years and is now routinely used to treat an estimated 40,000 patients per year worldwide. Furthermore, the use of hematopoietic stem-cell transplantation has been growing at a rate of approximately 30% per year (BORTIN et al. 1992). A limitation to the further increased used of this life-saving therapy is a better (cost effective, less invasive, widespread availability) source of the key therapeutic cells. A solution to this problem may be obtained if these cells could be effectively produced ex vivo.

Bone marrow is a soft amorphous tissue comprised of many different cell types, ranging from mesenchymal accessory (such as stromal) cell populations to stem cells and the other more mature cells of the hematopoietic system. The key process conditions for ex vivo bone-marrow growth were developed by mimicking the natural environment of bone marrow by providing slow, controlled nutrient perfusion and uniform levels of oxygenation of the stem and stromal cell components under precise conditions of temperature and medium composition. Consistent with the principles of tissue engineering, it is essential to provide culture conditions that are concurrently amenable for each of the many different cell types that are found in human bone marrow – conditions found inside bone-marrow cavities in the body. Using this approach, human stem cells found in bone marrow were able to survive in culture for an extended period for the first time, as well as replicate to produce more stem cells and mature progenitor cells, all as part of the marrow growth process (SCHWARTZ et al. 1991; KOLLER et al. 1993). This result is in direct contrast with alternative approaches that were taken in an attempt to grow human stem cells, based on conventional biological experimentation principles (Fig. 2). In these examples, hematopoietic stem/progenitor cells were isolated, e.g., via antibody-mediated cell selection based on their so-called CD34 cell-surface antigen, prior to a cell-culture process. In this procedure, the cultured stem cells do not grow, but die or irreversibly differentiate, and the cultures lose proliferative activity over a short period of time (HAYLOCK et al. 1992; SUTHERLAND et al. 1993; VERFAILLIE 1993). Presumably, critical heterogeneous tissue interactions have been eliminated by the removal of most of the stromal and accessory cell elements and by the employment of non-physiological culture conditions.

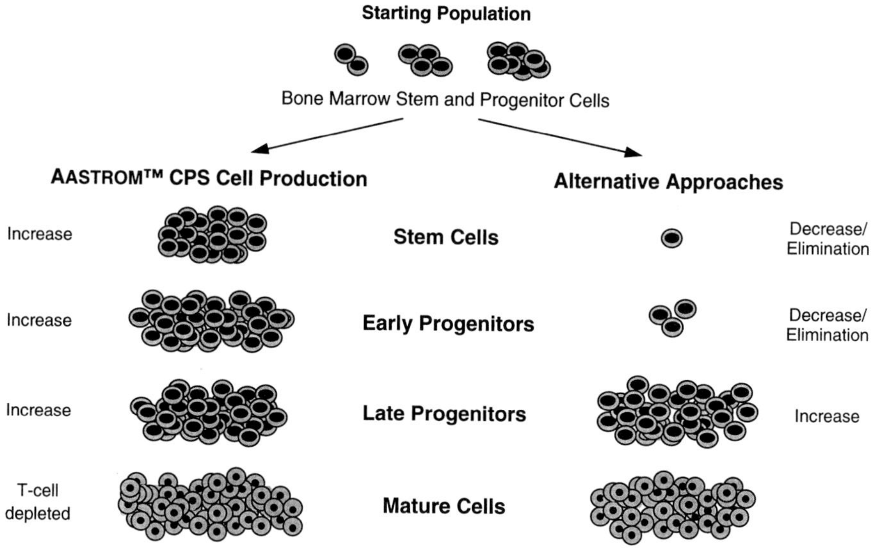

Fig. 2. Stem and hematopoietic progenitor cell production using a tissue-engineering approach versus alternative approaches based on $CD34^+$ cell purification. The AASTROM cell-production system (CPS) results in an increased number of stem, as well as early and late progenitor cells. Culture of purified $CD34^+$ cells typically results in an increased number of late progenitors, but decreased numbers of stem and early progenitor cells

Production of bone-marrow stem and progenitor cells is an excellent example of the merging interface of cell therapy with tissue engineering. This process was developed by scientists at the University of Michigan and AASTROM Biosciences, Inc. (SCHWARTZ et al. 1991; PALSSON et al. 1993; KOLLER et al. 1993). Its development for clinical application to the treatment of patients illustrates the interrelationship between cell- and tissue-production methodology and the medical-device and automated-system requirements to implement practical cell therapies.

C. Critical Requirements for Ex Vivo Cell Production

Cell- and organ-transplantation therapies have been practiced around the world in a diverse number of clinical and laboratory settings. Varying degrees of regulatory oversight have been applied, depending on a number of factors. In the United States, the Food and Drug Administration (FDA) has regulated certain products used in organ transplantation and has encouraged the practice of cell-therapy procedures under "Points to Consider" documents, which have, either explicitly or in effect, mandated good manufacturing practices (GMP) for the production of tissues and cells for therapy (FDA 1993). Recently, the FDA and its counterparts in other parts of the world have taken further steps toward comprehensive regulation of cell and transplantation therapies.

The inherent complexities of the numerous sequential laboratory processes used in cell therapies and the diversity of materials and reagents involved in these therapies likely makes their regulation challenging, but no less necessary. Efforts by manufacturers to reduce the inherent complexity and number of process steps required, and to more effectively provide for control of the processes and the materials used, will have a number of general benefits. These include: (1) more reliable and, thus, safer and more effective patient therapy, (2) inherently superior cost effectiveness, (3) greater practicability, and (4) greater worldwide availability of therapies to physicians and patients.

I. Process Reliability and Control: Automation

Perhaps the most critical of all issues to be addressed is the technical art inherent to most cell-culture processes. Just as it is probable to obtain ten different soufflés from ten different cooks working from the same recipe, site-to-site differences in cell culture are often as variable and subtle. For clinical cell-culture processes to move from the realm of limited experimental practice to routine and widespread clinical use, the technical art or sophistication must be eliminated such that the same cell product will be reliably obtained whether the cell-production process is performed in France or Indiana. This challenge can be best met by implementing a well-characterized, robust process that involves minimal human intervention and decision making.

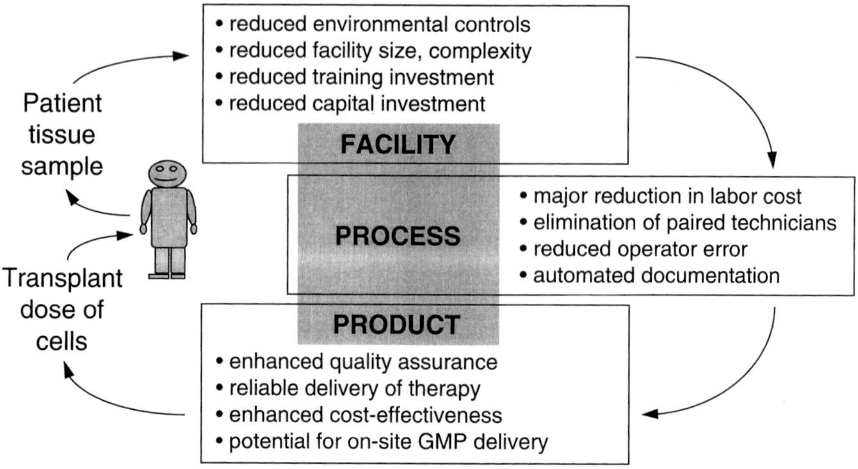

Fig. 3. Benefits of cell-production process automation. Process automation yields benefits in the facility design, cell-production process, and product quality

While the human variability factor is desirably eliminated from the processing of cells and tissues, the FDA and other regulatory agencies have appropriately required that human oversight be maintained for purposes of monitoring process control and product quality. Highly controlled and repeated operator training as well as thoroughly documented standard operating procedures (SOPs) and operator checklists may be used to address inherent variability in the practice of cell-production processes in the absence of process automation. However, in practice, these approaches may require the use of redundant personnel to check each others' work, and carry an inherently high expense.

Generally, if a laboratory process is under control and can lead to the same result repeatedly, then it can, and usually should, be automated if the manufacturer desires to replicate the process in multiple facilities. Significant cost savings may also result from automation in a single facility as well (Fig. 3). From a strategic point of view, the implementation of medical processes as automated medical devices may be desirable, because it eliminates variability due to human error or human initiative, reduces the need for highly skilled labor and, thus, reduces the cost of the process and the therapy, and makes the process amenable for widespread distribution without adverse consequences to otherwise successful clinical outcomes. In cell therapy, automation also meets a general desire by clinicians to provide better quality assurance to their patients.

II. Process Sterility: Closed Systems

With any cell-culture procedure, a major concern is the maintenance of sterility. When the product cells are to be transplanted into patients – which often occurs at a time when the patient is immunosuppressed or immunocompromised, as is the case with stem-cell transplants and organ transplants – the absence of microorganisms and viral contaminants is essential. Most laboratory cell-culture procedures are carried out under aseptic conditions, involving the practice of so-called sterile technique by trained laboratory technicians. Many of the bioreactor systems that have been developed for laboratory applications offer advantages over manual culture processes in that, once the culture is initiated, the culture chamber and fluid pathway are maintained in a sterile closed environment. However, even with these systems, the initial setup, inoculation and harvest steps, such as medium priming and collection of cells at the completion of the process, require non-sterile manual procedures and are at risk of contamination. In fact, laboratory cell-culture systems are only partly closed, i.e., they involve numerous aseptic connections, are inconveniently operated in a laminar air-flow hood to further minimize the risk of contamination and require pre- and post-processing steps requiring the use of open centrifuge tubes and other conventional laboratory technology.

If patient treatment is the end goal of a cell-culture process, it is far preferable for the process to be carried out within a system where the culture chamber and fluid path are functionally closed to the external environment, with sterile integrity maintained from the time the device is manufactured until it is ultimately disposed. Various medical technologies have been developed for use in blood banking, intravenous fluids administration and peritoneal dialysis, which may be adapted for use in the clinical cell-culture laboratory to facilitate "closing" of otherwise open processes. For example, tubing connection devices and microbial-filtration devices validated to achieve a sterility-assurance level of less than one potential contamination in a million connections are termed "sterile" under FDA guidelines. Better yet, are device systems that provide for pre-connection and subsequent pre-sterilization of tubing connections during manufacture, such that aseptic or open connections and fluid transfers are largely eliminated. With the successful implementation of such technology, expensive controlled-environment laboratory facilities may be eliminated as a requirement in cell and tissue therapy.

Gene therapy brings an additional set of requirements to the field of cell therapy, including the need to contain potentially infective viral gene vectors within a closed system. To a large extent, achievement of the requirements for cell-therapy-process sterility, as met by effective closed-system medical-device design, should result in achievement of the requirements for process containment. This is less true of traditional laboratory-based sterility-assurance measures, such as laminar flow hoods, where the risk of spillage and other containment lapses that are inherent in "open" devices tend to mandate even more burdensome protective measures, including glove boxes or whole body suits. The regulatory requirements for gene therapy remain at a comparatively early stage, reflecting, in part, the numerous additional associated issues of patent, and process safety. Effective process and system designs have the potential to enhance the widespread utilization of these exciting new forms of medicine.

III. Cell Recovery

In cell therapy, the product of the cell-culture process is the cells. Accordingly, efficient recovery and collection of the cells at the completion of the culture process is an important feature of an effective clinical cell-culture system. Recovery of cells from many cell-production devices is a challenge. At the end of the growth process, the cells are either packed into the interstices of a make-do dialysis cartridge, or they are suspended in many liters of culture medium. The former necessitates unreasonable and irreproducible physical force to dislodge cells – being neither reliable nor easily automated – and the latter case requires time, patience and a certain degree of good fortune – being neither reliable nor straightforwardly implemented within a closed system.

A superior approach for production of cells as the product is to culture cells in a defined, reasonable space, without physical barriers to recovery, so that the simple elution of the cell product results in a manageable, concentrated volume of cells amenable to final washing in a commercial, closed-system cell washer designed for clinical use, e.g., washing of blood cells prior to transfusion. An ideal clinical cell-production system would allow for the efficient and complete recovery of all cells produced, including both adherent and non-adherent cell types, where applicable. Furthermore, it should be possible to complete the harvest process without breaking the sterile barrier of the fluid path of the culture chamber.

IV. Optimization of Key Culture Parameters by Design

In any cell-culture process, there are several fundamental parameters that require periodic control. At larger scales, the process control requirements become even more critical, often requiring very frequent monitoring. For example, cultures must be provided with medium that allows for normal metabolic functions and growth, and typically this medium is delivered to the cell via a pumping mechanism (a bioreactor) or by a technician manually feeding or exchanging the medium on a regular basis. Use of an excess volume of medium can reduce the immediate control requirements, but in large-scale cultures this can be economically limiting – and in some cultures, for example the growth of bone-marrow stem cells, this approach is not feasible for biological reasons. In general, the rate of medium exchange determines the extent to which cells in culture are exposed to their own metabolic products, including waste products, and excessive exposure can defeat the ability of a cultured cell population to achieve the desired objective.

Attention to this important parameter led to the observation that bone-marrow stem-cell cultures require perfusion of fresh medium to maintain the ability to expand the immature stem and progenitor cell populations. Static culture methods using an excess of medium in situ caused, within a few days, irreversible differentiation of the early stem and progenitor cells to generate more mature cell populations (HAYLOCK et al. 1992; SUTHERLAND et al. 1993;

VERFAILLIE 1993). Such cultures act as though under stress and lose the capability for self regeneration. Likewise, too frequent a rate of medium exchange (corresponding to very rapid perfusion) causes such cultures to lose their ability to differentiate at all or to regenerate or expand the early stem and progenitor cells, presumably because important growth-factor products of the cultures are being removed too rapidly (SCHWARTZ et al. 1991; KOLLER et al. 1996).

The growth of cells or tissue also requires a source of oxygen, usually with the added presence of carbon dioxide to buffer the culture medium. Different cell types have different oxygenation requirements, and all cell cultures have differing oxygen-delivery requirements, depending on the cell density of the culture process (FLEISCHAKER and SINSKEY 1981; PENG and PALSSON 1996). Accordingly, a controllable and, ideally, flexible means for providing oxygen to the cells is a necessary component of the culture system. Under conditions of controlled medium perfusion, internal oxygenation may also be a critical requirement, since the common practice of using an external oxygenation circuit (the adaptation of a blood-oxygenator device) can require medium circulation rates between the oxygenator and the culture chamber that are too rapid to allow for effective growth of a desired cell population. In fact, most cells experience physiologically defined, very uniform perfusion and oxygenation conditions in their native in vivo environment.

Uniform distribution of the cell population and medium supply in the culture chamber are also important for the achievement of effective process control. This control is often achieved by use of a suspension culture design, which can be effective in cases where cell-to-cell interactions are not as important as cell-to-medium interactions. Examples of suspension culture systems include various stirred-tank designs (ZANDSTRA et al. 1994) and gas-permeable, non-adherent plastic bags (KLINGEMANN et al. 1993). Aside from use with mature blood cells, such as T cells, which circulate freely in the body in suspension, such designs are often deleterious for the growth of human tissues, as they do not support the development of three-dimensional structure in tissue. Analogous to the breeding of exotic animals, human stem cells are very discriminating in their reproductive behavior, requiring a carefully maintained combination of biological, chemical and physical parameters. The growth of bone marrow is precluded in environments favoring single-cell suspensions, because stem cells appear to require contact with the stromal and other accessory cells that are all part of bone marrow, in order to replicate (KOLLER et al. 1995; NAUGHTON 1995). Microcarrier particles can be added to stirred-tank bioreactors to encourage growth of adherent cells, but separation of the cells from the microcarriers at the end of a culture process can be problematic, and such technology limits the ability to achieve controlled perfusion and oxygenation conditions, as well as the efficient utilization of growth medium, which can often be very expensive. Furthermore, these carriers are not amenable to the growth of heterologous cell mixtures, such as bone marrow.

It is, perhaps, an obvious extension to note that the optimization of culture processes to produce specific cells and tissues for patient therapy is an included requirement for gene therapies, based on the infusion or transplant of such cells and tissues. An additional challenge presented by such gene therapies is the common need to introduce the gene vector or gene carrier into the cell or tissue of interest during the growth process, to ensure its satisfactory incorporation into the genome of the cell or tissue. Because of the concentration-dependent kinetics of gene transfer, as conventionally practiced, culture-device designs employing excess volumes of culture medium are at a distinct disadvantage in such integrated processes. They also impede the "pulsing" of gene–vector exposure, which is more readily applied in perfusion culture systems. In general, device designs that allow for effective introduction of gene vectors without disruption of optimized cell-growth processes will find more use in gene therapy than alternative designs.

V. Good Manufacturing Practices

The FDA and other regulatory authorities have increasingly turned to the application of GMP guidelines in order to bring quality assurance to the production of cells and tissues for human therapeutic use. These GMPs have their origin with pharmaceutical companies and other industrial concerns, which have come under regulation over the past century. The past decade has seen a dramatic refocusing of the regulation of blood-transfusion medicine and, more recently, the translation of similar standards to other cell- and gene-therapy clinical-research programs. This refocusing is intended to protect the public from exposure to infectious agents and ensure other aspects of product quality. Indeed, the FDA treats academic research laboratories generating experimental cell populations for transplantation as manufacturers of biological and related products, and has addressed its guidelines and regulations to all such manufacturers.

The essence of GMP is: (1) use of equipment, facilities and reagents that are validated to be capable of reliably producing the desired product without introduction of pathogenic substances; (2) process control that, through validation, demonstrates an ability to produce a product within desired specifications, and (3) final product and process documentation controls that prevent mix-ups or inappropriate release of a product before complete review of test and production records is conducted and accepted.

In a manufacturing environment that largely involves manual processes and human oversight, GMP is accomplished by selecting well-qualified personnel, providing them with extensive training, conducting extensive validation procedures, selecting the use of equipment and reagents that have, in turn, been produced under GMP, and developing a system of SOPs and extensive documentation that provides for quality assurance. In a multi-step biological laboratory process, such documentation may be many hundreds of pages, much of it representing forms to be filled out in their entirety during the manufacturing process. It is not uncommon for a therapeutic cell- or tissue-production process extending over several weeks to require in the order of 100 h of technician, supervisor, maintenance and quality-assurance personnel direct labor – more than half of it directed toward quality-assurance objectives rather than the direct handling of the biological product.

In an automated manufacturing environment, the same principles of process validation are used to demonstrate that the process is stable and capable of meeting specifications, but under the direction of automated instrumentation. Principles of statistical process control are then implemented to monitor the process to assure consistent conformance to specifications. If the cell-handling equipment is automated, it is feasible to integrate and also automate process monitoring and documentation to a significant degree, although final quality assurance (inspection of final records and release of product) should remain in human hands. As with many other industries, automation of biological processes has the potential to reduce labor inputs by as much as an order of magnitude, while at the same time improving various measures of quality assurance to a similar degree. Automated-process implementation, process control and process documentation have been widely adopted by the pharmaceutical industry, largely in order to cope with extensive regulatory and quality-assurance requirements (other elements of product cost are often relatively small in pharmaceuticals). In the cell-therapy field, such efforts to reduce cost and improve quality assurance will be even more critical, given that most processes are patient specific and, hence, involve a much greater degree of manufacturing.

In the current environment, as cell-culture processes become initially prominent in clinical care, much of the process control and record keeping depends on highly skilled, trained operators practicing manual technique and maintaining detailed records. The future of cell therapy lies in the availability of equipment and materials that are designed and manufactured for the intended purpose and automation that allows processes to be readily implemented, validated, controlled and documented. Only under these conditions can cell therapy be delivered to the market in a cost-effective manner.

D. Cell-Culture Devices and Procedures

In nature, tissue function and viability is dependent on the life-support process that is mediated by the whole organism. Nutrients, physiological salts, hormonal and other growth factors, and oxygen, are all brought to the tissue through the vascular system from other tissues within the organism. The waste products produced by the tissue, which can often be toxic, are carried away by the vascular system for disposal by other tissues within the organism. The other major components of tissue maintenance and repair are cellular; a pool of progenitor or stem cells that can replace cells that are lost or damaged, as well as the mesenchymal support or accessory cells that are critical for physiological tissue function (Naughton 1995).

The ex vivo reconstitution of human tissue requires these same elements to be managed by the culture devices and procedures. In other words, the different cell types required to reconstitute tissue function must be maintained in a physical and biological environment that is biocompatible and provides a means to control delivery of nutrients and oxygen to the cells, and carry away waste or other byproducts of the growing cell populations.

During the past decade, increasing numbers of medical researchers have sought to develop treatments based, in part, on the culture of human blood cells, including lymphocytes, monocytes, neutrophil precursors, and immature blood cells, including stem and progenitor cells. The evolution of technologies adapted or developed to meet these needs provides an excellent demonstration of the need for clinical systems for the production of cells for human therapy.

I. Traditional Cell-Culture Processes: Research Laboratory Environment

Traditional cell-culture technologies depend on controlled environments for cell handling. Cell-culture laboratories incorporate such features as laminar flow hoods, controlled access to the laboratory by gowned personnel and regular sterilization procedures to decontaminate laboratory surfaces. Personnel require extensive training to practice sterile technique, i.e., to avoid contamination of open containers and cell-transfer devices by contact with non-sterile materials. Despite these prophylactic measures, outbreaks of contamination in traditional cell-culture laboratories, e.g., by fungus, is a common occurrence – often with the impact of halting operations for days or weeks while the source of contamination is determined and resolved. Traditional cell-culture technologies further depend on incubation in an environment providing controlled gas mixtures and controlled temperature, usually satisfied by the use of commercial incubators ranging in size from large bench-top units to large floor-standing units.

Therapeutic requirements for numbers of blood cells (typically 1–100 billion per patient treatment) and limitations in maximum cell-culture density (typically one billion per liter of medium), together with space requirements for major laboratory hardware (hoods, incubators, refrigerators) and personnel activity, have resulted in considerable laboratory space requirements per patient therapy. Laboratory support operations, including preparation of media and the practice of various assays and other quality-assurance activities, expand these space requirements and associated capital investments and labor costs. Use of traditional cell-culture technology for patient therapy, thus, results in relatively high costs per patient treatment, reflecting the depreciation and ongoing expenses of laboratory infrastructure and personnel.

Ironically, such a laboratory environment is not conducive to the reliable and routine production of large numbers of cells for patient therapy, given its reliance on manual, highly skilled technique. Achieving GMP in such an environment is a daunting challenge, requiring the development and adherence to volumes of SOPs to eliminate inherent variability in laboratory practices. Technology choices, which underlie these limitations in the current practice of cell therapy, along with the economic limitations posed by direct labor inputs, are detailed below.

1. Culture Flasks and Roller Bottles

The earliest cell cultures were achieved in glass Petri dishes, which were largely supplanted by pre-sterilized plastic tissue-culture flasks in the 1960s and 1970s. Early attempts at large-scale culture of human cells for therapy in the mid-1980s involved the use of numerous glass roller bottles in a room-sized mechanized facility. Even today, most cell therapies have their genesis in plastic tissue-culture flasks, and process scale-up involves the use of more, larger, so-called T-flasks, which are fed with culture medium manually in a laminar flow hood.

As described earlier, one preferable objective is to provide a culture process that can deliver medium and oxygenation at uniform and controlled rates that mimic serum perfusion of tissue in vivo. In order to achieve these relatively slow delivery rates, a means of internally oxygenating the cells is often required. This is one requirement that is well met (ignoring other requirements) using the simplest of cell-culture processes – a culture dish or flask. Here, the surface of the culture is uniformly exposed to oxygen, and oxygen is available to the cells as needed. Similar conditions can be realized in a flat-bed bioreactor, with a gas-permeable/liquid-impermeable membrane a short distance from the cell bed (PENG and PALSSON 1996). This allows for a system to have variable medium-perfusion rates from static to high, with the oxygenation of the culture remaining constant and uniform. Such a design is also amenable to protection from environmental contamination.

2. Flexible Tissue-Culture Containers

Flexible tissue-culture containers, or culture bags, were developed in the mid 1980s in response to clinicians' desires to perform cultures of cells for human therapy in a reproducible and reliable manner across multiple laboratories and institutions. The use of aseptic-tubing-connections technology, used commonly in the medical-device industry (such as for blood collection and transfusion containers), rather than conventional sterile technique in laminar flow hoods, reduces the probability of contamination to less than one chance per 1000 connections. Flexible containers fitted with aseptic connectors were appropriated from blood banking, where blood-platelet concentrates are stored for several days in incubators inside gas-permeable/liquid-impermeable plastic containers (called "bags"), which permitted bicarbonate pH buffering of the platelets by the carbon-dioxide gas in the incubator. In the late 1980s, extensive trials of various forms of lymphocyte therapy were conducted using such culture bags, with low incidence of contamination. Today, these culture containers continue to find use in experimental cell therapies where oxygen-consumption requirements are minimal and non-adherent cells grow satisfactorily in static suspension culture. Such containers do not, however, support the growth of human stem cells, which require contact with a heterogeneous population of adherent stromal cells. Advantages of culture bags include relative simplicity of use, reduced skill-level requirements, and potential use without laminar flow hoods. However, to date, processes utilizing culture bags remain labor- and space intensive and are limited in their biological and, hence, clinical applicability.

3. Bioreactors

Semiautomated "platform" culture systems – typically referred to as bioreactors – have been available commercially for many years, and employ a variety of types of culture technologies. These are platform technologies in that they have been applied to the scale-up of numerous culture processes initially developed in purely manual culture systems. Of the different bioreactors used for mammalian cell culture, most have been designed to allow for the production of high-density cultures of a single homogeneous cell type. Typical application of these high-density systems is to produce, as the end product, a conditioned medium produced by the cells. This is the case, for example, with monoclonal-antibody production by hybridomas, or viral-vector production by "packaging" cell lines. Cell-therapy applications differ from those applications where the therapeutic end product is the harvested cells themselves.

These systems have made an important first step toward a usable clinical system, in that once setup and running, they provide varying degrees of automated regulation of medium flow, oxygen delivery, temperature and pH controls, and they allow for the production of large numbers of cells. While bioreactors, thus, provide some economies of labor and minimization of the potential for mid-process contamination, their setup and harvest procedures involve considerable labor requirements and open processing steps, most of which require laminar flow hood operation (some bioreactors are sold as large bench-top environmental containment chambers to house the various individual components, which must be manually assembled and primed). The labor required in setup alone can offset any in-process labor savings, for example. Furthermore, they are optimally designed for use with a homogeneous cell mixture, and not the mixture of cell types that exists within tissues, such as bone marrow, and many are best suited for production of protein products, e.g., from culture of hybridomas, rather than manufacture of cells and tissues as the end product.

Many bioreactors have a high medium-flow-rate requirement for operation. The reason for this feature is that the oxygenation mechanism is to oxygenate the medium outside of the growth chamber immediately before the medium is perfused into the growth chamber. Since a high-density culture will quickly deplete the medium of oxygen, the medium must have a short residence time in the chamber in order to be reoxygenated and recirculated back into the culture chamber. Furthermore, this process results in an absence of uniformity in oxygenation of the growing tissue, since cells proximal to the medium inlet see much higher concentrations of oxygen than do the cells proximal to the medium outlet. This results in different cells growing in different areas of the bioreactor – a major drawback if a uniformly heterogeneous tissue is the desired product.

An additional limitation is that many of the bioreactor designs, such as the various three-dimensional matrix-based designs (hollow-fiber cartridges or porous ceramics), can impede the successful recovery of expanded cells and/or tissues, particularly when cell growth is vigorous, and can also limit mid-procedure access to cells for purposes of process monitoring. For example, bioreactor devices based on hollow-fiber dialysis cartridges typically produce cells in the extracapillary space between the densely packed fibers. Recovery of viable cells from such devices is often problematic and never quantitative, usually requiring extensive physical agitation and rapid flushing with large syringes or pumps.

The various tradeoffs described have limited the utility of these systems and, in general, such bioreactors have not been used for human cell therapy to the same extent as the less automated, but more practical, culture-bag systems.

II. AASTROM Cell-Production System

There is increasing need for a cell-culture system that can provide for large-scale production of cells for therapeutic purposes, as well as be operated effectively in a clinical setting, similar to other blood-cell-processing instrumentation. AASTROM Biosciences, Inc. has developed a system with the goal of satisfying requirements for use in a clinical, GMP environment. The AASTROM CPS is at the final prototype stage, and not yet approved or available for sale. Although initially targeted for use with bone-marrow-derived stem-cell mixtures, the AASTROM™ CPS has also been demonstrated to be employed to efficiently produce cells of different lineages, such as blood T-cells, or mesenchymal fibroblasts.

1. System Description

The AASTROM CPS is targeted to meet the challenging, multiple requirements of a clinical cell-production system. It is a functionally sterile, closed system that eliminates the use of open labware and laminar flow hoods. The largely automated process provides a level of process control and reliability that previously has not been available in clinical cell-culture applications. The net result is a system with the capability of providing a cell product generated via a validated process under GMP.

The AASTROM CPS embodies a modular, closed-system process including a pre-sterilized, single-use disposable cell cassette operated by automated instruments. The instrumentation components of the system include an incubator unit and processor unit, along with a user-interface comprised of a computer-based system manager (Fig. 4).

a. Disposable Cell Cassette

The cell cassette provides a functionally closed, sterile environment in which cell production can occur. The single-use cartridge is provided fully assembled in a sterile package, opened just prior to use. The sterile fluid pathway contained in the cell cassette includes: the cell-growth chamber, a container for medium supply, a pump mechanism for delivery of medium to the growth chamber, a container for the collection of waste medium exiting the growth chamber, a container for the collection of harvested cells, interconnected tubing between the components, and sterile barrier elements throughout to protect the culture from contamination during use. The cell-cassette design

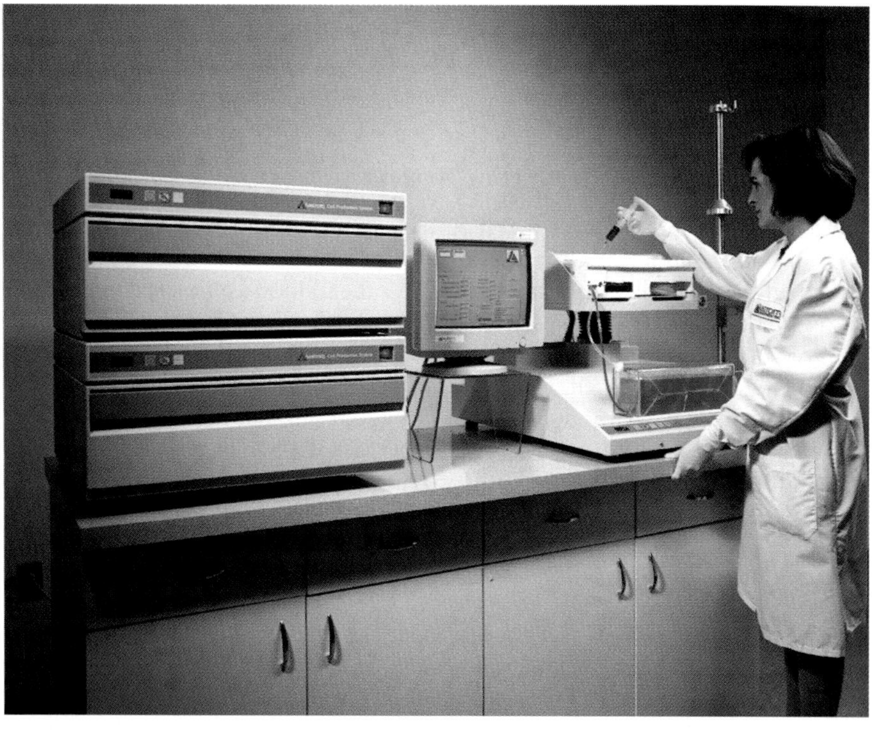

Fig. 4. Photograph of the prototype AASTROM cell-production system (CPS). The AASTROM CPS consists of the incubator(s) (*left*), system manager (*center*), and processor (*right*). A disposible cell cassette is being inoculated with cells as it sits on the processor platform

facilitates the introduction of culture medium, growth factors, and the introduction of the cells, in a contamination-free manner. The culture-chamber portion of the in-line fluid pathway contained in the cell cassette is where the key culture conditions are implemented. Since all fluid components are contained within the cell cassette, and the cassette is configured as a cartridge for one-step insertion into the control instrumentation, there is no risk of cross mixing patient materials when carrying out multiple patient expansions, and there is immediate turn-around time for the next use of the instruments.

The cell-cassette design is generic and adaptable to different medium conditions and the growth of either adherent cells (such as chondrocytes, fibroblasts, dendritic cells) or non-adherent cells (such as T-cell lymphocytes, CD34 progenitor cells). Furthermore, the culture chamber is configured for straightforward, complete removal of adherent and non-adherent cells in an automated manner. Finally, since all fluid components are self contained within the cell cassette, and the cassette is configured as a cartridge for one-step insertion into the control instruments, there is no risk of cross mixing patient materials during processing of multiple patient cultures. The cell cassette also facilitates immediate turn-around time of instruments for next use.

b. Incubator

A dedicated, small incubator controls the biological and physical environment and operations of each cell cassette necessary to support the cell-growth process. The incubator receives and self engages the disposable cell cassette, much like a VCR or video cassette. The incubator controls: (1) the flow of medium to the growth chamber, (2) the temperature of the growth-medium supply compartment, (3) the temperature of the growth-chamber compartment, and (4) the concentration and flow rate of gases delivered to the gas compartment of the culture chamber. The incubator also monitors various safety/alarm parameters to assure that the cell-production process is proceeding as expected and has the capacity for feedback control in the event that direct process monitoring is required. In the unlikely event of failure of an incubator, the cell cassette is readily transferred to another incubator without interruption of the culture process.

c. Processor

The processor performs the initial priming of the cell cassette with growth medium and, through instructions to the operator, the controlled inoculation of cells. The same unit also performs the removal (harvest) of the cells from the growth chamber at the completion of the cell-production process, sterilely transferring the cells to a pre-attached harvest container (analogous to a blood-transfusion bag). The heart of the processor is a mechanical platform which provides the cell cassette with automated tilting motions and valve actuations necessary to achieve priming and even distribution of cells for each procedure, and for final recovery of cells. The automated-system design provides for a high level of sterility assurance during each procedure by replacing the need for operator manipulation of culture devices within a laminar air flow hood.

d. System Manager

A user-friendly interface is a key component to a reliable, clinical system. This objective is accomplished in the AASTROM CPS through a central system manager, which employs a graphic CRT display. The system manager provides a convenient central-user interface and provides for redundant monitoring of each incubator in the network. Up to 50 incubators are monitored by a single system manager, providing the capacity to treat approximately 500–1000 patients per year (depending on the specific cell or gene therapy). The system manager can also perform procedure-scheduling tasks for the operators and provide a daily or weekly printed record of alarm events for quality-control and record-keeping purposes. The system manager also allows the operator to tailor a "batch-record" report of the entire cell-production process as required by GMP, stored either electronically or in printed form. The system manager is intended to facilitate GMP at any clinical facility.

e. ID Key

The ID key contains a non-volatile semiconductor memory device and clock and is affixed to the cell cassette at the beginning of a cell-production procedure. It is initialized by the system manager, utilizing data entered by the operator or obtained from a hospital information system with identifying patient information and specific protocol instructions. The ID key is accessed by the incubator or processor during each phase of the cell-production process to transfer protocol instructions and record the timing of major events and occurrence of alarms. The ID key provides reliable identification of the cell product, instructs the instruments for the cell-production process, prevents mix-ups and operator error, and stores the primary data for a complete process history record (effectively a "manufacturing batch record" for the cell product).

E. Applications for On-Site Delivery of Therapeutic Cell Production

The AASTROM CPS brings together two major breakthroughs in cell therapy: the ability to successfully grow human bone-marrow cells for supportive cellular therapy, and an automated-system concept that allows that and other processes to be reliably practiced at the point of use, in a conventional hospital-laboratory environment, with numerous features to facilitate GMP. The AASTROM CPS is, therefore, potentially useful in a number of cell- and gene-therapy applications.

I. Bone-Marrow Cell Production

Features of the cell cassette that make the AASTROM CPS particularly well suited for the growth of human bone marrow include: (1) unobstructed radial-flow fluid delivery allowing for substantially uniform exposure of cells to culture medium and fluid dynamic effects, (2) uniform internal oxygenation, to ensure all cells in the chamber are exposed to the same controlled oxygen levels, and (3) a large biocompatible uniform surface for adherent cell attachment and growth. The perfusion of the medium through the culture chamber occurs as a "single pass"; there is no recirculation. This provides a means for effective and controlled medium conditioning and removal of culture waste products. The internal oxygenation process circumvents the typical bioreactor requirement for external re-oxygenation achieved through rapid recirculation of medium. Furthermore, the chamber design is configured for simple and complete removal of both adherent and non-adherent cells.

With the successful production of the key bone-marrow cells in therapeutic quantities, a substitute for conventional bone-marrow transplantation or stem-cell therapy is enabled. This capability should result in potential advantages for the marrow donor in reducing the invasiveness of tissue procurement, along with various other clinical benefits. The use of ex vivo-produced bone-marrow stem cells and progenitor cells for stem-cell therapy is an excellent demonstration of how the lost function of a damaged or destroyed tissue can be repaired or restored with ex vivo tissue-engineered cells.

II. Other Cell and Tissue Production

The AASTROM CPS has been designed with the flexibility to accommodate additional applications in cell therapy via modifications to the culture process and, as necessary, to the culture environment. This may be accomplished by a family of cell cassettes, each uniquely optimized for a specific cell product. These include the production of lymphocytes, dendritic cells or various structural body tissues. The flexibility and system reliability achieved by using the ID key to instruct the processor and incubator enables the system to implement additional protocols in a well-controlled manner, readily amenable to validation by developers of new cell- and gene-therapy applications of the system.

Additional potential applications include the production of stem and progenitor cells from umbilical-cord blood, as well as the expansion of mesenchymal stem and progenitor cells to produce connective and solid organ tissues. The AASTROM CPS has recently been employed for the production of immature and mature lymphoid cells for the developing field of T-cell therapy. It is also being developed to implement the growth of these diverse types of tissues in the emerging field of gene therapy, where production of cells and tissues is a common requirement. The efficient perfusion design of the

AASTROM CPS may also facilitate the introduction of gene vectors into cells and tissues during cell-growth processes.

F. Summary

Cell therapies have shown great promise, but due to several limiting factors, have not yet realized their full potential. Once basic biological processes for ex vivo cell and tissue production are identified, the next requirement for therapeutic utilization is the need for clinical systems employing device technology and automation for process implementation. These clinical systems should ideally be amenable for routine use by the approximately 10,000 hospitals and clinics in the developed and developing world, which serve the patients intended to benefit from the transplantation of cells and tissues in native or altered form. The requirements for practical clinical systems for cell and gene therapy are not unlike the requirements for a validated, well-controlled, and reliable manufacturing process. The demands on clinical systems with the potential to achieve widespread therapeutic utilization and remain competitive over years of potential use are somewhat greater, and may turn on the effective automation of critical process and quality-assurance elements. As the field of cell therapy becomes increasingly established, the parallel need for efficient biological processes and effective, user-friendly automation should increasingly become equally important objectives for scientists, clinicians, entrepreneurs, and major corporations seeking to enhance the practice of medicine based on biotechnology and, more particularly, the therapeutic potential of cell therapy.

References

Bortin MM, Horowitz MM, Rimm AA (1992) Progress report from the international bone marrow transplant registry. Bone Marrow Transplant 10:113–122

Edgington SM (1992a) New horizons for stem-cell bioreactors. Biotechnology 10:1099–1106

Edgington SM (1992b) 3-D biotech: tissue engineering. Biotechnology 10:855–860

Emerson SG (1996) Ex vivo expansion of hematopoietic precursors, progenitors, and stem cells: The next generation of cellular therapeutics. Blood 87:3082–3088

FDA (1993) Medical devices; current good manufacturing practice (cGMP) regulations; proposed revisions; request for comments; proposed rule. Federal Register 58:61952–61986

Fleischaker RJ Jr, Sinskey AJ (1981) Oxygen demand and supply in cell culture. Eur J Appl Microbiol Biotechnol 12:193–197

Haylock DN, To LB, Dowse TL, Juttner CA, Simmons PJ (1992) Ex vivo expansion and maturation of peripheral blood $CD34^+$ cells into the myeloid lineage. Blood 80:1405–1412

Klingemann H-G, Deal D, Reid D, Eaves CJ (1993) Design and validation of a clinically applicable culture procedure for the generation of interleukin-2 activated natural killer cells in human bone marrow autografts. Exp Hematol 21:1263–1270

Koller MR, Emerson SG, Palsson BØ (1993) Large-scale expansion of human stem and progenitor cells from bone marrow mononuclear cells in continuous perfusion culture. Blood 82:378–384

Koller MR, Palsson MA, Manchel I, Palsson BØ (1995) LTC-IC expansion is dependent on frequent medium exchange combined with stromal and other accessory cell effects. Blood 86:1784–1793

Koller MR, Manchel I, Palsson MA, Maher RJ, Palsson BØ (1996) Different measures of ex vivo human hematopoietic culture performance are optimized under vastly different conditions. Biotechnol Bioeng 50:505–513

Koller MR, Palsson BØ (1993) Tissue engineering: reconstitution of human hematopoiesis ex vivo. Biotechnol Bioeng 42:909–930

Langer R, Vacanti JP (1993) Tissue engineering. Science 260:920–926

Mulligan RC (1993) The basic science of gene therapy. Science 260:926–932

Naughton BA (1995) The importance of stromal cells. In: Bronzino JD (ed) The biomedical engineering handbook. CRC Press, Boca Raton, pp 1710–1727

Palsson BØ, Paek S-H, Schwartz RM, Palsson M, Lee G-M, Silver SM, Emerson SG (1993) Expansion of human bone marrow progenitor cells in a high cell density continuous perfusion system. Biotechnology 11:368–371

Peng C-A, Palsson BØ (1996) Determination of specific oxygen uptake rates in human hematopoietic cultures and implications for bioreactor design. Ann Biomed Eng 24:

Schwartz RM, Palsson BØ, Emerson SG (1991) Rapid medium perfusion rate significantly increases the productivity and longevity of human bone marrow cultures. Proc Natl Acad Sci USA 88:6760–6764

Sutherland HJ, Hogge DE, Cook D, Eaves CJ (1993) Alternative mechanisms with and without steel factor support primitive human hematopoiesis. Blood 81:1465–1470

Verfaillie CM (1993) Soluble factor(s) produced by human bone marrow stroma increase cytokine-induced proliferation and maturation of primitive hematopoietic progenitors while preventing their terminal differentiation. Blood 82:2045–2053

Zandstra PW, Eaves CJ, Piret JM (1994) Expansion of hematopoietic progenitor cell populations in stirred suspension bioreactors of normal human bone marrow cells. Biotechnology 12:909–914

Subject Index

Printing: Saladruck, Berlin
Binding: Buchbinderei Lüderitz & Bauer, Berlin